AUDIO TESTS & MEASUREMENTS

HOW TO TEST ELECTRONIC COMPONENTS, AUDIOPHILE & GUITAR AMPLIFIERS AND LOUDSPEAKERS USING MODERN AND VINTAGE TEST INSTRUMENTS

Igor S. Popovich, B. Sc. (El. Eng.)

DISCLAIMER & COPYRIGHT NOTICE

Published by Career Professionals

P.O. Box 5668, Canning Vale South WA 6155, Australia

Revised edition, 2022

Bulk purchases

This book may be purchased in larger quantities for educational, business or promotional use. For special quantity discounts and savings please e-mail us at sales@careerprofessionals.com.au

National Library of Australia Cataloguing-in-Publication Data:

Popovich, Igor S.,

AUDIO TESTS & MEASUREMENTS: HOW TO TEST ELECTRONIC COMPONENTS, AUDIOPHILE & GUITAR AMPLIFIERS AND LOUDSPEAKERS USING MODERN AND VINTAGE TEST INSTRUMENTS

ISBN: 978-0-9806223-9-3

1. Electrical engineering 2. Electronics 3. Hi-fi

I Igor S. Popovich II Title

621.3

CONTENTS

CONTENTS, continued

WHY THIS UNIQUE BOOK AND WHO WILL BENEFIT FROM IT?

The aim of this book

After producing dozen or so DVD workshops about building, repairing, and upgrading tube amplifiers, preamplifiers, and tube testers, we produced two DVD programs on audio measurements, titled "Measurements in Electronics" and "Measurements in Tube Audio," which proved very popular."

Measurements in Electronics" covered the operational principles of various test instruments such as multimeters, ohmmeters, LCR meters, oscilloscopes, function generators, power supplies, vacuum tube voltmeters, attenuators, capacitance, and resistance decade boxes, and then demonstrated how to use such instrumentation to measure voltage, current, power, phase shift, resistance, capacitance, inductance, gain, attenuation, and other electrical parameters.

Its sequel, "Measurements in Tube Audio" DVD training course, focused on how to measure various parameters of tube amplifiers and preamplifiers: power consumption, gain, output power, THD, and IM distortion, S/N ratio, crosstalk, input and output impedance, damping factor, frequency range, internal AC & DC voltages and waveforms, phono stage RIAA accuracy and even some speaker measurements.

Most current books covering the T&M (Tests & Measurements) field were written to be used by first or second-year electrical engineering students and are thus too theoretical and way too broad in their coverage. Also, they focus more on high voltage tests and measurements in power engineering rather than electronics, meaning most of those topics are irrelevant to measurements in tube audio.

In the 1950s and 60s, there were quite a few books on the topic of electronic tests and measurements, mostly "popular" titles written in a friendly and easy-to-understand style, aimed at DIY constructors and amateurs who wanted to test and fix their own amplifiers, receivers, and other hi-fi components.

There were no digital multimeters or digital oscilloscopes in those days, not to mention virtual instrumentation and computer-based test and measurement platforms. Even the cheapest (single-channel!) analog oscilloscopes, for instance, were way too expensive to an average user. At the same time, top range equipment by brands such as Tektronix and HP cost as much as a good quality motor car!

So, we felt that there was a need for a book that would cover just enough theory so the reader can understand the engineering principles behind the instruments and the tests they perform, but without being too academic, focusing more on the application side of the T&M field. We hope that this book will fill this void and achieve that goal.

Who is this book for?

If you are an audiophile interested in the technical side of audio and the testing aspect of hi-fi equipment, this book is for you. Or, you could be a guitar player who likes modifying, repairing, or even building your own amps and speaker boxes. I feel that DIY enthusiasts and amplifier builders will form the bulk of our readership since tests & measurements are an essential part of the design & build process.

Even electrical engineering students should find lots of value in this book since T&M concepts and methods are explained simply and straightforwardly, emphasizing practicalities instead of overwhelming theory.

ABOVE: Dr. Slobodan (Bob) Popovich as a budding student of physics in 1958 in our native Yugoslavia, financed his university studies by building and repairing tube amplifiers, radios, and TVs

BELOW: Be careful what kind of book you give your kids to "read"; it may have a profound effect on their life! Here I am in 1963, "reading" my first book called "Practical Electronics," in German, no less!

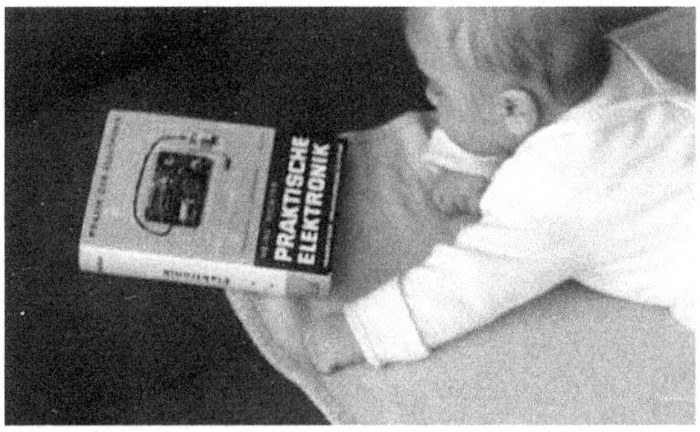

Our audio journey and T&M experience

My dad, Dr. Bob Popovich, a physics professor by day and an electronics guru by night, has been building receivers, guitar, and hi-fi amplifiers, all using tube technology, for more than six decades. The first smell I remember growing up was a melting solder, or rather the resin core inside it.

Upon completing my university degree in electronics in 1986, my practical education started. Over the last 30 years, Bob and I designed and constructed hundreds of different audiophile and guitar tube amplifiers and preamplifiers and serviced and improved countless more.

Tests & measurements were the most enjoyable side of amplifier troubleshooting and repairing. Even newly constructed amplifiers needed to be thoroughly tested and often modified to fine-tune their performance.

We used only the basic instruments, a digital multimeter in my and an analog one in Bob's case, plus an LCR meter, a function generator, and an oscilloscope. We've also designed and made a few simple test tools, jigs, and adapters, most of which you will learn about in this book.

Why include vintage test instruments but leave out virtual instrumentation?

I can already anticipate one criticism of this book: it will be the lack of coverage of virtual instrumentation and PC-based measurements. These days all you need is a computer with a good quality sound card and one of the dozens of available software packages, and you will have a powerful testing platform, not just a digital multimeter and oscilloscope, but also an FFT (Fast Fourier Transform) spectrum analyzer and a myriad of other virtual instruments.

There is nothing wrong with that approach; if you are serious about measurements, it is the easiest one to take. You don't have to buy half-a-dozen different instruments; you don't need to have bench space for them, or calibrate them. In other words, it is easy and convenient.

The first reason for not including virtual instruments here is that you probably know more about them than we do. In fact, we have never used them. The second reason is that real instruments, especially the vintage tube units, have that romantic and nostalgic feel about them, the beauty of lit-up tubes and moving coil analog meters, that magical feel of turning dials, something no sound card will ever give you.

However, the main reason is that these oldies can do what no PC sound card and software can - they can be used as learning platforms. Analyze their circuits and operation principles, and such knowledge will serve you well in your amplifier design & construction efforts.

Analyzing the design, wiring, and construction of a Tektronix tube oscilloscope, for instance, is much more educational than looking at tube amplifiers of the same era. Tektronix and other instrumentation giants such as HP, Beckman, and Fluke employed the best engineers, used the best components, and produced the best instruments of their time.

In comparison, Heathkits were poor cousins from the province but are still valuable, both educationally and operationally. Heathkit test & measurement gear was the most affordable of their era (the two decades, the 1950s-1960s). Millions of units were sold worldwide to amateur constructors and educational institutions. As a result, even today, they are the most common T&M brand sold on eBay and other online sources.

However, don't discard the vintage Heathkits based on their dated looks or technology; they are very cleverly designed instruments. Considering their relative simplicity, assuming they were correctly constructed (most were sold as kits), they perform remarkably well even by today's standards.

To use digital and virtual instrumentation properly, you need a broad and deep understanding of test & measurement principles and techniques. Older analog instruments illustrate such principles and techniques much better. Once you understand the hardware way of designing and constructing a test instrument, figuring out the software implementation is a natural progression.

Furthermore, many software methods are simpler and even more "primitive" than clever and resourceful hardware designs since "brute force" calculations enabled by fast microprocessors and heaps of memory can afford to use algorithms and measurement methods that would not work in the analog, hardware world.

Why is this book relatively thin?

You will notice small fonts and tight margins in this manual. To save on both paper (valuable trees) and printing costs, we typeset this book to minimize wasted space. Instead of 172 pages (albeit of a large, A4 size!), other books of this length, complexity and number of words run into 400-500 pages, which is insane. Ultimately, don't judge a book by its size, but by its contents and the usefulness of the information it contains!

Getting in touch with us

If you've liked the book and benefited from it, the best way to repay a favor is to recommend it to your friends and write a glowing online review on Amazon and similar websites. If you spot any errors or omissions, or should you have any constructive criticism of the book, I'd like to hear from you.

My e-mail is igorpop@careerprofessionals.com.au

I hope this book has answered at least some of your questions about the T&M field in general and audio measurements in particular. We wish you every success on your audio journey!

HOW TO READ THIS BOOK

One way to read a technical book like this one is to immediately go to a section or topic that interests you, and then to keep jumping back-and-forth to the related issues and chapters. This will, I suspect, be the way the more experienced readers will approach this book.

A more systematic approach is sequential, starting from the beginning and reading in order. This is what I would recommend. Although it seems more time consuming (since you will read about many issues you may already know a lot about or are not particularly interested in), paradoxically, this approach is often faster. You will not miss anything, and you will not waste time flipping forward and backward trying to clarify an issue that had been covered but which you've overlooked or perhaps not fully understood.

Whatever you do, don't treat this book as Holy Scripture. Underline or highlight the important parts, write your thoughts and ideas on its margins, sketch diagrams and circuits in its blank spaces.

Abbreviations used (in no particular order)

MAX	Maximum	AC	Alternating current	GND	Ground terminal
CV	Constant voltage (DC power supply)	CC	Constant current (DC power supply)	CF	Crest Factor (of a waveform)
MIN	Minimum	DC	Direct current	COM	Common terminal
DCV	Direct current volts	ACV	Alternating current volts	VOM	Volt-Ohm-Milliammeter
BNC	Bayonet Neill–Concelman, video & test equipment connector for coaxial cable	RCA	Unbalanced audio connector or Radio Corporation of America	SLO-BLO	Slow blowing fuse, delay fuse
IMD	Intermodulation distortion	THD	Total harmonic distortion	SS	Solid state
FFT	Fast Fourier Transform	CE	Common emitter	DMM	Digital multimeter
PP	Push-pull (amplifier)	CB	Common base	CF	Cathode follower
PP	Polypropylene (capacitor)	CC	Common collector	SPL	Sound pressure level
Gm	Mutual conductance (FET or vacuum tube)	MM	Moving magnet (phono cartridge)	PMMC	Permanent magnet moving coil meter
HV	High voltage	MC	Moving coil	NFB	Negative feedback
NTC, PTC	Negative and positive temperature coefficient (of a resistor or other component)	LOG	Logarithm, logarithmic scale or output, or taper (potentiometer)	GBW	Gain-bandwith product of a tube or amplifier
SET, PSET	Single-ended triode, parallel SET	LIN	Linear scale or taper (potentiometer)	EMF, CEMF	Electro-magnetive force, counter EMF
E/S	Electrostatic (field, interference or shield)	TR, VR	Turns or voltage ratio (of a transformer)	RIAA	Recording Industry Association of America
CCS	Constant current source or sink	IR	Impedance ratio (of a transformer)	XLR	Balanced audio connector (3-pins)
ESL	Equivalent series inductance (of a capacitor)	ESR	Equivalent series resistance (of a capacitor)	TPV	Turns-per-volt (of a transformer)
RMS	Root-Mean-Square (effective value of an AC signal)	DIN	Deutsches Institut für Normung (German Institute for Standardization)	FET, JFET, MOSFET	Field effect transistor, Junction FET, Metal Oxide FET
SG	Screen grid (in a tube)	CG	Control grid (in a tube)	SP	Suppressor grid
CT	Center tap (of a transformer)	dB	deciBel (1/10 of a Bel)	DF	Damping factor
CLC	Capacitor-inductor-capacitor filter	CRC	Capacitor-resistor-capacitor filter	ARRC	Average-responding RMS-calibrated meter
RFI	Radio frequency interference	AF	Audio frequency	RF	Radio frequency
CW	Clockwise (turn)	T&M	Tests & Measurements	CCW	Counterclockwise (turn)
LCR	Inductance-capacitance-resistance meter or circuit	CMRR	Common Mode Rejection Ratio	SNR or S/N	Signal-to-Noise ratio
IC	Integrated Circuit	FSD	Full scale deflection	EM	Electromagnetic
ZD	Zener diode	NOS	New Old Stock	PP	Peak-to-peak (AC signal)
LCD	Liquid Crystal Display	PF	Power factor	DUT	Device Under Test
FF	Form Factor (of a waveform)	TUT	Tube or transistor under test	CRO	Cathode ray oscilloscope
SPST	Single pole single throw (switch)	DPDT	Double pole double throw (switch)	BW	Bandwidth (frequency range)

SYMBOLS USED

MAGNETIC COMPONENTS

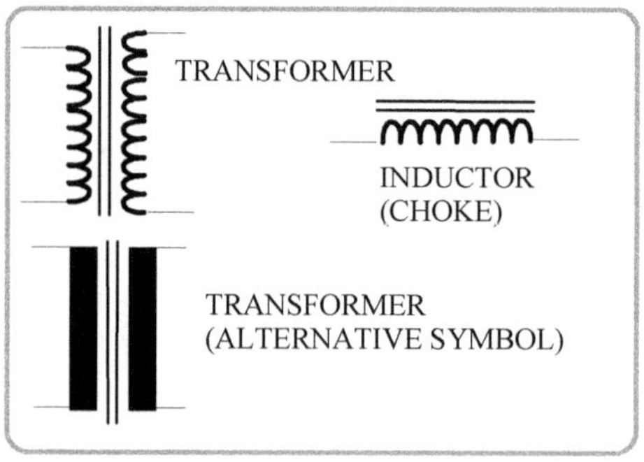

CAPACITORS

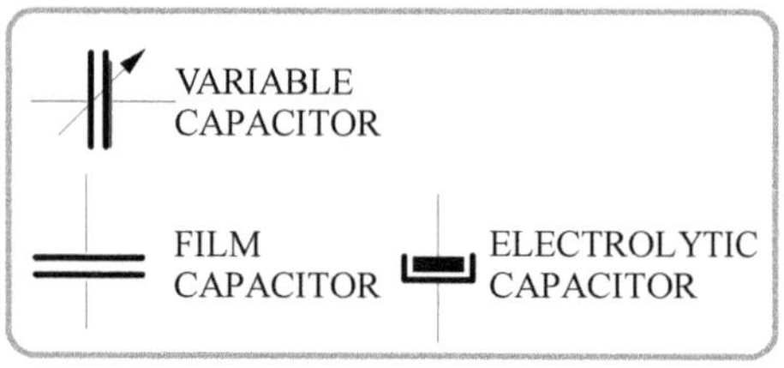

MISCELLANEOUS SYMBOLS

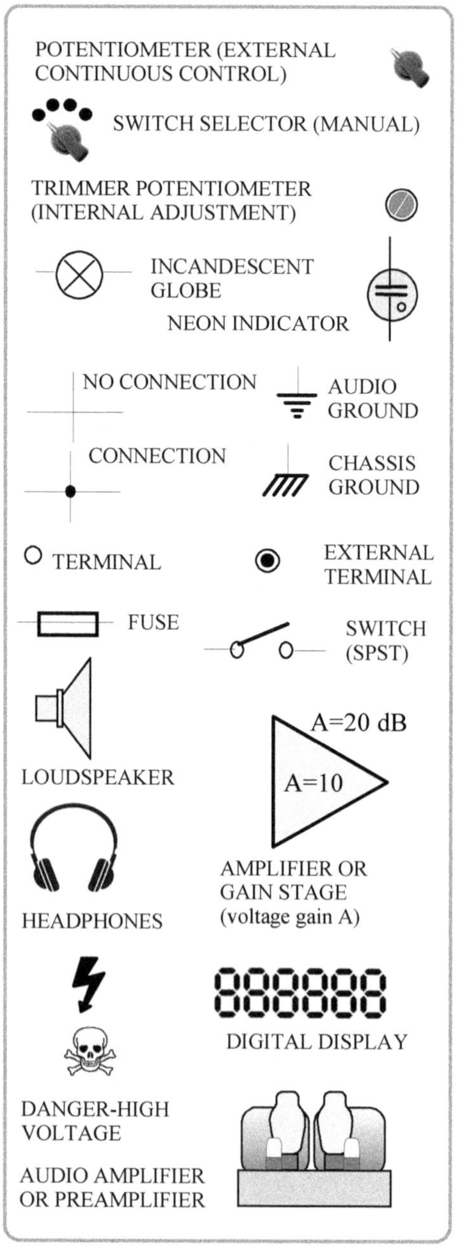

RESISTORS

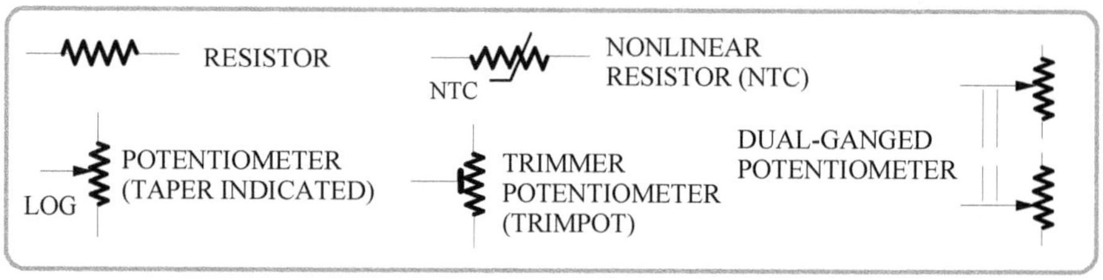

SEMICONDUCTORS

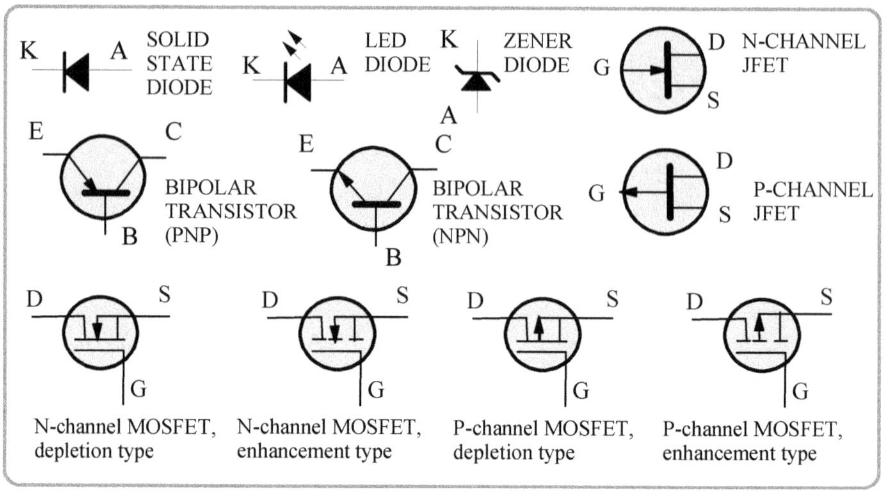

AC AND DC SOURCES

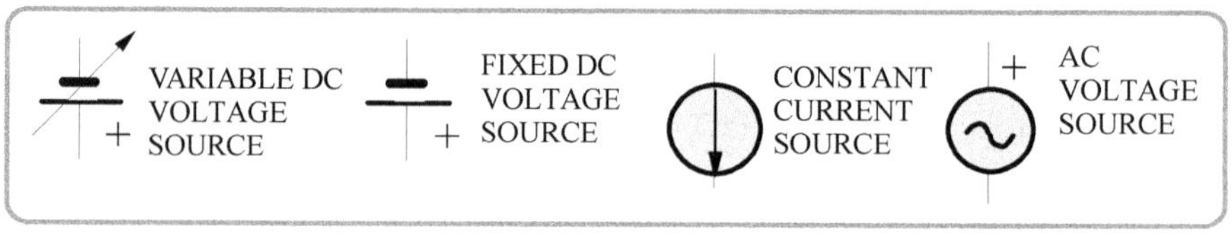

ELECTRON TUBES

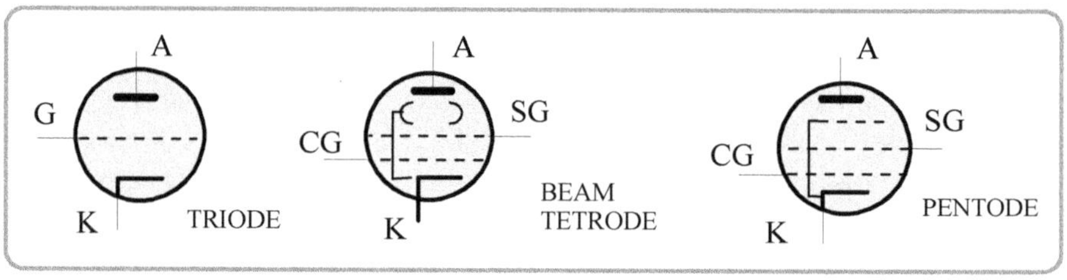

TEST INSTRUMENTS

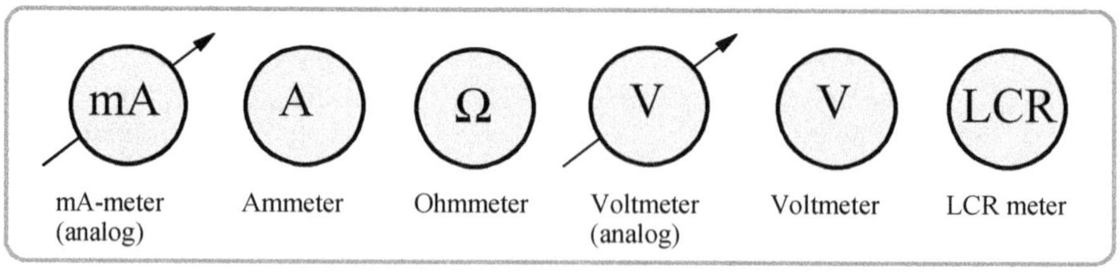

FRAMES

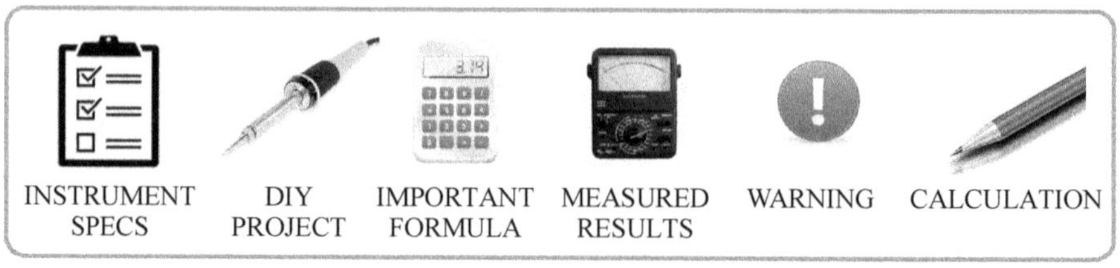

MATHEMATICAL AND OTHER OPERATIONAL SYMBOLS

- ≈ APPROXIMATE
- ⟶ SIGNAL OR CURRENT FLOW
- ❷ NUMBERED POINTS OF INTEREST (ON PHOTOS OR DIAGRAMS)
- || PARALLEL
- ∝ INFINITY
- ≪ MUCH SMALLER THAN
- ≫ MUCH BIGGER THAN

1 TEST INSTRUMENTS, ERRORS, LIMITATIONS & SAFETY ISSUES

- SAFETY RULES and PRECAUTIONS
- ABSOLUTE AND RELATIVE ERRORS
- PRECISION vs. ACCURACY vs. RESOLUTION
- INSTRUMENT LIMITATIONS
- DECIBELS AND LOGARITHMIC SCALES
- UNDERSTANDING T&M INSTRUMENTS' CIRCUIT DIAGRAMS
- MISCELLANEOUS JIGS, ADAPTERS AND TOOLS FOR AUDIO TESTS
- REPAIRING, RESTORING & CALIBRATING VINTAGE TEST INSTRUMENTS

> "Science starts with measurements, science makes no sense without measurements.
>
> Dimitri Ivanovich Mendeleev, Russian chemist and inventor, famous for formulating the Periodic Law and creating the periodic table of elements"

Testing audio equipment is not that different from testing other electrical and electronic gear

Building, upgrading, and repairing audio equipment in general, and amplifiers and preamplifiers in particular, require various tests and measurements at every stage of such undertakings. Once we have a design we want to build, we need to buy or gather components and test them before putting them together during the construction stage. Once we've completed building the amplifier or any other piece of audio gear, we perform continuity and ohmmeter tests ("cold" checks) before we hook up a dummy load (for power amplifiers) and energize the device (power it up).

If nothing smokes or explodes (!), we check DC voltages in various significant points. If all are within +/-10% of the nominal values, we connect a function generator to its input and run a sine or square wave signal through the amp or preamp while measuring the output voltage and observing its waveform on an oscilloscope. Then we start the fine-tuning stage, changing components' values and modifying circuits as needed.

Even if you have no desire to construct your own hi-fi or guitar gear, you may wish to check the performance or "fidelity" of the equipment you have or have just bought. In any case, understanding the basic test & measurement principles and methods used and the basic functionality of T&M equipment is a very useful skill.

If you are an electrician, an electrical or instrumentation technician, you can apply this knowledge to your daily work. The benefits of the knowledge gained here are transferable to tests & measurements in other applications and fields.

You could be an engineer or an electrical engineering student interested in music or hi-fi. Although a syllabus of most electrical and electronics degrees includes at least one course dealing with tests & measurements, the emphasis is usually on industrial and high voltage (power) tests, not on the audio field. In that case, you should find this book easy to understand, and this material should complement your existing knowledge well.

A bird's eye view of the testing & measurement process and setup

The simplified block diagram below shows a typical audio test setup, measuring the signal at the output of an audio amplifier, with a sine or square wave signal at its input. The dummy load is not shown, and the signal source isn't needed for all tests, so don't get stuck at the detailed level. Our aim here is to illustrate various subcircuits of the "test instrument," such as the analog processing block (even digital instruments have analog front ends), the digital signal processing hardware (if any), and the indicator or display. These can range from a tuning eye on simple vintage instruments, a null indicator such as headphones, tuning eye or moving coil meter in bridges, analog or digital meters and displays, cathode ray tubes on older oscilloscopes and spectrum analyzers, or LCDs on modern ones.

The critical aspect of any T&M setup is you, the user! You must decide what to measure, what instruments to use and how to do it. Once your testing is done, you have to scrutinize and analyze its results and decide if your results are valid, or if you've made a mistake, measured the wrong thing, used the wrong test instrument, or in the wrong (inappropriate) way!

As in any endeavor that involves some degree of danger (martial arts, sports, and outdoor activities, exercise, and such), we will start with the safety aspect of T&M. Please don't skip that part. Read it carefully, and memorize the checklist of electrical hazards and required safety measures forever. Happy reading!

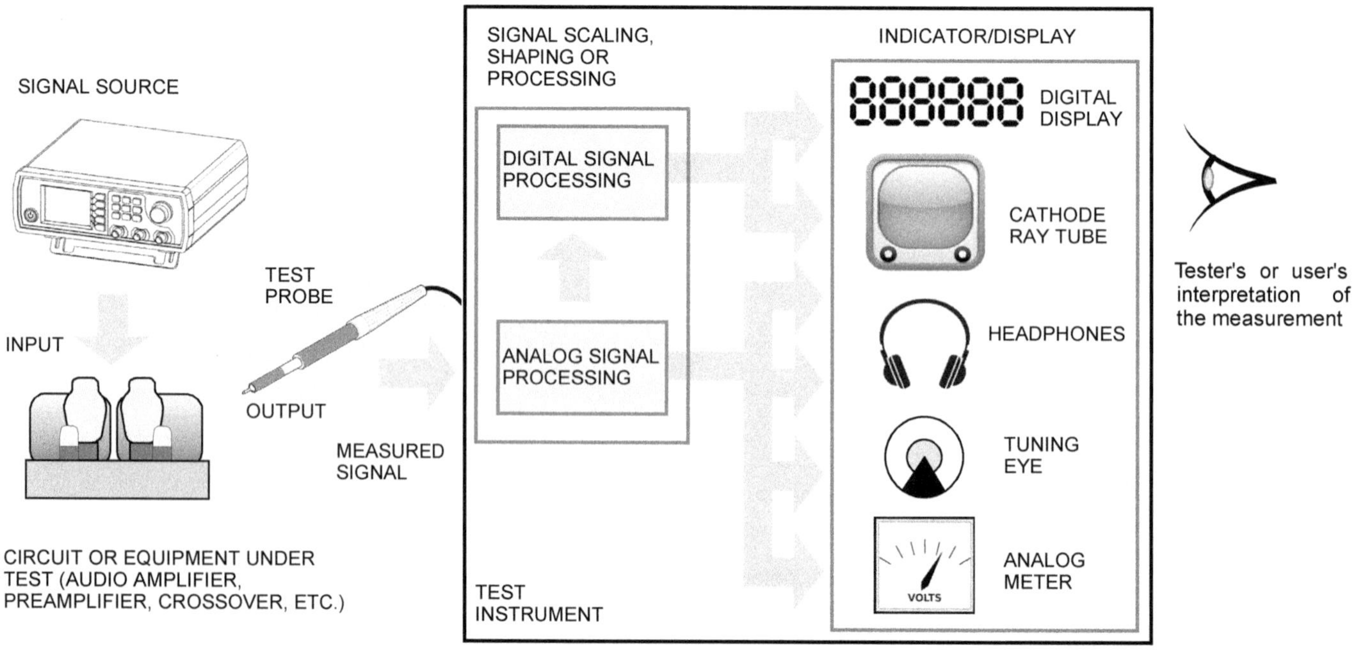

ABOVE: A simplified block-diagram of a typical audio test setup, in this case measuring the signal at the output of an audio amplifier, with a sine or square wave signal at its input.

SAFETY RULES AND PRECAUTIONS

Electric shock - how it happens and how to avoid it

Just as people can drown in 10cm of water and 10m deep water, they can get a lethal electric shock from 100V, which can be as deadly as 1,000V! The severity of an electric shock depends on many factors, such as the victim's age, gender, and physical condition. However, the main factor is the level of current flowing through the body.

The perception threshold in humans is around 1mA (milliAmpere, 0.001A). Currents up to 5mA produce tingling but not severe pain. Muscular contractions start at around 10mA, while 100mA of current would start the fibrillation of the heart muscle, preventing it from pumping blood and causing death unless the fibrillation is stopped.

Above 300mA, the heart's contractions are so severe that fibrillation is prevented, so if the electric shock is halted quickly, normal heart rhythm will probably resume. Thus, 100-300mA is the most dangerous range!

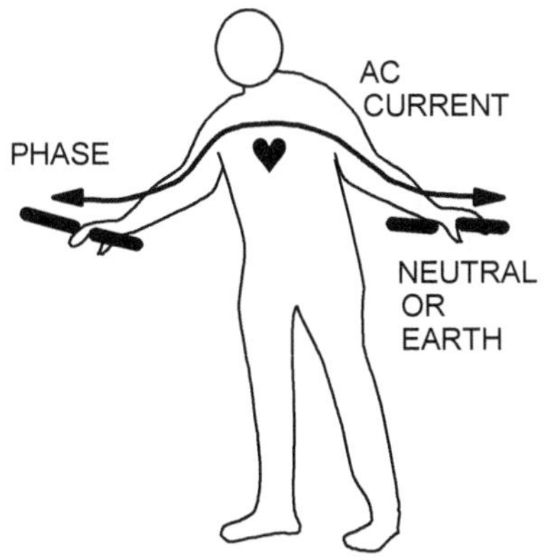

The worst situation: current flowing through the upper body (torso) and the heart

The lethal voltage level depends on the resistance of the current-conducting path through the body, which in turn depends on skin resistance and how the contact is made. Skin resistance, in turn, depends on its moisture level. Dry skin can have as much as 500kΩ of resistance, while the wet skin's resistance can be as low as 1kΩ! As electric current flows, it punctures and breaks down the outer layer of the skin (the epidermis), and the skin resistance falls rapidly. This is why it is crucial to break the contact with the live conductor as quickly as possible.

The most dangerous situation involves a voltage between two hands or arms, as illustrated when current flows through the upper torso where the heart is located. That is why you should only use one hand when measuring voltages, especially the high voltages present in tube amplifiers. The other hand should not be touching anything - keep it in your pocket!

Checklist: Electrical hazards and required safety measures

- Tube amplifiers and many test instruments involve voltages that may be lethal, so precautions must be taken to mitigate the risk of an electric shock when powering up, testing, and working on "live" circuits. One mistake may cost you an expensive transformer, a pair of tubes, or even your life!
- First of all, get an electrician to install in the switchboard of your house or workshop a device that goes under many names: ELCB (Earth Leakage Circuit Breaker), ground fault interrupter, or RCD (Residual Current Device). These are now mandatory by law in Australia, but electrical laws in many other countries are much more lax. These protective devices may save your or your children's lives (if they poke a knife into a toaster, for instance, or into a tube amplifier!)
- Unplug your amp from the wall outlet when not in use and while working on it. Once unplugged, discharge the power supply capacitors before working on the amplifier. Don't just use a straight wire for discharging; the spark and the high discharge current may damage the capacitor. Make a discharge cable with insulated crocodile clips at both ends and a 1kΩ 2W resistor in series to limit the discharge current.
- When working on a "live" amp, use only one hand. Don't touch the chassis or any other part of the amplifier with the other hand. If practical, wear thin cotton gloves.Wear shoes with rubber soles.
- Never stand barefoot on a bare concrete floor while working with electrical appliances or amplifiers. Use a rubber mat. Timber flooring and carpet (wool or polypropylene) are also good insulators. Use tools with insulated grips, and don't touch their exposed metal parts.
- Consider using an isolation transformer to power up your test bench. That will all but eliminate the probability of an electric shock. See the DIY project on page 13 on how to make your own isolation transformer from two general-purpose step-down transformers and save a fortune on a commercial isolation transformer.

Other hazards and precautions needed to mitigate them

- Your workspace and test bench should be clean, tidy, spacious, well organized, well lit, and well ventilated.
- Never work on amplifiers when you are tired, sick, absent-minded, upset, hot, sweaty, when you have food, sex, or anything else on your mind.
- Do only one thing at a time and don't do anything else: don't eat, drink, have your headphones on or talk on a mobile phone. Building, repairing, or modifying tube equipment is a demanding activity; it requires concentration and full awareness of all your senses. You must be able to hear when a capacitor starts frying, transformer insulation crackling or varnish smelling, or to smell a tiny whiff of smoke when a resistor starts burning!
- Never wear dangling jewelry, headphones, or anything else that may get caught inside the equipment you are working on and create either a mechanical or electrical hazard (short circuit).

- Tubes get hot very quickly. Never touch or handle hot tubes with bare hands. Use a thermally insulating glove.
- Tubes and soldering irons are fire hazards. Remove all flammable material from their immediate surroundings and always have a fire extinguisher ready. It must be of the type approved for electrical fires.
- Metalwork is another serious risk. Always wear eye protection and gloves when cutting, drilling, or grinding metal chassis or other parts. Secure the pieces you are working on (clamp them down). Have a First Aid Kit handy.
- Treat tubes gently; one moment of lapsed concentration may cost you hundreds of dollars. When pulling tubes out, do not jerk them left and right. Pull them straight out vertically
- Amplifiers are heavy; bigger ones weigh 20 to 30 kg! Exercise caution when lifting them to the workbench and whenever you move them around. Bend your legs (your knees), not your back!Keep children and pets away from the amplifier and your working area.

A critical safety upgrade for vintage amplifiers and test instruments

Most amplifiers, tube testers, capacitor checkers, signal generators, and other audio or test equipment built before the 1970s in the USA used a two-pin mains plug. There was no earth pin, so these units were not earthed (or grounded) at all. Furthermore, these 2-prong plugs are reversible, so instead of the neutral being connected to the metal chassis, you can end up with the live 115 VAC on it. In many cases, the fuse and the power switch are in different mains circuit branches, as illustrated below in the Scott 299D amplifier.

Film capacitors are almost always connected between the phase or neutral and the chassis. Over the decades, most of these capacitors become leaky, some even completely short out, which allows the mains voltage to end up on the metal chassis. These days only X2-rated and approved capacitors can be used in such critical positions (the short circuits cannot develop). These leaky capacitors are a severe hazard and must be removed. Our example, the Scott 299D amplifier, had two of those, C211 and C212.

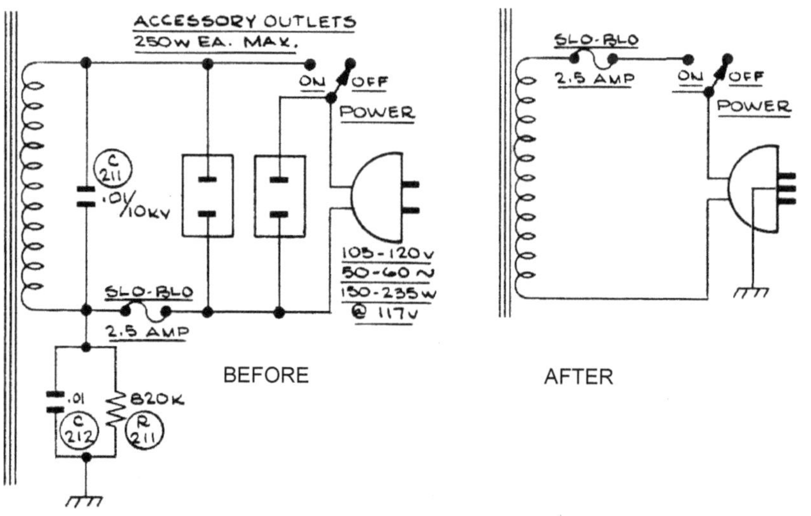

The mains circuit of Scott 299D amplifier in its original state (ABOVE LEFT) and after safety modifications (ABOVE RIGHT).

Before you start repairing or upgrading anything, your first job is to perform this safety conversion. Bin the two-core mains cable, the auxiliary outlets, and the two-pin plug, and replace them with an approved 3-core mains cable and plug.

Remove any capacitors and resistors connected between phase and neutral and ground, such as C211, C212, and R211 on the Scott 299D diagram.

Connect the earth lug via an insulated green or green-yellow wire to a dedicated bolt with good galvanic contact with the metal chassis. Of course, this upgrade does not apply to double-insulated or "floating" instruments. In fact, earthing (grounding) such instruments may render them inoperative or even cause unwanted short circuits.

The grounding problem in Dynaco ST70 amplifier

When testing the original Dynaco ST70 (whose chassis was not grounded due to a 2-pin power plug used), the 10Ω resistor between the negative terminal of the audio input and the chassis started smoking. We measured $11V_{AC}$ between the chassis and the "earth," enough to push more than 1 A of AC current through that loop. The dissipation on that poor 1 Watt rated resistor was P=V*I = 11*1.1 = 12 Watts!

The audio generator grounded the other end of the resistor and caused the circulating AC current to flow. Solution? We removed the offending resistor, grounded the chassis, and all was well.

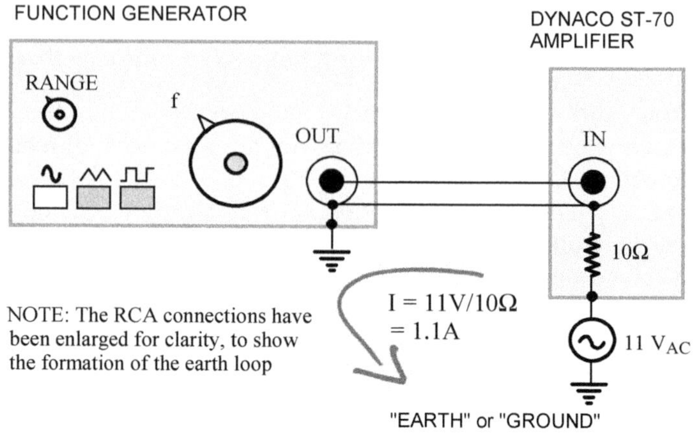

The test equipment grounding problem

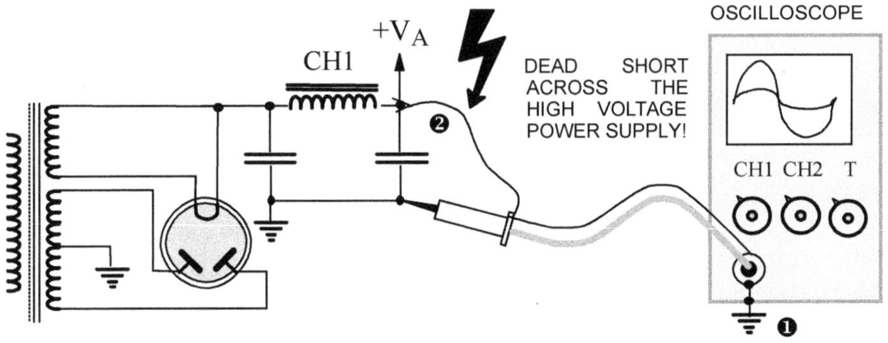

Most instruments have single-ended inputs/outputs, meaning the negative side is grounded (1). These are connected to the shield and the clip of the test leads (2).

Say you want to observe the ripple voltage waveform on the high voltage line ($+V_A$). If you accidentally reverse the probes and connect the positive probe (the tip) to the ground and the clip (the probe's ground) to $+V_A$, you will cause a short circuit!

The principle behind isolation transformers' operation

Isolation transformers galvanically isolate the primary and secondary windings. Touching the secondary LS (Live) or NS (Neutral) terminal would not result in electric shock since these are floating, not "referenced" to ground or earth, so the path through the human body cannot be closed and circulating current cannot flow.

As indicated by the arrows, one fault in the secondary (load) side, such as a short circuit to earth (or the metal case, which is earthed), is tolerated. It will not endanger the user since it will simply "ground" (or "earth") either a secondary phase (LS) or neutral (NS). However, the isolation properties of the transformer would be lost since that would "reference" the secondary winding to the ground.

To have two faults simultaneously (as in the diagram) is highly unlikely, and even if it does happen, it will short-circuit the transformer's secondary and the secondary fuse (4) would blow, or the primary circuit breaker (3) would trip.

Notice that only the transformer is "floating"; its metal case (enclosure) is always grounded or earthed (5)!

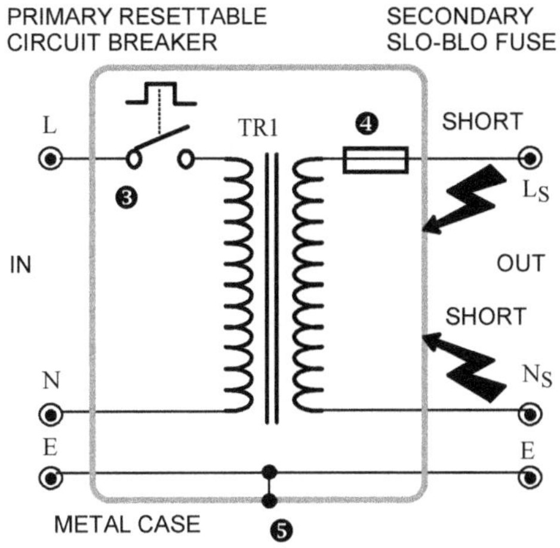

ABOVE: Two possible short circuits on the secondary side, usually cased by a shorted load (not shown).

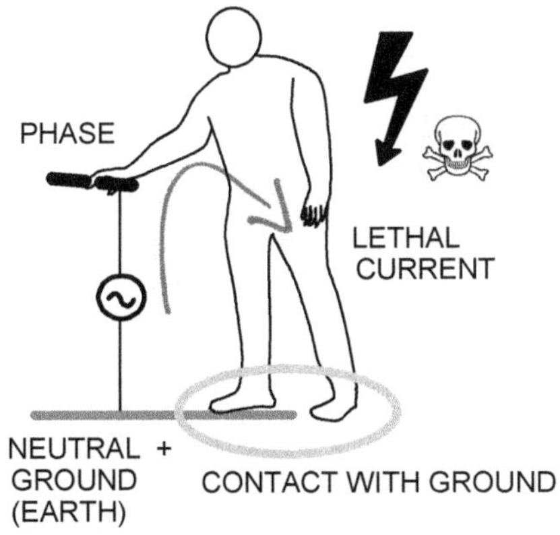

LEFT: Ordinary mains installation with neutral and earth bonded together (at the switchboard). Once the body closes the circuit by touching a live conductor, lethal currents flow!

RIGHT: Floating system when isolation safety transformers are used. Neither neutral nor phase are referenced to ground so touching a live conductor is harmless!

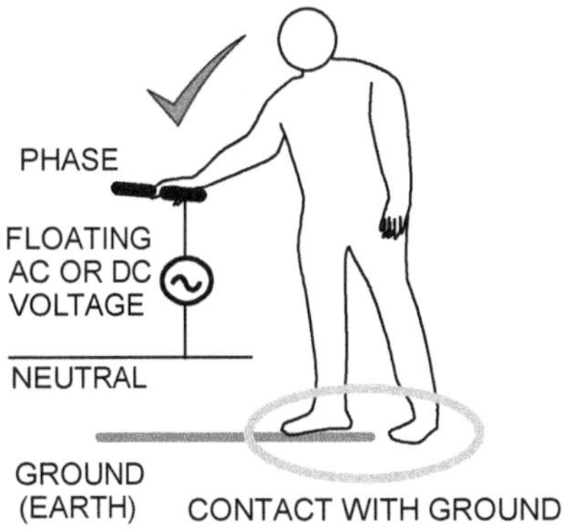

MAKE YOUR OWN ISOLATION TRANSFORMER **DIY PROJECT**

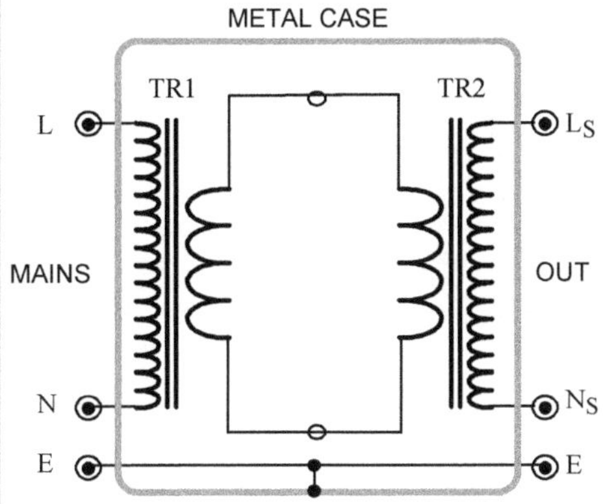

Isolation transformers are expensive. If you are on a limited budget, get two step-down transformers of a suitable power rating and connect them back-to-back (secondary of TR1 to the primary of TR2). Transformers can be any type, EI, C-core, R-core, or toroidal.

With one secondary (12V, 24V, 48V, etc.) and identical transformers, you will only get a 1:1 voltage ratio. With multiple secondaries, you can get a step-up and step-down feature, depending on how you connect the low voltage secondaries.

For general workshop use, 500VA or a higher rating is needed. Remember to earth the metal chassis (enclosure), as illustrated. You can include an AC ammeter and AC voltmeter on the secondary side to monitor the voltage or current delivered to the load.

ABSOLUTE AND RELATIVE ERRORS

Absolute and relative errors of calibrated dial scales

Many vintage and some modern (or rather "currently produced") T&M instruments have a scale on a dial, which can be a variable capacitor (many LCR bridges), a precision potentiometer, or even a variable transformer as in this case. Sold in Australia, this made-in-China "variable voltage regulator" is rated at 2kVA and designed for $220V_{AC}$ voltage input. Notice that even its name is misleading, it does not regulate anything. It enables the user to manually vary the output AC voltage. The output voltage will still follow the mains voltage fluctuations present on its input!

What is the absolute (in Volts) and relative error (in %) if it's connected to the nominal Australian $240V_{AC}$ mains voltage and set at $220V_{AC}$ (to power up a hi-fi amplifier designed for European 220V standard)?

With 220V input and the dial set at 220V, the output voltage would be 220V (assuming that the Variac's scale is calibrated correctly and there are no other errors).

However, we have a 240V input, so at the same 220V setting, the output will be proportionally higher, or 240V.

The absolute indication error is thus actual indication or "dial" D_{ACT} minus nominal dial D_{NOM}, or $\varepsilon = D_{ACT} - D_{NOM} = 220-240 = -20V$ and the relative error is $\varepsilon_R = (I_{ACT} - I_{NOM})/I_{NOM} = (220-240)/240 = -20/240 = -0.0833$ or -8.33%!

Note: "I" here stands for indication, not for current. Both errors are negative since the actual indication on the dial is lower than the nominal figure (the real voltage present).

With 240V in, to get 220V output we need to set the dial at $D = 220V*(1+\varepsilon_R) = 220*(1-0.0833) = 201.67V$!

The answer could be deduced in another way, by using a simple ratio, $V_{REAL100\%}/V_{MARKED100\%} = V_{REAL}/V_{MARKED}$ or $240/220 = 220/V_{MARKED}$, so we get $V_{MARKED} = 220*220/240 = 201.67V$!

ABOVE: Since its scale was marked based on the assumption of $220V_{AC}$ voltage at its input terminals, the indication of this variable transformer will be wrong at any other input voltage.

Absolute & relative errors of test instruments in general

An absolute error is always in the same units as the measured quantity, volts for voltage, ohms for resistance, etc. The relative error is the absolute error divided by the actual or nominal value and is often multiplied by 100 and expressed in percentages. Both errors can be positive (when the measured or indicated value is higher than the actual value) or negative.

ABSOLUTE ERROR

ε_A = MEASURED VALUE - TRUE VALUE

RELATIVE ERROR

ε_R = (MEASURED VALUE - TRUE VALUE)/ TRUE VALUE = ε_A/TRUE VALUE

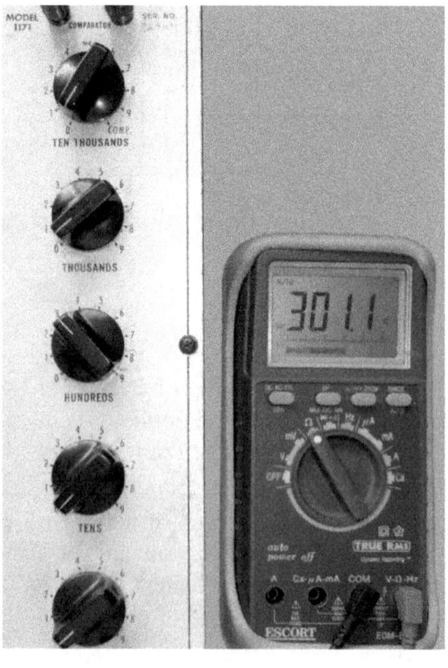

ABOVE LEFT: The selector switches on this Eico decade resistance box indicate 300Ω, yet DMM measures 301.1Ω. For most test purposes, this level of accuracy would be acceptable.

ABOVE RIGHT: An internal trimmer potentiometer inside this Eico 930 high voltage power supply enables "DC Volts" meter calibration, in this case against a digital multimeter as a reference. It is impossible to distinguish 1.6V on a 300V analog scale.

PRECISION vs. ACCURACY vs. RESOLUTION

Accuracy is not the same as precision

In the measurement field, terms precision and accuracy are often misunderstood and thus used interchangeably and often incorrectly. Accuracy describes the difference between the measured and the actual value of the quantity (parameter) measured, the degree of "agreement" between the two.

Precision describes how well identically performed measurements agree with each other. So, for a measurement to be accurate, it also must be precise; accuracy implies precision, it is a necessary condition. The opposite, however, is not true; an instrument can be very precise but grossly inaccurate. Precision does not guarantee accuracy!

The shooting analogy (right) may help. Who is a better shooter, A or B? More precisely (no pun intended), who is a more precise shooter?Well, A is more accurate; his shots are much closer to the actual value or the "bull's eye"! However, his shot pattern is more scattered; therefore, he is less precise! Shooter B is less accurate, all his shots are much further away from the "true value" or the target than A's, but he is more precise, his shots are less scattered. Instead of two shooters aiming at the same target, A and B could also be two instruments measuring the same signal or parameter.

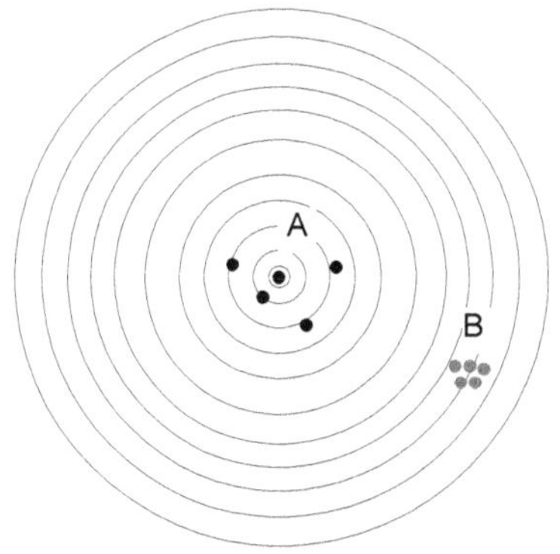

ABOVE: Who is a better shooter, A or B? By the same analogy, which test instrument is more accurate, A or B? Which is more precise?

Scale resolution and accuracy of analog meters

The *accuracy* of any test instrument depends on so many factors (the testing principle used, calibration, component drift, mains voltage fluctuations, hum and interference, regulation and stability of its power supply) that it is hard to ascertain it even for an experienced and knowledgeable user. What we can relatively quickly determine is the *resolution* of the instrument's scale. No matter how accurate the measurement may be, it cannot exceed the resolution of the scale!

Let's agree that one division is the smallest unit of resolution. We can guess if the needle is somewhere in between divisions, but that is another story, called guesswork, not measurement. The scale of the vintage TV-10/U military tube tester made by Hickok has four mutual conductance (Gm) ranges, 3,000-6,000-15,000-30,000 micromhos.

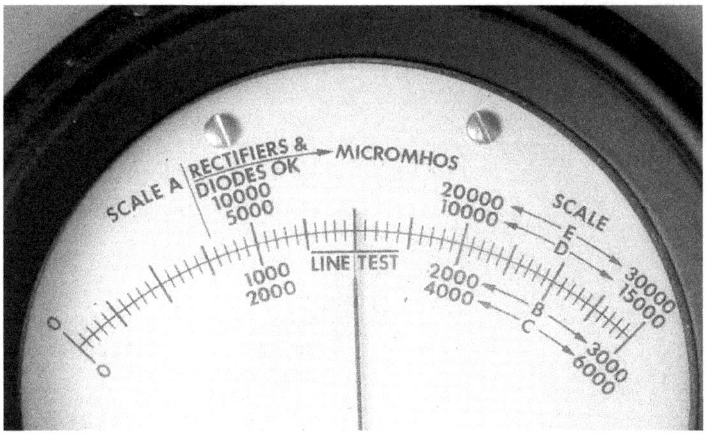

What is the resolution of each scale (the Gm value of one minor division)? On the 3,000 micromhos scale, there are four major divisions between 2,000 and 3,000 micromhos. Each major division has five minor divisions, 4 x 5 = 20 in total. 20 minor divisions are equivalent to 1,000 micromhos, so 1 mindiv = 50 micromhos.

On the 6,000 scale, 20 mindiv is equivalent to 2,000 micromhos, so the resolution is 100 micromhos! Using the same reasoning, we get that the resolution of the 15,000 scale is five times lower than that of the 3,000 scale, 50x5=250 or 5,000/20 = 250, and the resolution of the 30,000 scale is 500 micromhos (10,000/20 = 500).

Nonlinear scales, increasing uncertainty and decreasing accuracy

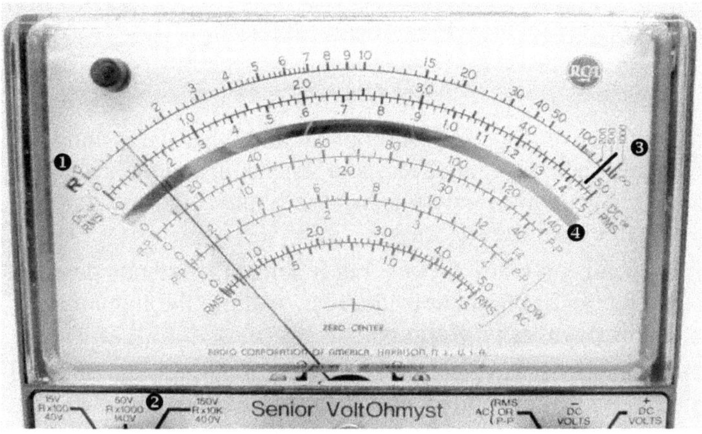

On most analog multimeters the resistance or "ohms" scale, marked on the RCA Senior VoltOhmyst "R" (1), is highly nonlinear - the scale resolution decreases from left to right, and so does the accuracy of the resistance measurement.

With the range switch (2) in Rx1,000 position, we would read this indication as 1.0x1,000 = 1,000Ω. However, with the indicator in position (3), what is the resistance measured? All we can say with reasonable certainty is "somewhere between 200 and 300 kΩ". The rest would be pure speculation. It seems closer to 200, so probably 235kΩ? Thus, it is always better to change the range switch so the indication is on the left-hand side of the scale.

Catastrophic or gross errors

Both analog and digital indicating instruments open the possibility of the tester making a gross or catastrophic error. For instance, going back to the RCA Senior VoltOhmyst electronic multimeter (previous page), one possibility is using the wrong scale. Say we are measuring an AC voltage with the range switch on $140V_{AC}$ range (2), but instead of reading the upper 0-140V scale (4), we take a reading on the lower $0\text{-}40V_{AC}$ scale. The true reading is $13V_{AC}$, but the lower scale would make us think the value is around $4V_{AC}$!

The other possibility is reading the indication on the correct scale but multiplying it with a wrong "range" factor. For instance, we correctly read "1.0" on the "R" scale, but for some reason, absentmindedly, multiply it by 100 instead of 1,000 and get 100Ω instead of 1,000Ω.

INSTRUMENT LIMITATIONS

We've already mentioned a few limitations of every test instrument: the resolution of its scale or display and its overall accuracy. This applies not just to simple analog instruments with a moving coil meter but to all instruments, even digital ones, as we will see soon when discussing digital multimeters.

There are other limitations, which apply to specific instruments, uses, or environments. These are not always published for general purpose T&M instruments such as multimeters but are of interest to their use in harsh and marine environments, high altitudes and temperatures, industrial and similar demanding applications.

An IP (Ingress Protection) rating system has been devised for industrial applications. This applies not just to electrical enclosures, motors, and similar industrial apparatus but also to test instruments.

The complete description of the IP ratings and associated tests can be found in IEC (International Electromechanical Commission) Publication 529. A simplified summary table will suffice here for our purposes (right). After all, most of your audio measurements will be performed in clean, dust- and water-free environments.

	SOLIDS (1st figure)		WATER (2nd figure)
1	Protected against a solid object larger than 50mm (e.g. a hand)	1	Protected against vertically falling drops of water (limited ingress permitted)
2	Protected against a solid object larger than 12.5mm (e.g. a finger)	2	Protected against vertically falling drops of water when tilted up to 15 degrees from the vertical (limited ingress permitted)
3	Protected against a solid object larger than 2.5mm (e.g. a screwdriver)	3	Protected against sprays of water up to 60 degrees from the vertical (limited ingress permitted)
4	Protected against a solid object larger than 1mm (e.g. a wire)	4	Protected against splashes from all directions (limited ingress permitted)
5	Dust-protected, limited ingress of dust that will not interfere with operation of the equipment or 2-8 hours.	5	Protected against jets of water (limited ingress permitted)
6	Dust-proof, no ingress of dust at all for 2-8 hours.	6	Water from heavy seas or from powerful jets shall not enter the enclosure in harmful quantities.
		7	Protected against immersion in water for 30 minutes (depth of 15-100cm)
		8	Protected against immersion in water under pressure for long periods of time

ABOVE: The IP (Ingress Protection) ratings guide

Dynamic range

A cheap multimeter may be able to measure DC voltages between 1mV (0.001V) and 300V, while the measurement range of a more complex (and expensive) digital multimeter may extend from 0.1mV to 1,000V. We will talk more about the measurement range of various instruments throughout this book.

Frequency range

Humans, as do most animals, have optical and acoustic sensors that can (together with optical and other nerves and the brain itself as a processing unit) measure electromagnetic signals such as light (eyes) and mechanical signals such as changes in sound pressure (ears). These "instruments" can only detect waves within a specific limited frequency range. Between the two, which has a broader frequency range, the human eye or the ear?

The range of human hearing declines (narrows) with age and depends on many factors, but it generally extends down to about 20Hz and up to 20,000 Hz (10 kHz). This spans 20-200-2,000-20,000 or four decades, making the human ear a relatively wide-spectrum acoustic instrument that detects mechanical waves (vibrations).

In comparison, a typical human eye will only respond to electromagnetic wavelengths ("visible light") between 380 and 740 nm (nanometers, or 10^{-9} meters), which, when recalculated as frequency, corresponds to a very narrow 430–770 THz (TeraHerz, 1 THz = 10^{12} Hz or one trillion Hertz) frequency band, less than one octave (2:1 ratio of frequencies)!

Anything under ("infrared") or above that ("ultraviolet") is invisible to the human eye, which is thus a very high frequency narrow-band electromagnetic sensor, compared to a low frequency and wideband mechanical detector called ear!

Returning to T&M instruments, cheaper multimeters may only accurately measure AC signals up to 2 or 3 kHz, while the frequency range of elaborate DMMs (digital multimeters) may extend to 30 kHz or higher. Likewise, cheaper oscilloscopes can display waveforms of periodic signals with frequencies of up to 20Mhz, while the frequency range of more complex and expensive ones may extend to 400MHz or even higher.

Measuring the frequency range of AC voltmeters and multimeters

To make audio measurements using your multimeter as an AC voltmeter, its bandwidth must be wider than the bandwidth of a typical audio amplifier. Since the frequency range of many tube amps reaches 30, 40 or even 60 kHz on the upper side, many el-cheapo multimeters (and some expensive, famous brand ones!) will have a snowball's chance in hell to measure anything at such high (ultrasonic) frequencies.

This simple test will enable you to measure the upper-frequency limit of your voltmeter or multimeter with high accuracy. It uses a comparison method, comparing the meter's reading with the amplitude display on an oscilloscope. Modern oscilloscopes have a guaranteed accuracy of at least 20 or 40 Mhz, way above the audio range of measurements.

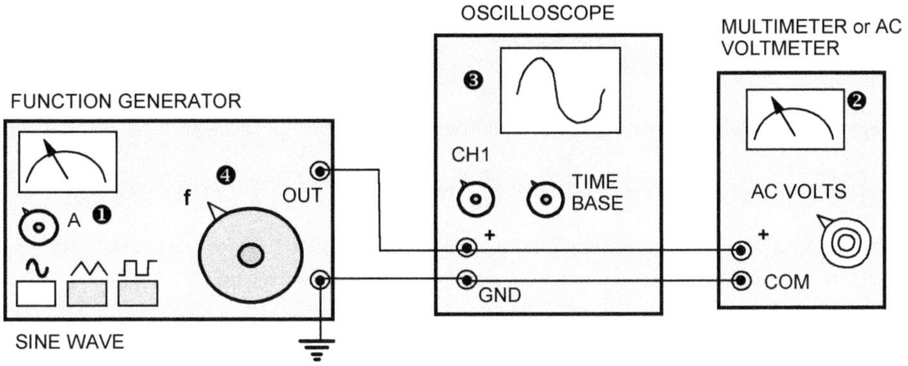

LEFT: Measuring the frequency range of an AC voltmeter or multimeter (analog or digital) by an oscilloscope comparison method

Start with the frequency of 1 kHz on the function generator (sine wave) and adjust its amplitude (1) to get the reading of 2.83 V_{RMS} on the AC voltmeter or 4 V_{PP} (peak-to-peak) on the scope (3). Increase the audio oscillator's or function generator's frequency (4) until the voltmeter's reading drops to 2.0 V_{RMS} (effective or RMS value) or 2.8 V_{PP} (if your meter is, for instance, a peak-to-peak type VTVM). The voltage dropped to approximately 71% of its value (0.707 to be exact) at 1 kHz, and by definition, that is f_U or the upper -3dB frequency of the voltmeter.

Why 2.83V or 4.00V? These voltages were chosen simply because of the easy calculation. 71% of 2.83V is 2.0V and 71% of 4.0 is 2.8! You can choose any voltages you like; just go up in frequency until the output voltage drops to 71% (-3 dB) of its original value at 1 kHz.

If you want to determine the lower -3dB frequency, reduce the frequency back to 1 kHz and recheck the indication on the voltmeter and the scope, it should return to the original 2.83 V_{RMS} indication on the voltmeter and 4.0 V_{PP} on the scope. Reduce the oscillator's test frequency until the amp's output voltage drops to 2.00 V_{RMS} or until the peak-to-peak voltage on the CRO drops to 2.8 V_{PP}. This is the lower -3 dB frequency of the voltmeter, its f_L.Ensure that the amplitude of the generator's output signal is constant and that the f_U of the oscilloscope is much higher than the f_U of the multimeter/voltmeter under test. Since most modern service type scopes have a bandwidth of at least 20 MHz, that is almost always the case.

Measuring the input resistance of an AC or DC meter or multimeter with another

Ideally, the input resistance of a voltmeter and many other instruments (oscilloscopes, etc.) would be infinite, so they would not affect the tested circuit. Ammeters, in contrast, would ideally have zero internal impedance (resistance). Real instruments vary widely in their internal resistance values, so it is very important to understand the impact of such a limitation or imperfection on the accuracy of your measurements. We will talk more about this "loading effect" soon. If unsure about the actual internal resistance of your AC or DC voltmeter or a multimeter, simply connect another multimeter on "Ohms" (use it as an ohmmeter) and measure such resistance.

Eico 680 is a vintage transistor and circuit tester. As a DC voltmeter, it only has two ranges, 5V and 50V, and no sane person would use it as such these days, but since it was the only analog non-electronic meter we had in our workshop, it's perfect to illustrate this experiment.

The left photo (next page) shows a DMM (used as an ohmmeter) measuring Eico's internal resistance on its 5V voltmeter range (around 100kΩ). The resistance on the 50V range was around 1MΩ. These are quite respectable (high) results for a simple passive voltmeter because the analog meter used in Eico 680 is quite sensitive, 0.05mA FSD (Full Scale Deflection). Remember, an ideal voltmeter would have infinite resistance. In other words, no current would flow through it so that no current would be diverted from the tested circuit, and so there would be no loading effect.

The sensitivity formula confirms these results. To get 0.05mA through Eico's meter (full-scale indication) with 5V coming in, the resistance of the voltmeter (internal resistances switched in) must be $R=5V/0.05*10^{-3}A = 100{,}000\Omega$! The photo on the right shows Eico 680's internal resistance when used as an ammeter (2.5Ω). Again, the resistance of an ideal ammeter would be zero.

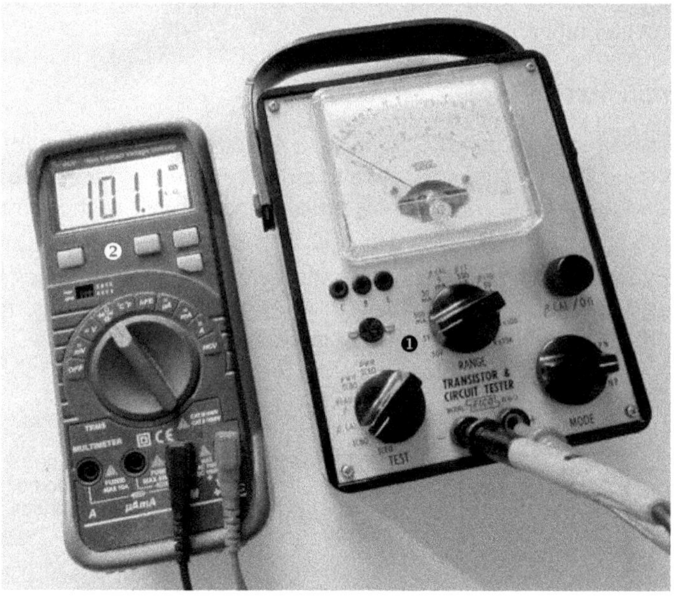

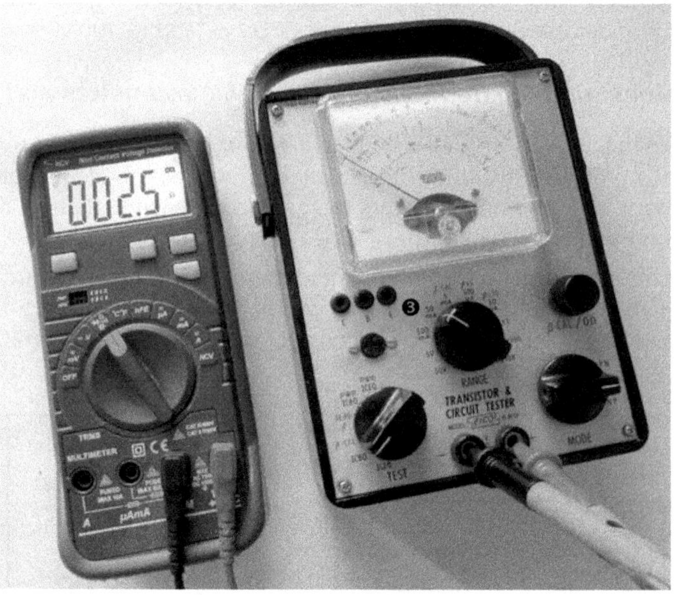

ABOVE LEFT: Measuring internal resistance of an analog meter on its 5V voltmeter range (1) using a digital multimeter (2)

ABOVE RIGHT: Measuring internal resistance of an analog meter on its 50mA ammeter range (3) using a digital multimeter

DECIBELS AND LOGARITHMIC SCALES

We have already mentioned the amazing wideband instrument called the human ear, whose dynamic frequency range is four decades wide (20 Hz to 20,000 Hz). To be able to cope with such a wide frequency range, instruments of this type (both "natural," such as the ear, and "man-made") respond not in a linear but a logarithmic manner to such stimuli.

Many multimeters and other instruments have linear and logarithmic scales, so it's of paramount importance to understand logarithms and logarithmic scales.

The proper and improper use of decibels

Decibel or "dB" is a unit for a power ratio or gain: **$P= 10*\log(P/P_0)$** [dB], where P is the power produced by a system or a device (an amplifier, for instance) and P0 is the referent power chosen for a particular purpose, comparison or measurement. Various referent power levels have been used over the decades, which created confusion among users. We will talk about that in a minute.

If we substitute the power of an amplifier $P=V^2/R$ into the formula above we get $P=10*\log(P_{OUT}/P_{IN}) = 10\log[(V_{OUT}^2/R_{OUT})/(V_{IN}^2/R_{IN})] = 10\log[(V_{OUT}/V_{IN})^2(R_{OUT}/R_{IN})] = 20\log(V_{OUT}/V_{IN}) + 10\log(R_{IN}/R_{OUT})$ [dB]

If, and only if $R_{OUT}=R_{IN}$, **$P = 20\log(V_{OUT}/V_{IN})$**.

We will adopt this simplification since the last factor, $10\log(R_{IN}/R_{OUT})$, almost always gets omitted (it's not even mentioned in nine out of ten books!). Although improper from the strict dB definition point-of-view, such usage of dB works, providing it is used consistently, meaning as long as everyone (mis)uses it in the same way! Again, it works because a dB is a relative unit that compares two levels, either input vs. output, or input or output versus a third or referent level.

CALCULATION

A tube amplifier has an input impedance of 47kΩ and supplies an 8Ω load. Its voltage gain is 15 times (A=15). We will assume an ideal case, that its output impedance R_{OUT} is zero. What is the voltage gain of this amp in dB?

These days most people would simply calculate $P = 20\log(V_{OUT}/V_{IN}) = 20\log 15 = 23.5$ dB! That is certainly convenient but, strictly speaking, it is not correct because this simplification is only correct when $R_{IN}=R_L$, and in our case, the input and output resistances are very different!

The correct answer is $P=20\log(V_{OUT}/V_{IN})+10\log(R_{IN}/R_L)= 20\log 15+ 10\log(47{,}000/8) = 23.5+37.7 = 61.2$dB

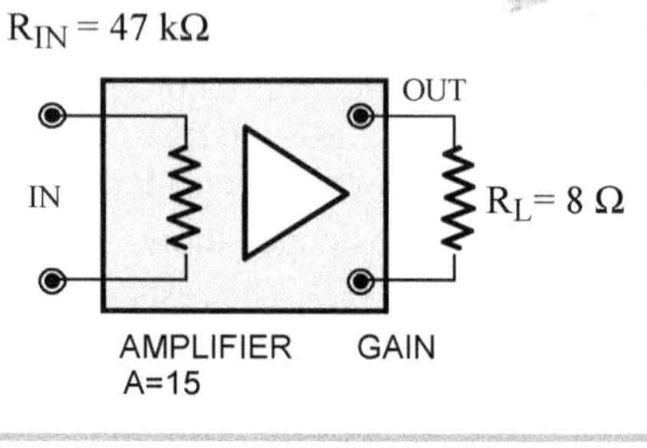

Voltage, current and power amplification through the audio chain

For a 2A3 triode amplifier supplying $5V_{RMS}$ to an 8Ω load, the output power is $P_{OUT} = V^2/R = 25/8 = 3.125W$. The AC (signal) current through 8Ω load is $I_L = V_L/R = 5/8 = 0.625A$. For a higher power tube amp supplying $40V_{RMS}$ to an 8Ω load, the output power is $P_{OUT} = V^2/R = 40^2/8 = 200W$, while the current through the load is $I_L=V_L/R = 40/8 = 5A$.

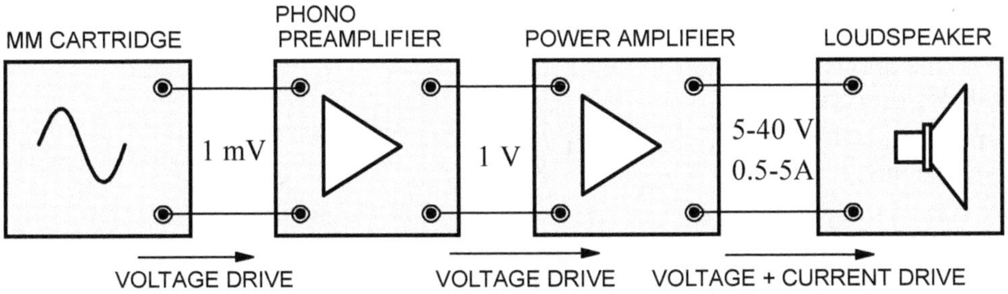

LEFT: Typical values of voltage and current signals in an audio chain

All amplification devices upstream of the power amplifier provide only voltage amplification, while the power amplifier amplifies both the voltage and current.

We can now proceed with analyzing a multistage amplification chain (illustrated below). We have an audiophile system comprising of a signal source, a turntable with a moving coil cartridge (not shown), an MC step-up transformer, a MM (moving magnet) phono preamplifier, and a power amplifier. The amplification factors are given for each component, and all signals' RMS values are specified. What is the overall amplification factor in dB?

There are two ways to do this. You can calculate the gain of each stage or device in dB and then simply add them up: A_{TOTAL} [dB] $=A_1+A_2+A_3 = 103.5$ dB , or, you can calculate the overall gain of the system $A_{TOTAL}=A_1A_2A_3 = 15{,}000$ and then convert it into dB: A_{TOTAL} [dB] $= 20\log A_{TOTAL} = 20*\log150{,}000 = 20*5.176 = 103.5$ dB.

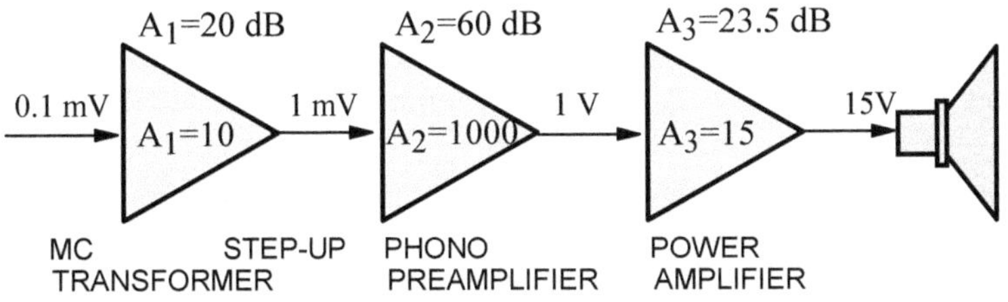

$A_{TOTAL}=A_1A_2A_3 = 150{,}000$

A_{TOTAL} [dB]$=A_1+A_2+A_3 = 103.5$ dB

Different "types" of decibels and dB scales

There are two dB units, dBm and dBV. dBm (also called dBmW) has a communication origin, referenced to the 1 mW of power that a sine source of 0.775 V_{RMS} would dissipate on a 600Ω resistive load. dBV is the unit we have been talking about so far, used for voltage ratios such as amplification and attenuation factors. It is referenced to the level of 1 Volt.

See how the two dB scales have a different zero on this analog meter face, from a harmonic distortion analyzer (which can also be used as an electronic voltmeter): 0 dBm is at 0.775 V_{RMS} (7.75 on the lower scale 0-11) and 0 dBV is at 1.0 V_{RMS} (10 on the same scale).

The referent power level for dBm in instrumentation use is $P_0=0.001$ W, so P [dBm] = 10log(P/0.001).

To convert dBV to dBm we need the same referent impedance (R), in this case, 600Ω. Remember, $P=V^2/R$!

Taking the log of both sides of that equation log(P) = 2*log(V) - log(R) or dBm - 30 = dBV - 10log(R).

Finally, dBV= dBm - 30 + 10log(600) = dBm -30 +27.78 so **dBV= dBm-2.218.**

From the scale we can see that approx. 2.2 dBm = 0dBV, so our calculation is correct.

0 dBm is not the same as 0 dBV!

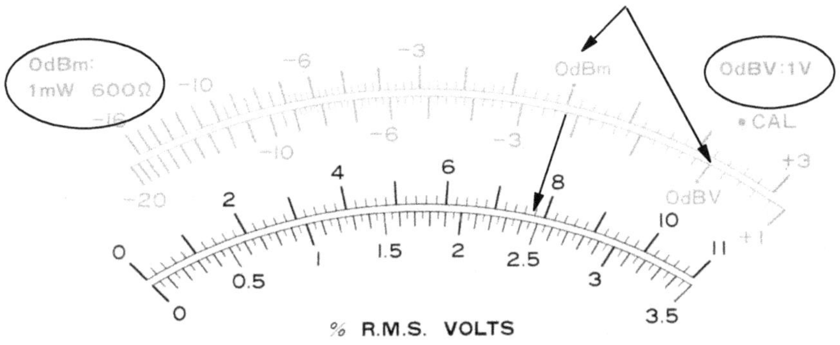

dBV-dBM CONVERSION

dBV= dBm-2.218

The logarithmic power curve

Let's return to the half-power a.k.a. -3dB frequencies. To understand the logarithmic nature of the amplifier's power drop at those frequency limits we need to study the logarithmic power curve, the logarithmic equation in a visual form

$P = 10\log(P_2/P_1)$ [dB]

The ratio P_2/P_1 is on the horizontal axis (the abscissa) and the decibels of that ratio are on the vertical axis (the ordinate).

When $P_1=P_2$ their ratio is 1, and log1=0 (because $10^0 = 1$). Any number to the power of zero equals one (point A).

As P_2 becomes larger than P_1, the log curve is in the positive territory, and as P_2/P_1 drops under 1, the curve goes into the negative values.

In point E, P_2 is twice as large as P_1, which is +3dB. In point D, P_2 is half as large as P_1, which is -3dB. Now you understand why -3dB frequencies are called half-power frequencies. In point C, $P_2/P_1=10$, and log10=1, so P_2/P_1=10dB. In point B, $P_2/P_1 = 0.1$ so P_2/P_1= -10 dB.

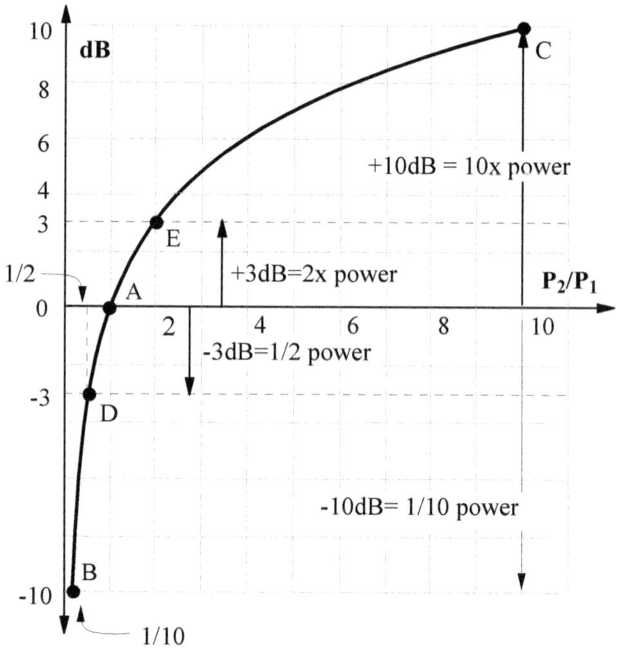

The "dB graph"

An audio (input, output, or interstage) transformer is a bandpass filter, as is a whole audio amplifier. An RL-type first-order model can approximate the low-frequency behavior of these devices. First-order systems have a roll-off slope of 20dB per decade or 6dB per octave. A decade is 10x lower or higher frequency (a ratio of 10:1), an octave is a ratio of 2:1 (double or half the frequency).

The "dB graph" (below right) shows the low-frequency response of an audio transformer or a whole amplifier in decibels. X-axis is linear, expressed as a ratio of frequency and -3dB frequency f_L. Y-axis is attenuation in dB, relative to A_0 (midband level), a log scale.

Very few manufacturers specify their amps' and transformers' frequency range as -1dB, since it looks bad, others use -2dB frequencies, but most stick to "standard" -3dB figures (they look the best). -2dB points used to be more common in the 1960s, much less so now, with -3dB being the most common spec today. This graph can help you convert these disparate specs to the same level so you can compare them!

The high-frequency behavior of these devices can also be approximated by the RL-type first-order model, but only in some cases. Often parasitic capacitances also feature prominently in the model, so now we have three elements or an RLC circuit with its resonant frequency and a corresponding rise and peak in the amplitude - versus - frequency characteristic.

RIGHT: By using multiplication factors, this universal dB graph enables you to convert and compare attenuation figures expressed in different dB levels

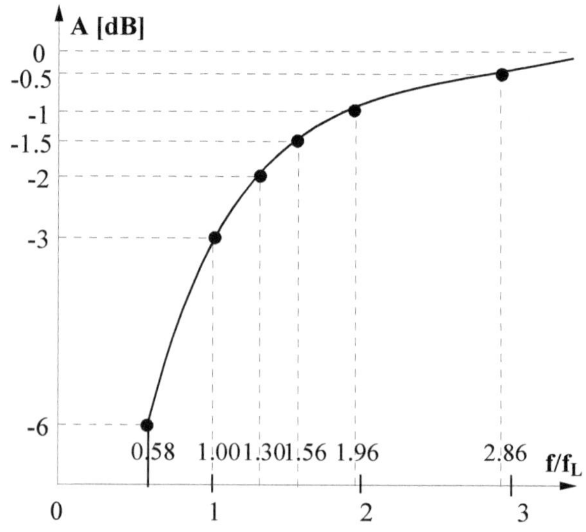

LOW FREQUENCY ATTENUATION

$A_{LF}= 20*\log(V_{IN}/V_{OUT})=10*\log[(1 + (f_L/f)^2]$

For RL high-pass filter model, attenuation at low frequencies in dB is $A_{LF} = 20*\log\sqrt{[(1 + (f_L/f)^2]} = 10*\log[(1 + (f_L/f)^2]$

CALCULATION

A vintage Australian-made Ferguson "Medium Fidelity" push-pull transformer offers three choices of primary impedances (7, 8 and 10 kΩ). It has the following specs: P=15 W, 40-30,000 Hz (-2dB), L_P=50H (primary inductance), and L_L=29 mH (leakage inductance).

"Medium Fidelity" most likely refers to a relatively high f_L of 40 Hz and a relatively low f_U of 30kHz. 40Hz is the -2 dB frequency, which is $1.3*f_{3dB}$ (from the "dB graph), meaning that the -3 dB frequency would be $f_{L\text{-}3dB}$ = 40/1.3= 30.8 Hz. The upper -3dB frequency $f_{U\text{-}3dB}$ can be estimated in the similar fashion, $f_{U\text{-}3dB}$ = 1.3*30,000 = 39 kHz.

S/N (Signal-to-Noise) ratio

An important parameter of any amplifier, preamplifier, or in general, any electronic system (including electronic measuring instruments!) is called signal-to-noise ratio, and its very name indicates that it is a relative indicator, 20 times the logarithm of a ratio of the signal voltage V_S. and noise voltage V_N at its output: **SNR = 20log (V_S/V_N) [dB]**.

The lower the noise voltage V_N, the higher the S/N ratio SNR and the quieter the amplifier.

Say we have a 300B amplifier and a load of 8Ω. Without any signal at its input (inputs short circuited to ground), we take a true-RMS multimeter and measure 1 mV_{AC} voltage on the speaker terminals. This is the hum or noise that the amp is generating itself. We want to specify its S/N ratio referenced to a level of 1 Watt at the output. How do we do it?

1 Watt of power dissipated on 8Ω load means the voltage on the load is $V_S = \sqrt{(PR)} = \sqrt{(1*8)} = 2.83$ V. Substituting 2.83 for V_S and 1 mV for V_N into the formula we get SNR = 20log (2.83/0.001) = 20*log2,830 = 20*3.452 = 69 dB.

Audio equipment manufacturers use every method at their disposal to make their equipment look better, and their test figures seem more impressive. One trick is with various ways of specifying the S/N ratios. They can declare S/N ratios at a certain "standard" power level, such as 1 Watt, or at the amp's rated power output. Let's look at a typical example.

WHICH AMPLIFIER HAS A HIGHER S/N RATIO?

ANALYSIS

The Fizzy Fuzz amp's S/N ratio is declared as 83 dB at the rated power of 100 Watts on its 4Ω output, while the Dindle Doondle amp's (also rated at 100 Watts) S/N ratio is only 69 dB at 1Watt and 8Ω impedance. Which amp is quieter (has less hum)?

Most audiophiles would conclude that 83dB is better than 69dB, but is that really the case?

We cannot compare the figures directly, since they are referenced to two different power levels and two different output load impedances, so we must convert one of the figures into the other's reference/impedance level, or we need to convert both S/N figures into volts of noise. If we express V_N from the S/N formula we get V_N = V_S/INVLOG(SNR/20)

For Dindle Doondle amp V_N = V_S/INVLOG(SNR/20) = 2.83/INVLOG(69/20) = 2.83/2,818 = 1.0 mV

For the Fizzy Fuzz amp we have to first find the signal voltage that will produce 100 Watts on 4Ω and that is $V_S = \sqrt{(PR)} = \sqrt{(100*4)} = 20$V. Now we can calculate the noise voltage as V_N = V_S/INVLOG(SNR/20) = 20/INVLOG(83/20) = 20/14,125 = 1.42 mV!

Despite having a more impressive- looking S/N figure (83 versus 69 dB), the hum voltage on the Fizzy Fuzz's output is higher. It is in the manufacturer's interest to declare the S/N figures referenced to the rated power level and not to 1 Watt because the S/N figures then look more impressive to the untrained eye!

LIN (linear) and LOG (logarithmic) scales

Four types of X-Y scales are used in electronics. The independent variable (X-axis) and the dependent variable (Y-axis) can be presented using a linear (LIN) or logarithmic (LOG) scale, resulting in four possibilities: LIN-LIN, LIN-LOG, LOG-LIN, and LOG-LOG. LOG scale is used when the range of the variable depicted is wide, as in audio frequency range, which is studied from 1Hz to 100,000 Hz, or 5 decades. Each decade is a factor of 10:1.

The example below illustrates the amplifier's gain, where a linear horizontal scale would be impractical since it would be impossible to read the gain from the curve at the frequency extremes, where most of our interest lies.

In some books, you will see the frequency curve of a 1st order system (a simple RC filter or an amplifier stage) with a straight "end" and in others with a curved end. The straight end means the vertical scale is logarithmic (LOG), and it thus has no end; there is no vertical zero "0" marked! A curved end means the vertical scale is linear (LIN) with a definite end, zero at the horizontal axis!

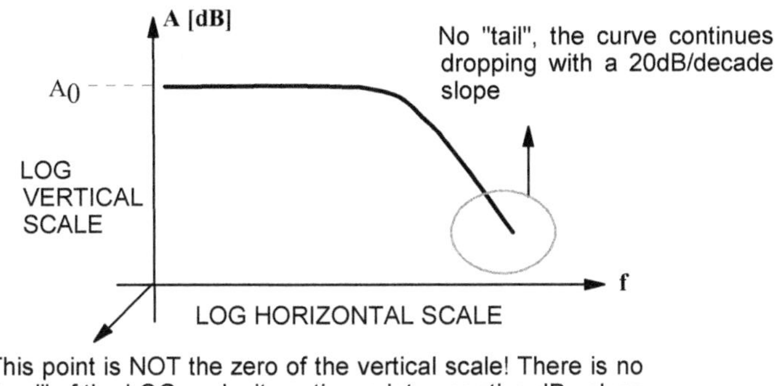

This point is NOT the zero of the vertical scale! There is no "end" of the LOG scale, it continues into negative dB values of A (A lower than 1 but larger than 0)

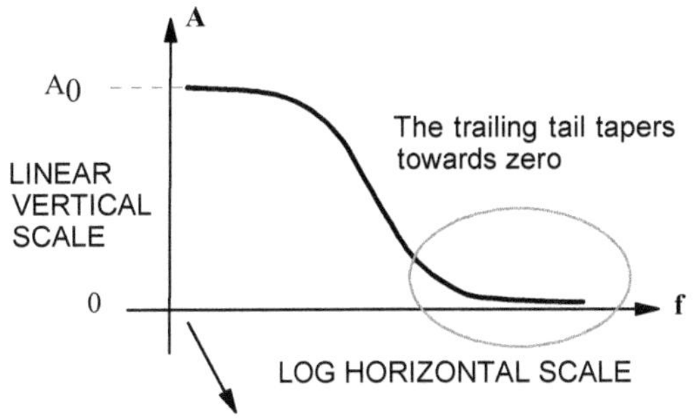

There is a definite end at ZERO, gain A cannot be lower than zero!

UNDERSTANDING T&M INSTRUMENTS' CIRCUIT DIAGRAMS

In the 1950s and 60s, we would have to devote a whole chapter to this important topic. The instruments were simpler and often of kit variety. The user would buy the kit and put it together per the supplied instructions. However, manufacturers used their own drawing layouts and symbols in those days, and there was very little uniformity. Luckily, there is generally no need to read circuit diagrams of modern instruments. Many are entirely digital and software-based anyway, so the following discussion is for those who feel inclined towards vintage test gear.

How to read switching arrangements on T&M instruments' circuit diagrams

Despite clear standards, at least on the national if not the international level, different circuit diagram symbols have been used by amplifier and test equipment manufacturers. While understanding various versions of symbols for resistors and capacitors, for instance, isn't difficult, switches and switching arrangements, so crucial in T&M instruments, still cause confusion. So, let's spend some time clarifying at least a few issues through a couple of examples.

The partial diagram on the right is part of the Sprague Tel-Ohmike TO-5 Capacitor Analyzer circuit. The bank of eight P/B (push-button) switches (called S1, treated as one switch), marked A to H, is the same as on the drawing. Only switches C to F are illustrated here. All switches are 2-position type, i.e., they can either be OUT or IN. Also, some switches could be 1-pole, others can be of a 2-pole or 3-pole variety. A "pole" is a separate set of contacts not connected to each other on the actual switch, although some may be connected through the external circuit. Also, just because a switch is "OUT" does not mean that it's switched "OFF" or that its contacts are disconnected. Using Switch "D" as an example, we can identify four contacts or switch "blades."

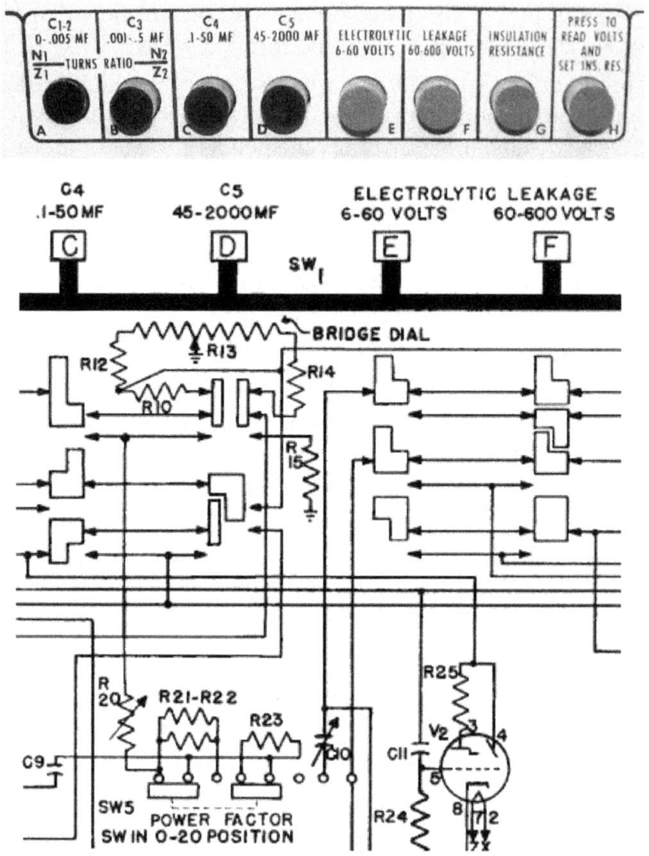

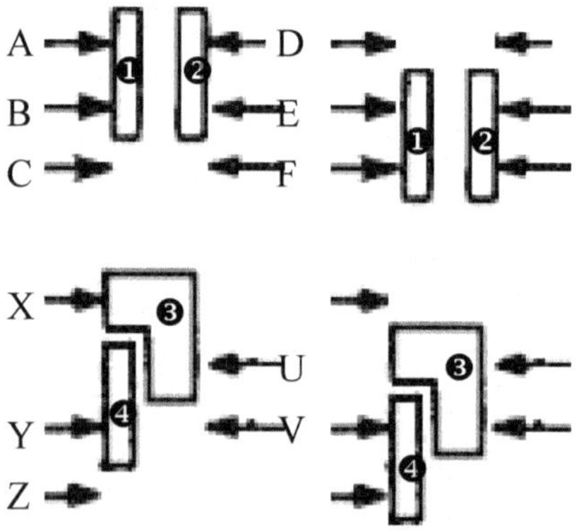

FAR LEFT: Pushbutton "D" OUT

LEFT: Pushbutton "D" pressed (IN)

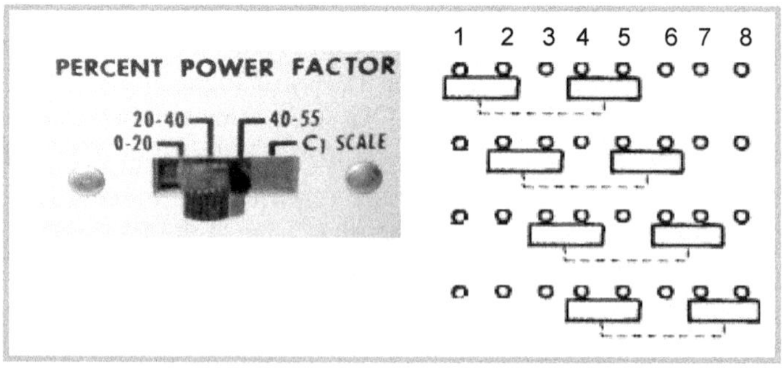

They are drawn as various shapes as they are in real life. Usually, four sets of contacts mean that the switch is a 4-pole type. Contacts #1 and #2 in the OFF or OUT position (left) connect points (lugs) A & B and D & E. Contact #3 bridges lug X to lug U, while contact #4, although making contact with lug Y, doesn't do anything, meaning it does not connect that lug Y to anything else.

Once the P/B is pressed, blade #1 disconnects lug A from lug B and connects B to C; ditto for contact #2, which now connects E and F, while lug D is disconnected. Contact #3 now bridges lugs U and V, disconnecting lug X, and contact #4 connects lugs Y and Z. If you feel you need some practice, study the other three switches and figure out their operation.

Moving to SW5 on the same drawing, the 4-position sliding switch ("Percent power factor") has two interlocked sets of contacts, meaning the two contacts move together at all times. With lugs numbered from 1 to 8, in the first (leftmost) position, contact A connects lugs 1 & 2, while contact B connects lugs 4 & 5. In the next position marked "(20-40"), both on the front panel and on the drawing), contact A now connects lugs 2 & 3, while contact B connects lugs 5 & 6.

Similar mechanical interlocking can be seen on the next page; switch S7 has two poles or two separate switches S7a and S7b, each with four positions. The interlocking is indicated with a thin dotted double line.

Switch S1 is self-explanatory; only one of its 11 lugs is connected to its "output" or central contact (switch shown in position 1). This arrangement is widespread in tube testers as a filament (heater) voltage selector.

Switch S2 connects 9 out of its 10 lugs together at all times, effectively isolating lugs 1 to 9 one by one from the rest, another relatively common type in certain tube testers.

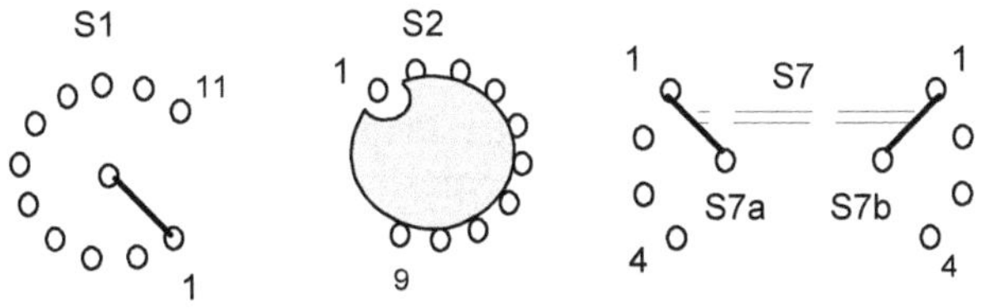

Tube tester switching example

At first glance, this switching matrix from Triplett 3423 tube tester seems a complete mess, so the first step in such cases is to relate the schematic to the actual control panel of the instrument. Here we see three rotary switches, D-GRID, E-SCREEN, and F-PLATE, each selecting only one tube pin (pin 1 on all sockets, pin 2 on all sockets, etc.) plus X or Y positions.

Nothing is connected in the position marked "0", so the grid, screen, and plate voltages are not brought out to any pins.

The drawing shows all three switches in position "1" (pin 1).

In position "2," the point "0" would be connected to pin 2, etc. Switch "D" or "grid" switch is different. Being a 2-pole switch, its lower pole S18-1 is identical to switches E and F, but its upper pole shows the same arrangement as the switch S2 (top of the page).

In position "1" as drawn, contact 1 is disconnected, and all other tube pins (2, 3, ... X, Y) are shorted together. Thus, it is an isolation switch used for "SHORTS" testing of a tube, so shorts between pin 1 and any of the other pins are detected, then shorts between pin 2 and all other pins (including pin 1 again) in position "2" and so on.

The final example is the bank of eleven 3-position toggle switches, marked 1-10, X, and Y.

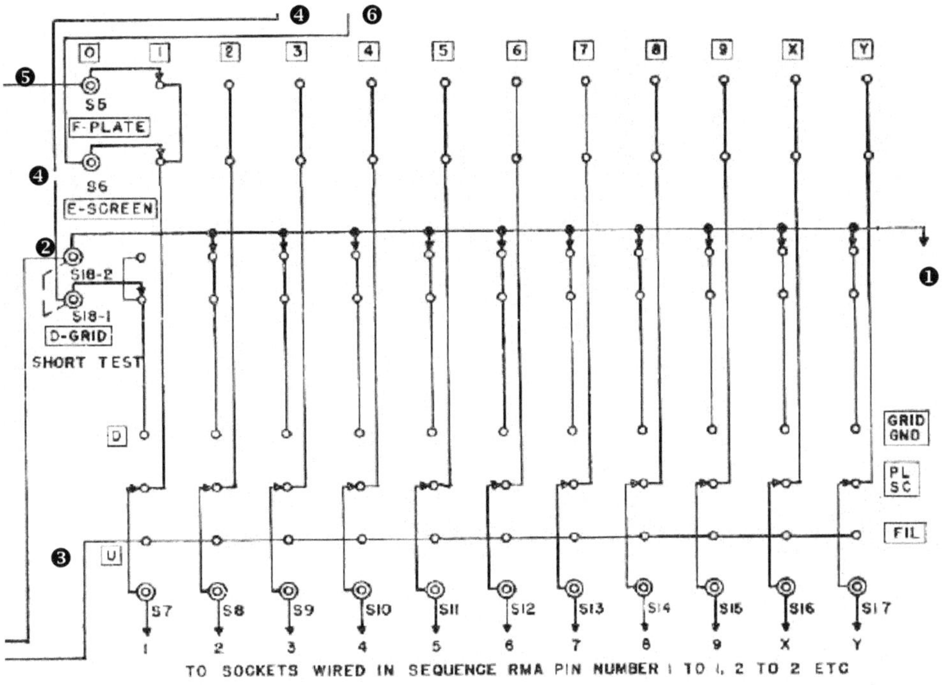

ABOVE: Partial circuit diagram of Triplett 3423 tube tester, showing the pin switching arrangement

BELOW: The central section of Triplett 3423 tube tester's control panel, with 11 lever switches (nine tube pin switches 1-9, plus X- and Y-switches), the Value-Gas switch, G-switch (main test circuit selector) and three switches that allocate pins to control grid (D), screen grid (E) and plate (F)

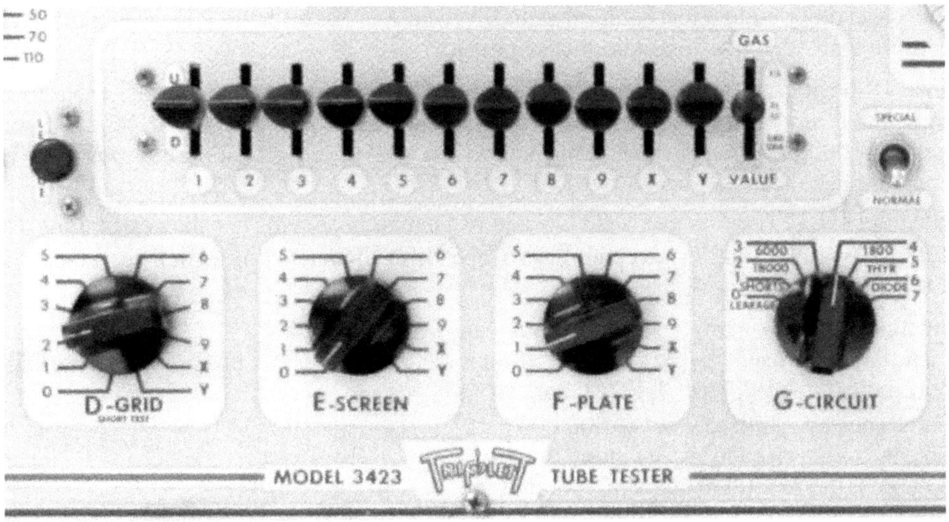

On the schematic diagram, they are marked S7 to S17. In the central position (P-SC, meaning plate and screen), as seen on the photo and the diagram, the pins from all tester's sockets are connected to the corresponding pins on the three already mentioned switches D-GRID, E-SCREEN, and F-PLATE.

In the "U" or UP position, any tube pin can be connected to the "GRID GND" bus (1), which is always connected to lug "0" of the D-GRID switch (2) and to the common side of the filament voltage supply (power transformer secondary). Grid DC bias and 4 kHz test signal come in through connection (4), plate voltage through connection (5), and screen voltage through (6). In the "D" or DOWN position, any tube pin can be connected to the "FIL" or filament bus (3), the filament voltage supply tap on the power transformer's secondary winding, as selected by the A-FILAMENT switch (not shown).

MISCELLANEOUS JIGS, ADAPTERS AND TOOLS FOR AUDIO TESTS

Tube socket test adapters

These test sockets or similar adapters make it quick & easy to measure DC voltages and AC signals on tube pins without opening the amp up and exposing yourself to dangerously high voltages and hot tubes. Each pin on the tube socket is brought out to its own test point, so measuring the voltage between pins or each pin and GND is a breeze. Simply plug the male adapter into the tube socket and then plug the tube into the jig; power thc amp up, and you are ready to measure!

You can make such a jig for any socket you need, 4-pin, miniature 7-pin, 9-pin Magnoval, RimLock, Loctal, etc. Instead of miniature pins, you can use standard 4mm plugs and sockets, but then your plastic housings need to be bigger.

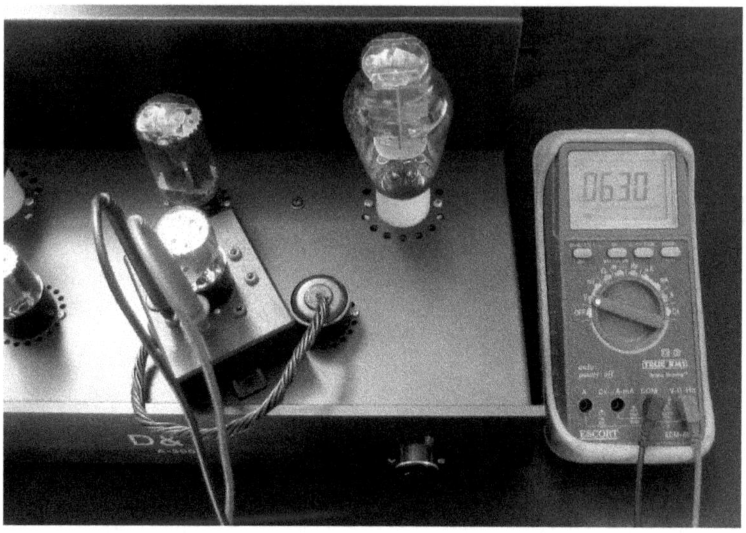

ABOVE RIGHT: These DIY test sockets are the most helpful tool of all! Build one for all commonly used socket types (7-pin mini, Noval, Octal, Magnoval, RimLock, etc.) Pin 1 is marked with a plastic sticker on the Noval (miniature 9-pin) jig.

ABOVE LEFT: The octal test jig plugged into a 300B tube amp, heater voltage on pins 7 and 8 of a 6SN7 preamp tube being measured.

R-, C- and RC-substitution boxes

You can fix, build, and ear-tune amplifiers and audio gear without substitution boxes by swapping fixed resistors and capacitors, but these small jigs make it all so much faster and easier. Laboratory decade resistors and capacitors (meaning very precise, i.e., with low tolerance values) can be pretty expensive. However, there are still vintage units by Heathkit, Eico, and other budget test equipment makers of the 1950s and 60s. Their level of precision (usually 1%, sometimes 0.5%) is more than adequate for audio measurements.

Two 18-position rotary switches to select one of the common resistor values (15Ω to 10k on the left and 15k to 10M on the right bank), and a changeover switch to select one or the other ("Lo" or "Hi"), and you have Eico's RTMA resistance box, model 1100. It is often cheaper to buy a used one than to buy all the parts and make it yourself. Model IN-12 is the Heathkit equivalent.

Heathkit IN-17 (next page) is a 6-decade resistance box, with resistances ranging from x1 (ohms) to x100,000 ohms or 100kΩ. Heathkit IN-27 (next page) is a 3-decade capacitance box with three 11-position rotary switches, marked x.01mF, x.001mF and x100mmF.

Heathkit and many other American manufacturers used "m" improperly. It does not indicate "milli" or 1/1,000 part of a Farad, but "μ" ("micro") or 1/1,000,000 part! So, the first decade is in steps of 10nF ("nano"), the second in steps of 1nF, and the smallest values are in steps of 100pF ("pico") or 0.1nF!

Except for the two coupling capacitors at the output of phase inverters in push-pull amps, where you'd need two of these boxes, one box is enough for voicing tests on guitar amps or one channel of a stereo hi-fi amp. Change the coupling capacitor's value from, say, 10nF to 22n to 47n and listen to the changes in the amp's tone.

ABOVE: Eico RTMA resistance box, model 1100

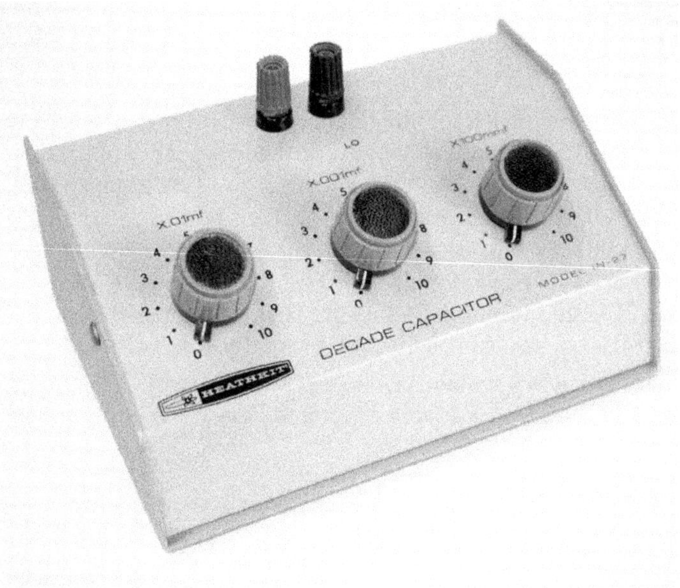

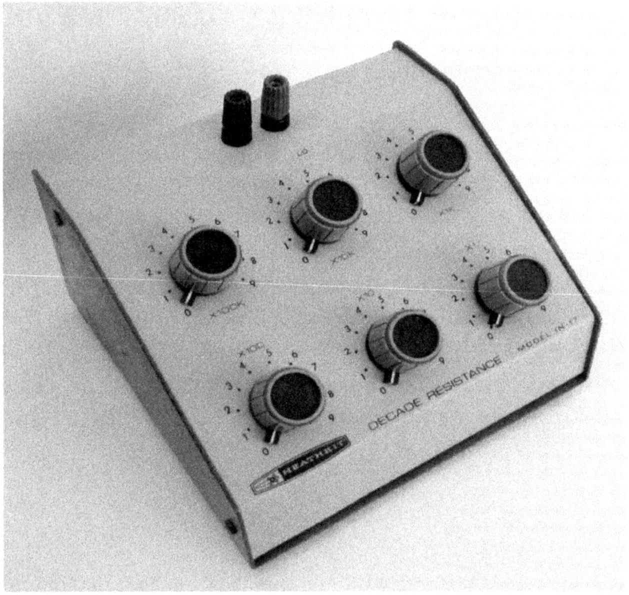

Dummy loads for amplifier testing

High power resistors used instead of loudspeakers for amp testing convert audio power into wasted heat, which is why they are called "dummy" loads. One such made-in-China wire-wound resistor (100 Watts, 8Ω) is pictured below.

Alternatively, resistors of lower power rating can be connected in any combination (series, parallel, or series + parallel) to get the resistance and power rating you need. 7 Watt and 10 Watt wire-wound resistors are cheap and commonly available. Two of many options are shown below in the DIY PROJECT frame. With some values, you will not end up with precisely 8Ω, but that is fine as long as you know the resistance of your dummy load and can alter your calculations accordingly.

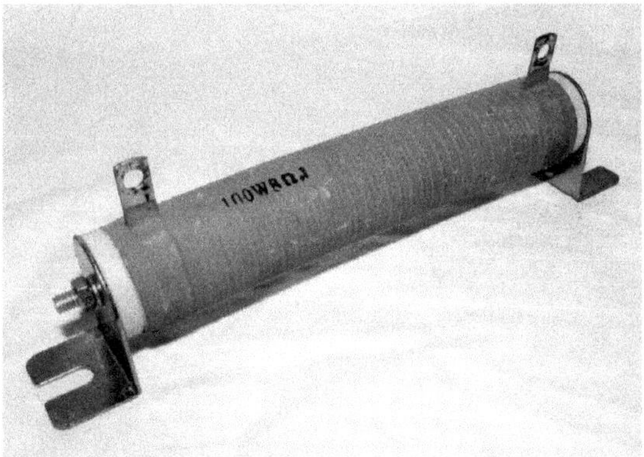

MAKING A DUMMY LOAD FOR AMPLIFIER TESTING

DIY PROJECT

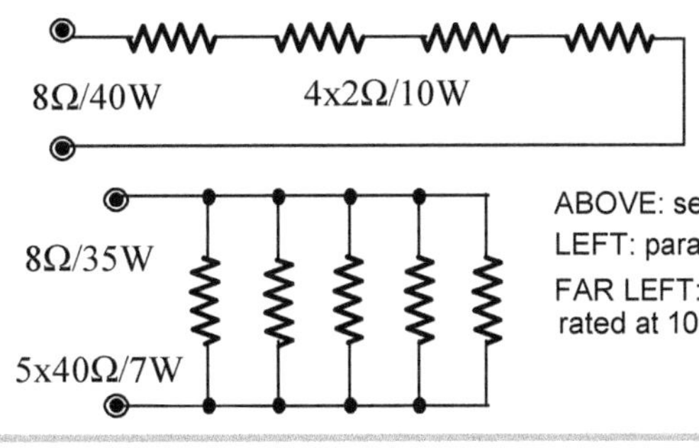

ABOVE: series arrangement
LEFT: parallel connection
FAR LEFT: 8 Ω wirewound resistor rated at 100 Watts

BOB'S BLACK BOX, A SIMPLE POWER METER (WATTMETER)

DIY PROJECT

Bob's Black Box (or BBB for short) is a simple wattmeter, or, more precisely, a VA-meter (apparent power). The load current flows through the 10Ω resistor. The voltage drop measured by an AC voltmeter between two test points, TP1 and TP2, is proportional to this current. The load current is I_L=V/10, and by multiplying this current by the mains voltage, we get the power draw in VA (apparent power).

The diagram shows an Australian power outlet, but any international type can obviously be used.

There is a fuse in the active (line) feed for safety reasons, the test points are in the neutral line, and a double-pole switch is used for both line and neutral connections.

The neon indicator glows when the mains voltage is present, and the on-off switch is closed.

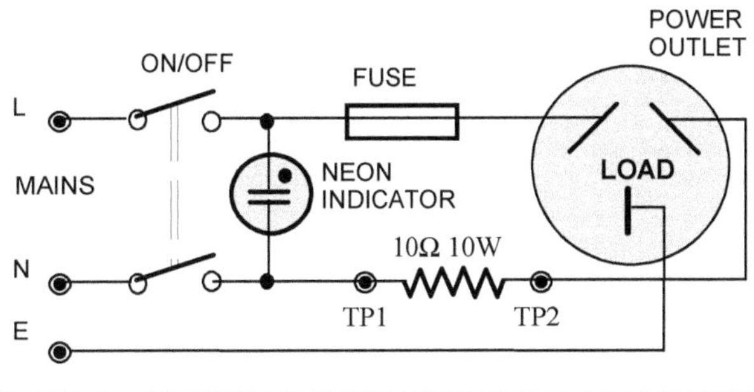

REPAIRING, RESTORING & CALIBRATING VINTAGE TEST INSTRUMENTS

Replacing leaky film capacitors

Even after 50+ years of service, some vintage film capacitors have low D-factor (0.004, for instance) and don't need replacing unless the application requires absolutely no leakage. Other brands' film capacitors are so leaky (D=0.526!) that you don't have to waste your precious time testing them one by one. Many instruments have 6, 10, or even more film capacitors, all of the same make, so if one tests bad (high leakage), replace all of them.

Capacitor testers and other instruments that use high DC test voltages will have leaky film capacitors of $900V_{DC}$ - $1{,}200V_{DC}$ rating. Instead of messing around with two or three 400 or 450V capacitors in series (plus equalizing resistors in parallel with each!), mains-rated film capacitors such as the two X2 rated examples below are cheap replacement option, as are small motor start caps. These two types are rated in AC volts, but you can easily convert that to DC volts.

An AC voltage is a sine wave with a peak-to-peak value which is its RMS value (AC rating) multiplied by 2.82. So, a $450V_{AC}$ motor start capacitor is suitable for at least 1,270 V_{DC}, and a $275V_{AC}$ capacitor pictured below can withstand at least 775 V_{DC}!

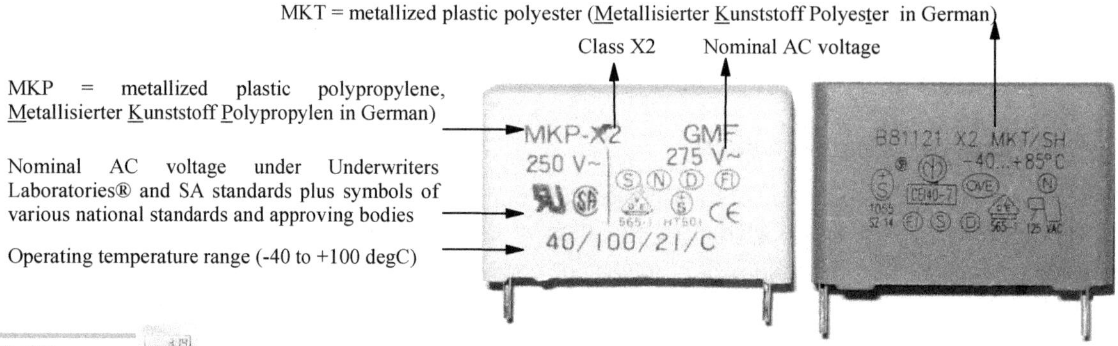

DC VOLTAGE RATING OF MOTOR-START & OTHER AC-RATED FILM CAPACITORS

$V_{DC} = 2.82 * V_{AC}$

ABOVE: Two examples of X2-rated (for mains filtering applications) film capacitors, in metallized polypropylene film and metallized polyester film technologies.

Replacing selenium rectifiers

Selenium rectifiers (1) age, regardless of their use. So, even NOS (New Old Stock) rectifiers sitting on a shelf for 40-50 years should not be used as replacements. Gradually, their forward voltage drop increases, and the reverse leakage current increases. Their DC rating reduces as they age; they run hotter and often even catch fire. The smoke is highly toxic.

The bigger the size of the plates, the larger the current capacity of the selenium rectifier. The more plates in a stack, the higher the rated voltage of the selenium stack. Each plate can take about 30V, so by counting the number of plates, you can determine the voltage rating of the selenium diode.

When you replace selenium rectifiers with silicon diodes, the DC voltage will increase by 5 to 20 Volts. The forward voltage drop of silicon diodes is much lower, only around 0.6V.

This increase may be useful or may present a problem, depending on the circuit and the application. If such a voltage rise is unwanted, simply add a suitably sized (in terms of power rating and resistance) resistor in series with the silicon diode(s).

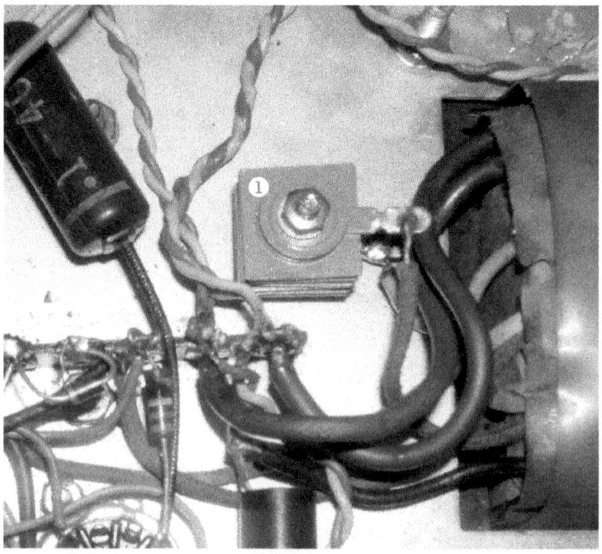

RIGHT: Selenium rectifiers should be pulled out and replaced without even testing them, and so should 50+ year old electrolytic and even film capacitors.

Replacing multi-section electrolytic capacitors

Considering their low capacity, modern multi- section can-type elcos (mounted on top of the chassis) are costly, $50-$70 on eBay! Their only advantage is the preservation of the original amplifier looks.

The same aim (original looks) can be achieved by retaining the faulty multi cap (disconnected from the rest of the circuit) and adding a few single radial or axial caps (at less than $1 each) under the chassis.

That's what we did on this vintage power supply (right). The three new elcos were mounted on the added terminal strip, and the original multi-section elco (2) was left in place.

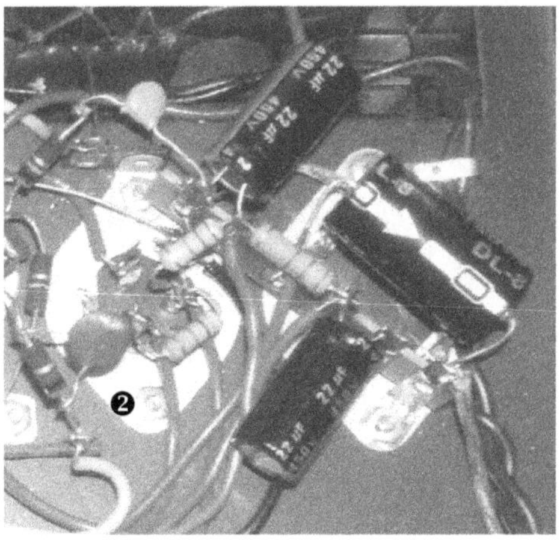

The art of salvage: how to source quality parts

If you plan to keep fixing, modifying, or building audio equipment or test instruments, the best long-term investment is to build up a large stash of various used parts and components. Get as many vintage TVs, radios, tape recorders, oscilloscopes, and any other piece of electronic gear you can find. Sunday markets, garage and deceased estate sales, swap meets are just some sources, not to mention online sources and industrial equipment auctions.

A large stash of quality parts will save you not just many hours you would otherwise spend online or going through electronic shops looking for a specific component, but also hundreds, even thousands of dollars. Compared to their trade prices (what manufacturers pay), modern components sold in electronic shops to DIY enthusiasts are grossly overpriced. Many are of poor quality, bought cheaply, and sold at huge markups (hundreds and even thousands of percent).

When you look at a Tektronix tube oscilloscope or other quality tube test instrument, you can see it was designed and built properly. In comparison, most vintage tube amplifiers from the same period look and feel cheaply made (and they were). The components used in test instruments were first class.

Just look at all tubes, trimmer capacitors, trimmer resistors, carbon composition resistors, precision resistors, tube sockets, potentiometers, extension shafts, porcelain soldering strips, and quality hookup wire inside this tube oscilloscope (pictured below)!

Most of this gear will be dusty & dirty, so clean them up with compressed air. Small compressors are available in auto-parts shops. Make sure you wear a good quality dust mask; the dust may not be just full of bacteria and viruses but also may contain toxic particles.

Alternatively, use a water hose. Don't soak the transformers; a gentle hosing of the surface dust is all that is needed. Then, dry them in the sunshine or a very slow oven, 50-70 degC.

Don't just snap the component leads off with your cutters; try to unsolder them first. That way, the reused components will have slightly longer leads which may prove critical in your project. There is nothing worse than when the leads of a particular resistor or capacitor are a few millimeters short, so you have to spend ten or more minutes searching for another one in your stash.

ABOVE: The internals of a Tektronix tube oscilloscope, a treasure trove of the highest caliber components. Even after 40-50 years, most are of superior quality to currently produced garbage.

Unfortunately, in some vintage gear the component leads were first wrapped around a terminal or tube socket lug and then soldered, so removing them is a tedious and frustrating task.

Buy a large flat tip soldering iron (100 Watts) for this task; most regulated temperature 40W soldering irons aren't powerful or hot enough for rapid desoldering.

Not all vintage parts are worth salvaging. None of the electrolytic capacitors will be suitable for reuse, and many film capacitors will be leaky. Selenium and copper-oxide rectifiers are also to be thrown out. Most other components, such as tubes, sockets, terminal strips, wire, mounting hardware, extension shafts, etc., will be fine.

THIS PAGE WAS DELIBERATELY LEFT BLANK

2 SIGNAL SOURCES, TRACERS, POWER SUPPLIES AND FILTERS

- AC POWER SUPPLIES & ISOLATION TRANSFORMERS
- LINEAR DC POWER SUPPLIES
- HIGH VOLTAGE POWER SUPPLIES
- DIY PROJECT: 0 - 600 V_{DC} HV POWER SUPPLY
- FUNCTION GENERATORS
- LOW DISTORTION OSCILLATORS
- SIGNAL TRACERS
- ELECTRONIC FILTERS

> "One accurate measurement is worth a thousand expert opinions."
>
> Grace Hopper

AC POWER SUPPLIES & ISOLATION TRANSFORMERS

The way it used to be: Sencore AC "Powerite" PR57

These vintage units often come up for sale on eBay; it seems that Sencore had sold quite a few in its day. It is a low-power variable transformer (1), a 1:1 isolation transformer (2), and a safety (leakage) analyzer in one compact benchtop unit. The leakage probe plugs into the front of the unit (3). The magnetic circuit breaker (4) doubles as an on-off switch and protects the primary, with a fuse on the secondary side (5).

B&K Precision still makes and sells similar AC power supplies, models 1653A (2A capacity) and 1655A (3A capacity).

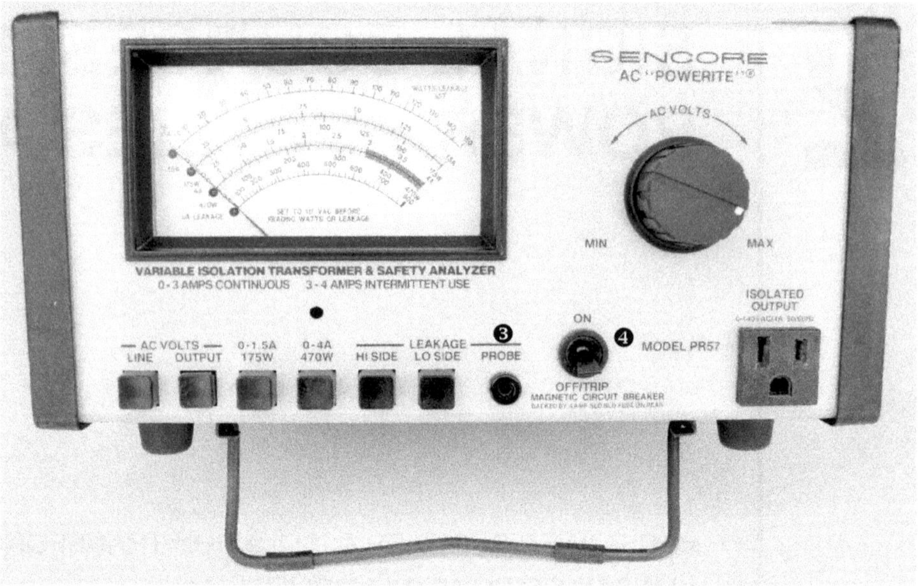

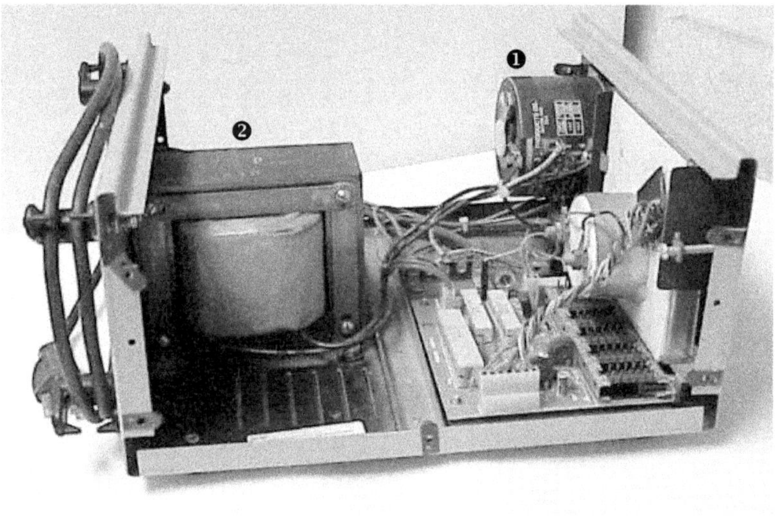

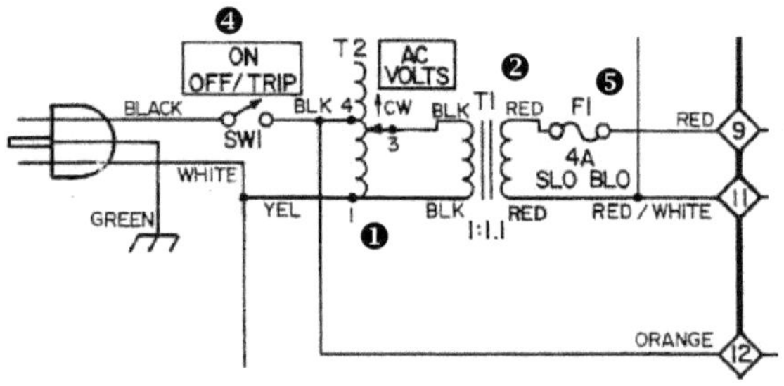

ABOVE: How the Variac and the isolation transformer are connected inside PR57

LEFT: The internal view of the unit

Case study: Unsafe in more than one way

The back of this DC power supply, made in China, illustrates a few (at best) perplexing and (at worst) potentially lethal issues.

Unless an instrument is double insulated as indicated by the appropriate symbol (and this one isn't!), it must be grounded (earthed) using a 3-core power cable and a 3-pin power plug. Here a 2-pin plug (Euro- or US-style) is used (6), meaning the metal chassis is not earthed, making the instrument illegal.

The sticker (7) says, "To avoid electric shock the power cable protective grounding conductor must be connected to ground." There is no such conductor here.

The fuse holder that can be unscrewed by bare hands is now illegal in Australia and many other countries (8). The approved fuse holders have an X-slot and can only be opened up using a Philips-head screwdriver.

Finally, the "CE" sticker (9). Most Chinese products haven't had any risk assessment, safety evaluation, or testing done, so they haven't been certified against the relevant standards. Various websites discuss this topic and illustrate the differences between the real CE logo and the fake ones. Those guys didn't even correctly scan & copy the letter proportions!

LINEAR DC POWER SUPPLIES

A power supply inside audio and T&M equipment is invariably a DC (direct current) supply, albeit of fixed voltage output and a constant load (the amplifier's audio circuit). The benchtop, stand-alone DC power supplies have a variable voltage output, which, once set to a specific value by the user, they maintain through negative feedback.

First & foremost, a DC power supply performs AC to DC conversion, converting mains AC voltage of a particular frequency (50 or 60Hz) to a steady DC voltage of the required level. This could be a low voltage used for tube heaters, typically 6.3 or 12.6V, a negative bias voltage, typ. -10 to -60V, or a high anode voltage, typ. 200-1,000V.

Apart from the desired voltage, every power supply must have a required current capacity; for instance, it must supply $400V_{DC}$ anode voltage at 300mA, which would be the current drawn by the load (audio section of the amplifier).It must also provide adequate energy storage. Energy can only be stored in the filtering chokes and capacitors, not in resistors, so superior power supplies use high inductance chokes and large capacitor banks.

It must be efficient in its energy conversion from an AC to DC supply, i.e., its internal losses should be as low as possible. Low losses mean less heat dissipation, less vibration, no buzzing of transformers, and a cooler and quieter running amplifier. Internal losses are modeled by the internal resistance of a power supply, which is zero in an ideal case. Actual power supplies have a finite internal resistance. An ideal DC supply provides a constant output voltage regardless of the load and its current demands (current draw).

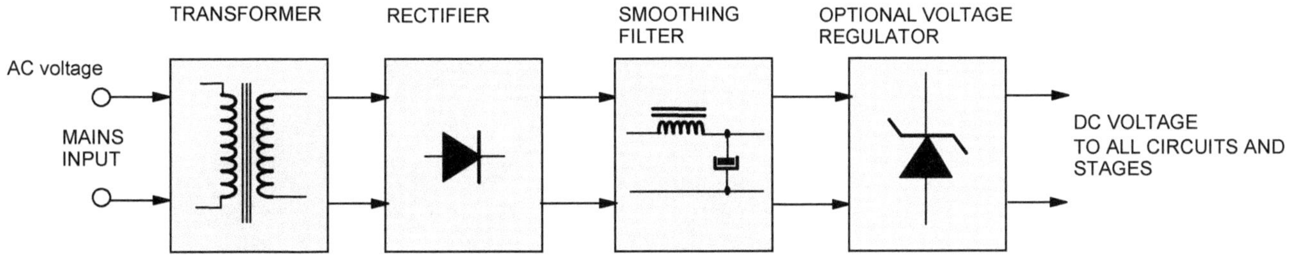

ABOVE: Block diagram of a typical linear power supply

Since the output DC voltage is obtained by rectification and filtration from an AC waveform, some remnants of that AC, called ripple, remain. While an ideal DC supply would provide a perfectly horizontal line, the output voltage of actual power supplies fluctuates above and below some steady DC value V0. Thus, the smaller the ripple, the closer the DC supply comes to an ideal one.

An ideal DC power supply would have no internal resistance or impedance. One consequence of the internal resistance of an actual power supply is the voltage drop that load current will cause on such a resistance, thus reducing the output voltage (the voltage divider effect). So, the output of an actual power supply will drop (sometimes called a voltage "droop") with increasing load current.

The factor that describes how close an actual power supply comes to an ideal one in that regard is called voltage regulation.

Power supply regulation

The ideal DC power supply would hold its output voltage constant at the V_0 value regardless of the load (increased load current). Due to the power supply's internal resistance R_I, load current causes a voltage drop on that resistance, so the output (load) voltage will decrease. The higher the input impedance of the power supply, the steeper such a droop will be.

Regulation R is a measure of how close power supplies get to the ideal: $\mathbf{R=(V_0-V_L)/V_0 *100\%}$. The lower the internal resistance of the power supply, the better the regulation and the steadier the output voltage.

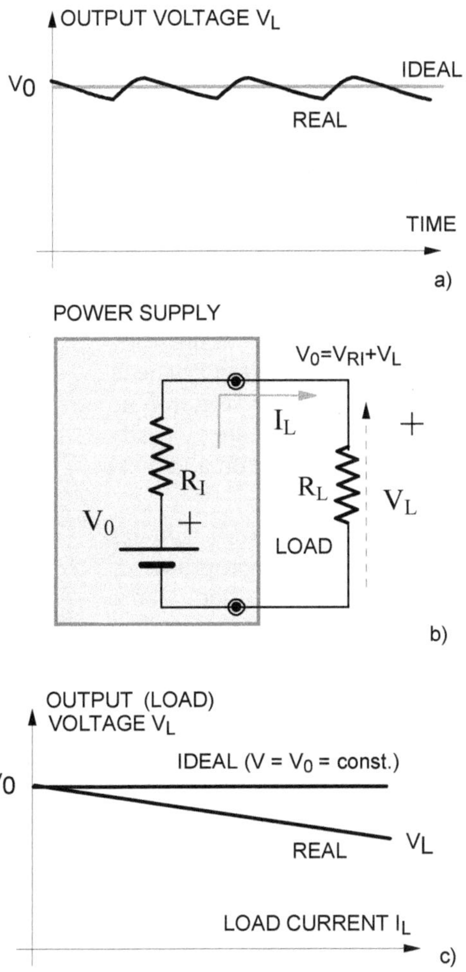

RIGHT:

a) The output voltage waveform of an ideal and real DC power supply

b) The model of a real DC voltage source with internal resistance R_I

c) The output voltage versus load current characteristics of the ideal and real power supply

Commercial low voltage DC power supplies

This Australian-made 3-channel DC power supply, made in the mid-1980s, has been serving us for many years. The left channel (1) can supply 0-8V at up to 4A, while the middle (2) and right channel (3) are identical, 0-25V_{DC} at up to 1A. Those channels can be used independently or track each other (4).

All channels can be used in a fully floating mode, or their positive or negative side can be grounded by wire jumpers across the corresponding binding posts (5). The channels can also be stacked in series, so Ch2 and Ch3 "on top" of each other can then provide up to 50V_{DC} at 1A.

All channels feature overcurrent fold-back protection circuits, whose limits can be adjusted by the "CURRENT" knobs.

The inside view reveals a large PCB (6). The circuitry for each channel is separate, so it's straightforward to understand the circuit and troubleshoot it, even without the circuit diagram.

The large power transformer (7) can supply 4A per channel, so the DC outputs are very conservatively rated. The limits can be raised by readjusting the blue trimpots on the PCB.

Each channel has its own heatsink (8) with a couple of 2N3055 power transistors. Should they require replacement, the large filtering electrolytic capacitors (9) are easily accessible, mounted on a vertical steel divider.

Modern DC power supplies (right) are often microprocessor controlled and offer more features, such as a constant voltage (C.V.) or constant current (C.C.) mode.

Channels can operate in series, in parallel, and independently. Digital readouts are used almost exclusively.

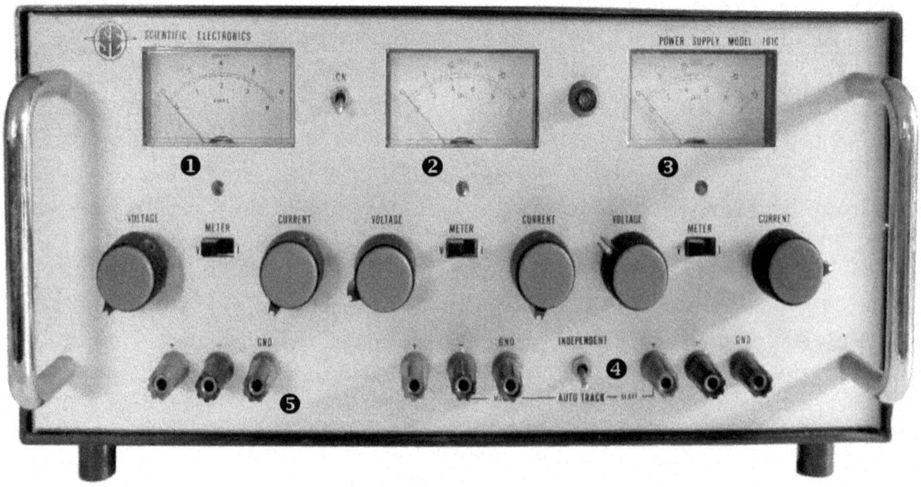

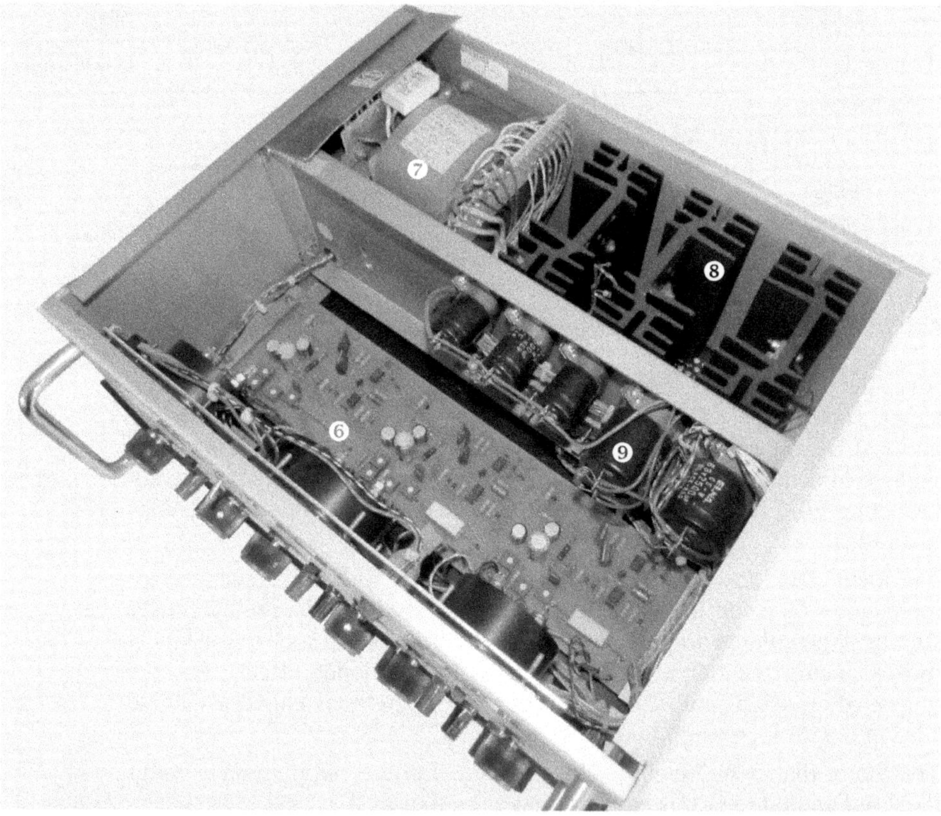

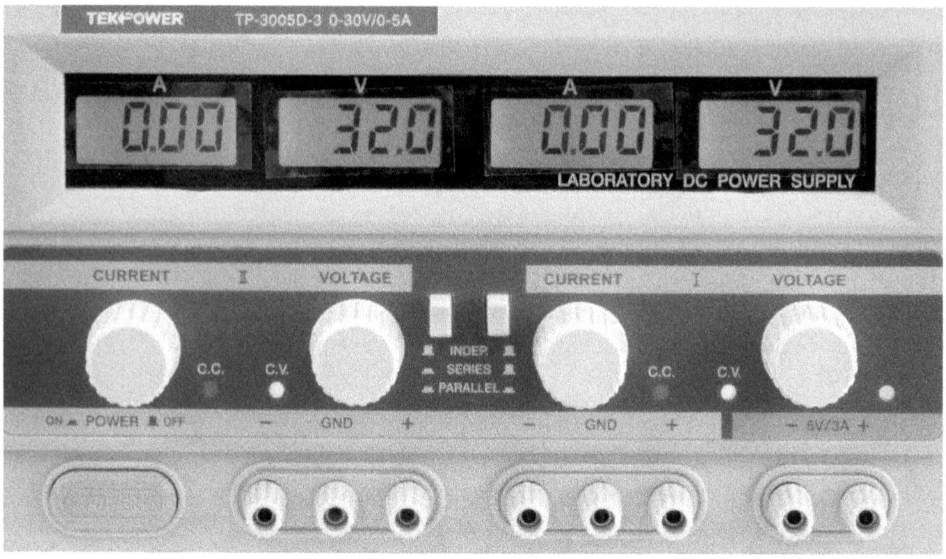

HIGH VOLTAGE POWER SUPPLIES

Vintage commercial HV regulated power supplies

Vintage high voltage regulated power supplies are available for sale online, but most are limited to 125 mA of current and $400V_{DC}$, with maximum power delivered to a load of only P = 0.125*400 = 50VA. That could be enough for low-powered single-ended designs with 2A3 or 300B triodes but is inadequate for larger SE and most high-power push-pull circuits.

These vintage HV power supplies use the same series regulated design with two triode-strapped 6L6 tubes in parallel, so it doesn't matter which one you get, Eico, Precision (PACO), Knight, or Heathkit. The most common by far are various reincarnations of Heathkit's "Regulated H.V. Power Supply."

Later units used solid-state diodes instead of tube rectifiers and Zener diodes instead of VR (Voltage Regulator) tubes. They were also sold under the name Heath Zenith and Heath Schlumberger. Those later models feature transformers with dual primaries (115-230V), while the other three brands mentioned were only made for 115V mains voltage.

The bias voltage is adjustable from 0 to -150V, but since an ordinary potentiometer was used, it is impossible to adjust it precisely; as soon as the knob is moved a tiny bit, the voltage increases or reduces two or three volts. Thus, these crude instruments aren't suitable for precise measurements. For bias supply, you are better off using a precisely regulated low voltage DC supply 0-25V or 0-50V if you are testing triodes requiring a high negative bias, such as 300B and 2A3.

If you plan to do lots of circuit experimentation and testing, you should build a better and more powerful HV power supply. You can go as high in voltage and current capabilities and other functionality as you need or want. Our design can go up to $600V_{DC}$ and 300 mA (180 VA). Before we look at its design, let's see how these regulated power supplies work.

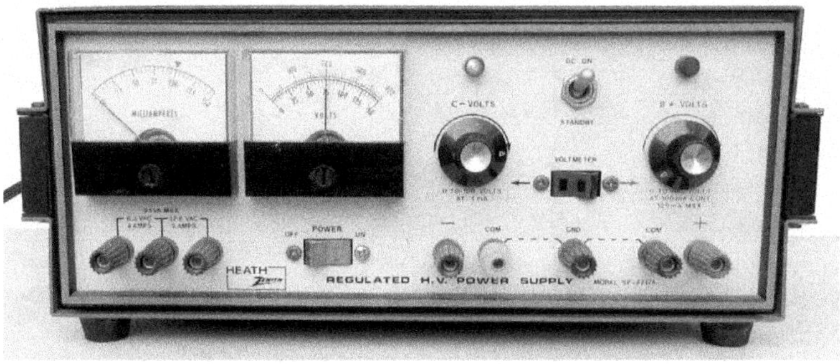

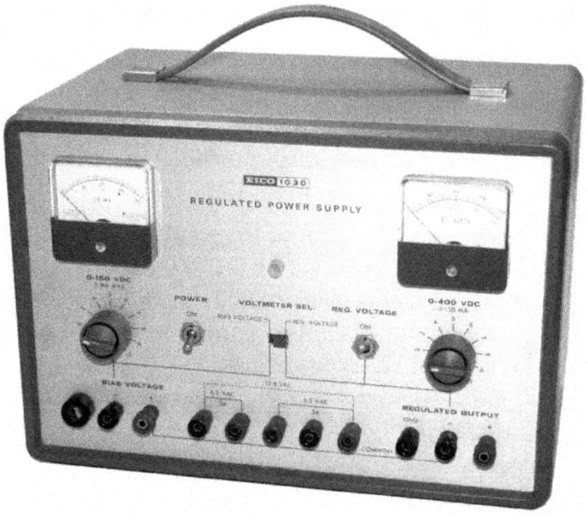

ABOVE: Vintage tube-based high voltage regulated power supplies such as Heathkit IP32, Heath Zenith IP2717 (ABOVE) or Eico 1030 (RIGHT) are useful vintage workshop instruments, not just in amplifier design but can be are heart of your own tube testing and matching setup. However, they are limited both in voltage/current capabilities and the precision of their adjustments.

RIGHT: An internal trimmer potentiometer enables "DC Volts" meter calibration, in this case against a modern digital multimeter as a reference.

BELOW RIGHT: The operating principle behind series voltage regulators

The series pass tube as a variable resistor

A variable resistor connected between the input and output terminals would be able to neutralize any changes in input voltage or output load current by adjusting its resistance, thus changing the voltage drop across its terminals. A higher input voltage means a larger voltage drop on that resistor and a constant output voltage. A larger load current means lower resistance, lower voltage drop, and a constant output voltage.

However, a resistor cannot do that by itself, so a servo-circuit employing an error detector and a negative feedback mechanism is needed, with a power tube acting as a variable resistor. Series regulators have a power tube in series with the output, and since it must pass the total load current, it is also called a "pass tube."

The whole regulator is a negative feedback DC amplifier that keeps the output voltage steady by opening and closing the pass tube as needed, just like the cruise control on your car keeps the speed constant by adjusting (opening or closing) the throttle.

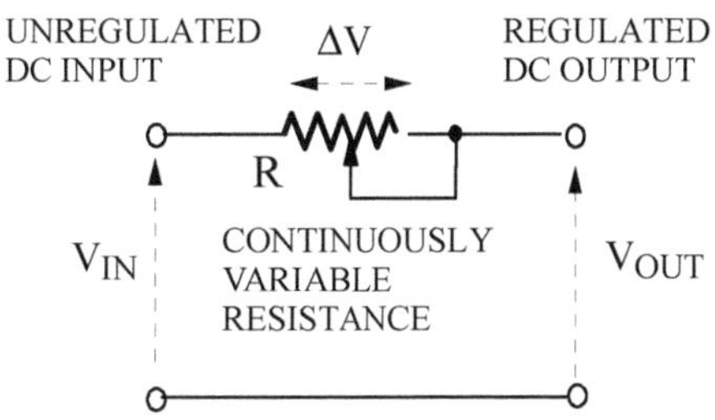

Series regulator with a triode error amplifier

Triode V1 is a DC error amplifier. Part of the output voltage is taken from the voltage divider network of two fixed resistors and one trimpot and fed to the triode's grid. The voltage is $V_{G1} = V_{R2} = KV_O = R_2/(R_1+R_2)V_O$ The gain of this amplifier whose load is R_X is $A_1 = -\mu_1 R_X/(R_X+r_I)$

The change of the anode voltage of the error amplifier drives the grid of V2, the series pass tube, so $V_{G2}=A_1V_{G1} = -\mu_1 KR_X/(R_X+r_I)V_O$. The change of the plate current of V2 is $\Delta I_{P2} = KA_1Gm_2\Delta V_O$.

The ratio of the change in output voltage to the change of output current is the output impedance of the voltage regulator: $Z_{OUT}=\Delta V_O/\Delta I_{P2} = 1/(KA_1Gm_2)$, where A_1 is the amplification factor of the error amplifier, Gm_2 is transconductance of V_2 and the constant $K=R_2/(R_1+R_2) = 50/147 = 0.34$.

For the 200 mA regulator with 6080 as a series pass tube and 12AX7 (both triodes in parallel) as error amplifier, illustrated on the right, we have $A_1= -\mu_1R_X/(R_X+r_I) = -100*470/(470+33) = -100*0.93 = -93$.

Since Gm_2=13 mA/V, and $K=R_2/(R_1+R_2$, the output impedance is $Z_{OUT}= 1/(KA_1Gm_2) = 1/(0.1875*93*0.013) = 4.4\Omega$!

The 33nF capacitor between the grid of the error amplifier and the output terminal is called a speedup capacitor. It makes factor K=1 for ripple and sudden surges, increasing the amplification at higher frequencies and thus improving the regulating speed of the voltage regulator.

The 470kΩ anode resistor of the error amplifier (R_X) is connected to the unregulated side, which is not ideal, the performance would be improved if it was connected to the other (regulated) side of the pass tube, but then there is a danger that the whole circuit would hang up and not start regulating.

Zener diodes are noisy and should be bypassed by a film capacitor and an elco, stabilizing the voltage further. Notice that we used 13mA/V for Gm_2, although 6080 triode has Gm of 6.5 mA/V, since V2 is actually two 6080 triodes connected in parallel (one physical tube with two triodes inside), so the mutual conductance Gm is doubled.

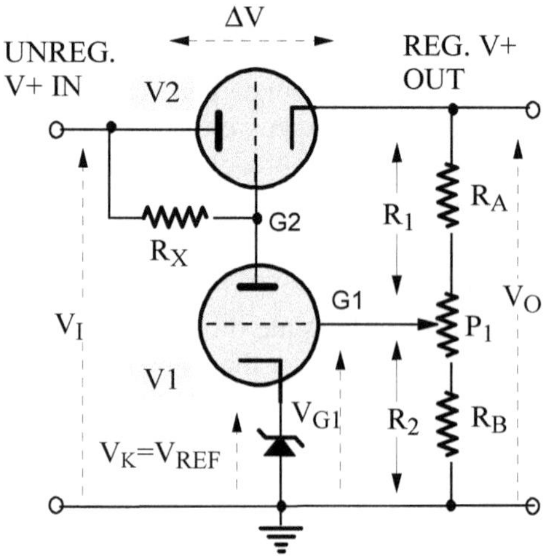

ABOVE: Basic series voltage regulator with a triode error amplifier and Zener diode reference.

BELOW: Practical implementation of a simple series high voltage regulator with paralleled 6080 duo-triode as a pass tube and 12AX7 error amplifier

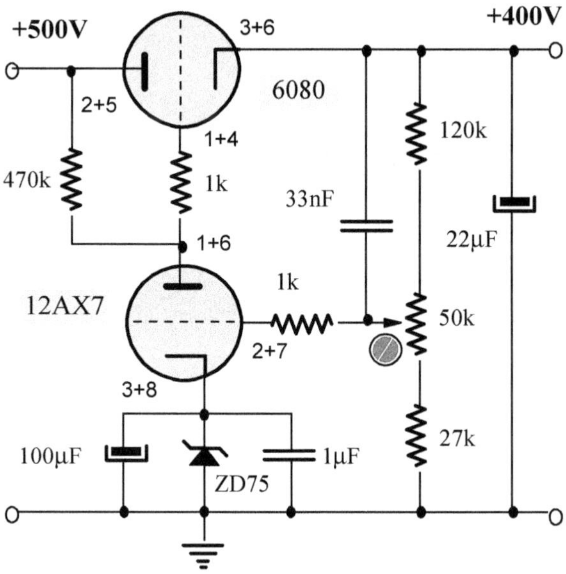

DIY PROJECT: 0 - 600 V_{DC} HV POWER SUPPLY

Our design

It is now apparent how improvements in performance could be made. Although a triode with the highest gain was used for the error amplifier (12AX7 with μ=100), to reduce the impedance of the power supply, we could use its two triodes in cascade, for an overall amplification factor of 88*88 =7,744!

The gain of each stage would be 88 and not 93 due to the higher internal resistance of a single triode. That would lower the Z_{OUT} by the same factor (88 times), bringing it under 54 mΩ!

We could also use a pentode since their μ is much higher than that of triodes or a pass tube with higher Gm. One such tube is E130L, with Gm=27.5 mA/V, more than double that of two 6080 in parallel. However, we will use a much more common KT88 for the pass tube and a 6AU6 pentode for the error amplifier. Both perform very well in this design, last a long time, and are easy to find and cheap to replace when they do fail.

Should you need a higher current capacity, you could use one of the recently developed power tubes, KT120 or even KT150, but then again, why not parallel two cheaper KT88 tubes and double the current output? That would also double the mutual conductance of the pass tube and lower the output impedance of the power supply even further.

The voltage doubler (next page) was used simply because we had a power transformer with a 240V_{AC} secondary available. Feel free to use a 2-diode rectification (with a center-tapped secondary of 2 x 480V_{AC}) or a solid-state bridge rectifier (480V_{AC} secondary without a CT).

Likewise, the dual power supplies producing +260V and -260 also use voltage doublers and CRC filtering.

As in any tube amp or instrument design, the power supply topology will depend on the parameters of the available power transformer.

Three Zener diodes are used to keep voltages on the 6AU6 electrodes at fixed levels. The cathode is at -250+150 = -100V and the screen grid is at -100+44V= -66V! The very high anode resistance of 470kΩ ensures a high amplification factor. The 6AU6 anode output drives the grid of the series tube via a 1k grid stopper resistor. The 6AU6 pentode is not critical; EF86 or a similar pentode can be used.

Switching between the bias and the anode voltage metering is annoying and often inconvenient, so separate voltmeters are used for the two DC voltage outputs.

The pass tube (KT88) is working as a pentode, which increases its internal impedance and the internal impedance of the whole power supply, a price that had to be paid for achieving a higher output voltage than what would be possible in a triode connection.

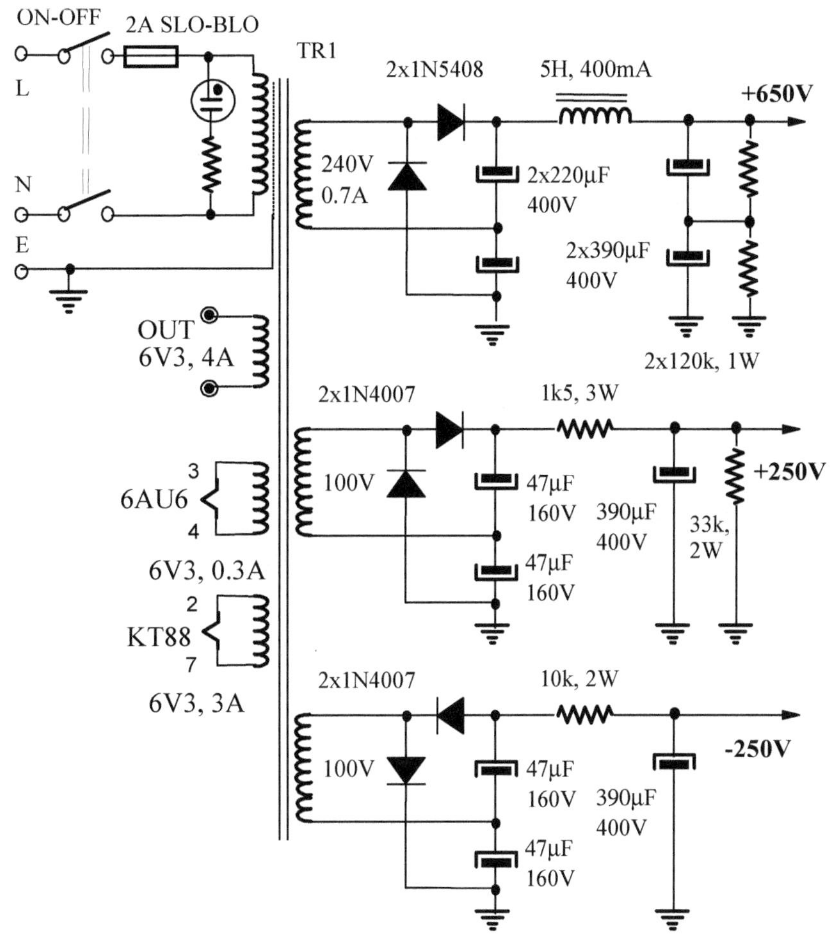

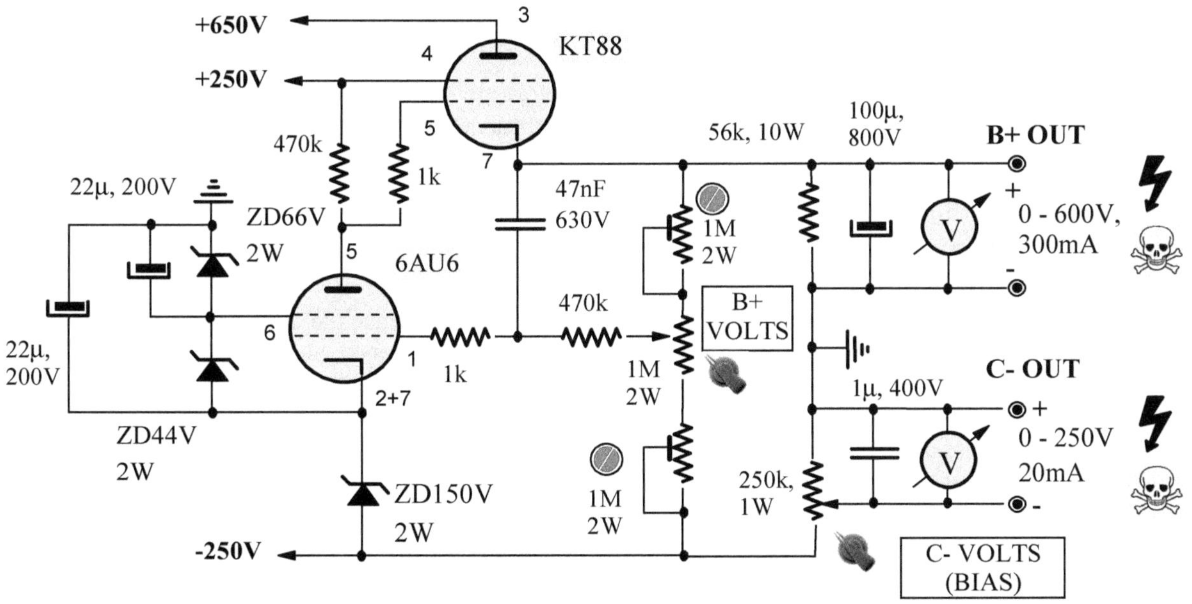

TUBE PROFILE: 6AU6

- Sharp cutoff pentode
- 7-pin mini socket (B7G base) , heater: 6.3V, 300 mA
- Maximum ratings: V_{AMAX}=330V_{DC}, V_{SMAX}=330V_{DC}, P_{AMAX}=3.5W, V_{HKMAX}=200V
- As a triode (at V_A=250V, V_S=250V, V_G=-4.0V): I_A=12.2mA, Gm=4.8 mA/V, μ=36, r_I=7.5kΩ
- As a pentode (at V_A=250V, V_S=150V, V_G=-1.0V): I_A=10.6mA, I_S=4.3mA, Gm =5.2 mA/V, μ=5,200

FUNCTION GENERATORS

Controls of a typical function generator

Oscilloscopes display input and output voltages of an amplifier and voltages in any other point of interest in any amplifier stage. However, something has to provide these test voltages of a variable frequency, variable amplitude, and waveform, and these instruments are called function generators. Alternative names used are "audio oscillators" (these only provide sine and sometimes square waves of audio frequency, while function generators' range extends much higher, into MHz region) and for those with a frequency sweep capability, "sweep generators." More on those very soon.

General-purpose function generators usually aren't of a low distortion kind and aren't suitable for precise distortion measurements but are fine for daily testing and repair work. For instance, Sampo FG1617 and FG1627, the models this illustration is based on, have a specified sinewave distortion of "less than 2%". For THD measurement of tube amplifiers, the distortion of such a source should be less than 0.05%.

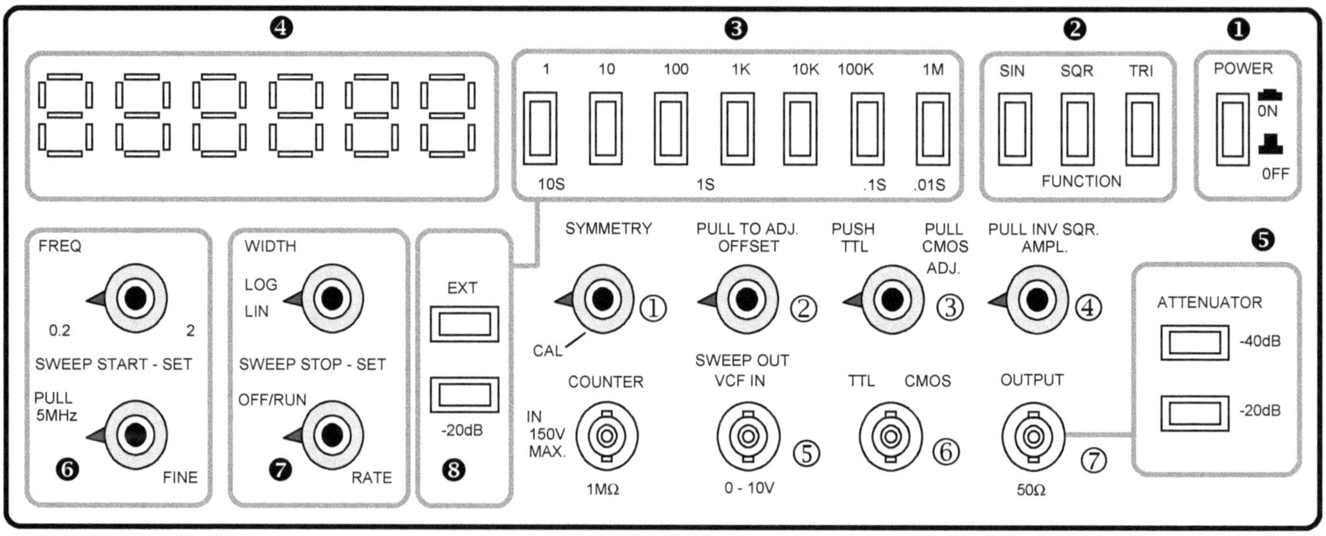

ABOVE: The controls of a typical function/sweep generator, based on Sampo FG1627

A term "signal generator" is also used, but strictly speaking, signal generators are high frequency sources for radio work, not audio frequency oscillators. The main functional groupings of a typical function generator are (❶ to ❽):

1. Power switch
2. Waveform (function) selector push-buttons, sine, square, and triangular (sawtooth) wave
3. Frequency select range push-buttons
4. Frequency and counter display (digital)
5. Amplitude attenuator for output voltage, -20, -40 or -60dB (both buttons pressed). Since the output voltage goes up to 14 Volts, that is scaled down to 14/100 = 140 mV on -20dB range, 1.4mV on -40dB range, and 0.014 mV when -60dB attenuation is selected (great for testing moving coil phono preamps and step-up transformers).
6. Coarse frequency adjustment (also sets the start frequency in the sweep mode) + Fine frequency adjustment, 1/10 of the course range
7. The top potentiometer adjusts the width of the sweep; the switch selects LOG sweep (push) or LIN sweep mode (pull). When the knob is pulled out, it sets the sweep stop frequency. The "RATE" adjusts the sweep rate from 25ms to 5 seconds. OFF/RUN starts and stops the sweep mode.
8. Selects external input in frequency counter mode, the -20dB push-button attenuates the amplitude of external pulses to be counted, used if the amplitude of those pulses are over 10V

The other controls (① to ⑦):

1. In CAL. position the signal is symmetrical, rotating the knob changes the symmetry
2. DC offset adjustment - when pushed in DC offset is zero
3. TTL/CMOS Selector - TTL or CMOS amplitude level of the output signal
4. Output amplitude adjustment + pull for square wave signal inversion
5. Input BNC terminal for VCF control signal input. In internal sweep mode the output of the sweep signal
6. CMOS or TTL level square wave output BNC terminal
7. Main signal output

Sweep generators

Thurlby Thandar Instruments model TG230 sweep/function generator is in the same class as Sampo FG1627. On the debit side, there is only the main frequency dial (3) and no fine adjustment, and only -20dB attenuation of the output signal (4), while Sampo also has -40dB and -60 dB options. The "Display Select" pushbutton (9) selects the signal frequency, output amplitude or DC offset to be displayed, a feature most function generators do not have, so their amplitude and offset have to be measured by an oscilloscope.

It has an internal and external AM (amplitude modulation) capability (8) for those repairing old AM radios and receivers, but that is of no interest for hi-end audiophile measurements.

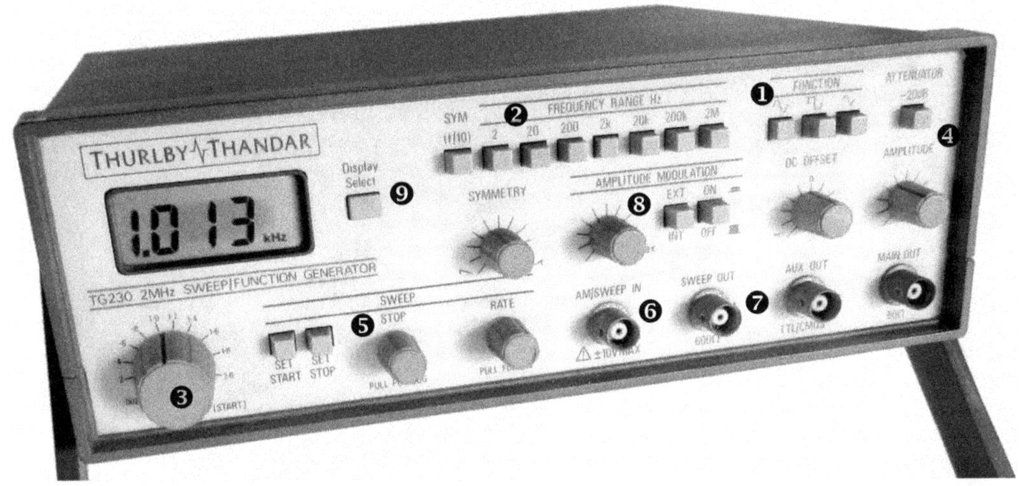

Analog function generators are often available for sale online from electronic and T&M surplus websites and on eBay for under US$50.

Metrix GX240 is a very similar sweep generator of French origin. Markings (1) to (5) correspond to the equivalent functions on TG230. The VCF input (6), equivalent to "AM/Sweep In" on TG230, and "Sweep out" BNC socket (7) are at the back.

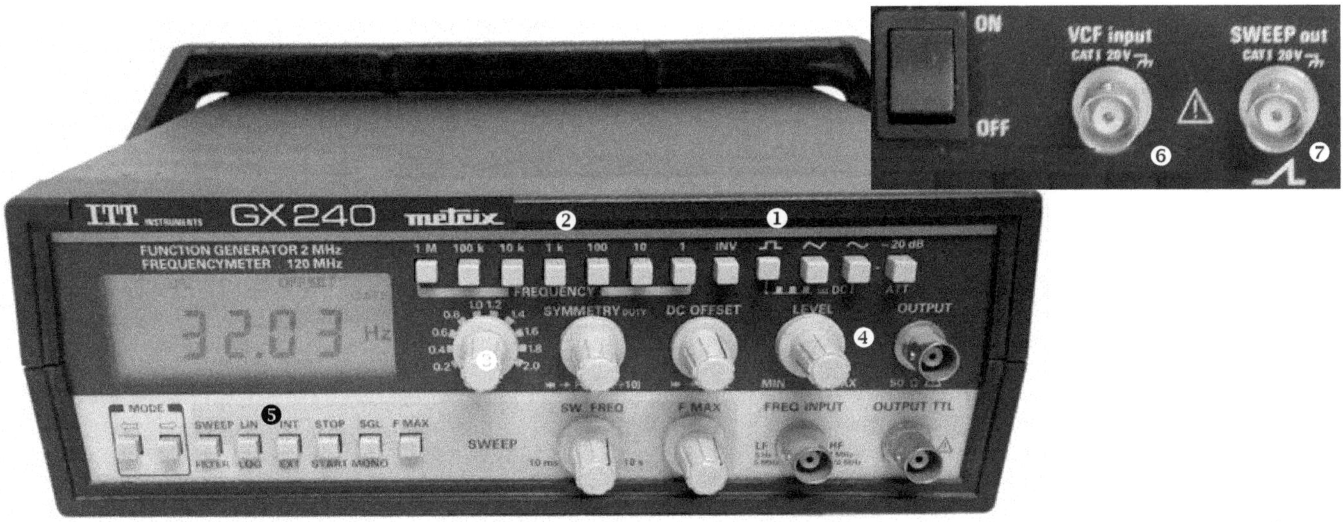

Tube audio oscillators

Made in Japan, Tech TE-22D is a typical example of mid-1960s service-type tube-based audio oscillators designed to compete with more expensive USA-made models such as HP 200CD. Only sine waveform is generated. The distortion is moderately high (0.2-0.5%, depending on the frequency range used), so not the best choice for THD and IM tests, but instruments of this class are simple, easy to repair, and intuitive to operate, perfect for quick audio tests and repair jobs.

There are two Noval vacuum tubes, 6BM8 triode-pentode, 6AQ8 duo-triode, a single rectifier diode, variable capacitor for frequency tuning, front switches, and potentiometer, plus a few precision resistors and capacitors is all there is to it (photo on the next page).

Thurlby Thandar TG230

- Sine wave distortion: under 0.5% on 100, 1k and 10k ranges, less than 1% on 10 and 100k ranges
- Sweep range: 1000:1 within each range
- Sweep rate: adjustable, 20ms to 20s
- Ext. sweep input sensitivity: 0 - 3V for 1000:1 sweep
- Max. allowable input voltage: ±10V
- Sweep linearity: better than 1%
- Sweep out 6V ramp from 600Ω
- AM depth: variable 0 to 100%
- AM frequency: 400Hz (int.), 0-100kHz (external)
- External sensitivity: cca $2V_{PP}$ for 50% modulation
- Square wave DC range: ±10V from 50Ω
- Symmetry: variable between 1:9 to 9:1

The On-Off switch is in the left bottom corner (1) and two output binding posts to its right (2). The two-position switch that selects square or sine wave output (3) is mounted on the same shaft as the frequency range selector (4), with four ranges, 20-200 Hz, 200-2,000Hz, 2kHz-20kHz, and 20-200kHz. The large dial (5) is the frequency tuning knob, while another dual switch is the output level attenuator (6), marked 1/1, 1/10, 1/100, and 1/1k, and potentiometer (7) that adjusts the output level continuously. The 1/10 attenuation is -10dB, 1/100 is -20dB, 1/100 30dB and 1/1,000 is 40 dB attenuation.

The max. output voltage is around $6V_{RMS}$ on 1/1 range, so on the finest range of 1/1,000, it is adjustable from zero to 6V/1,000 = 6mV, the level useful for testing moving magnet phono preamplifiers.

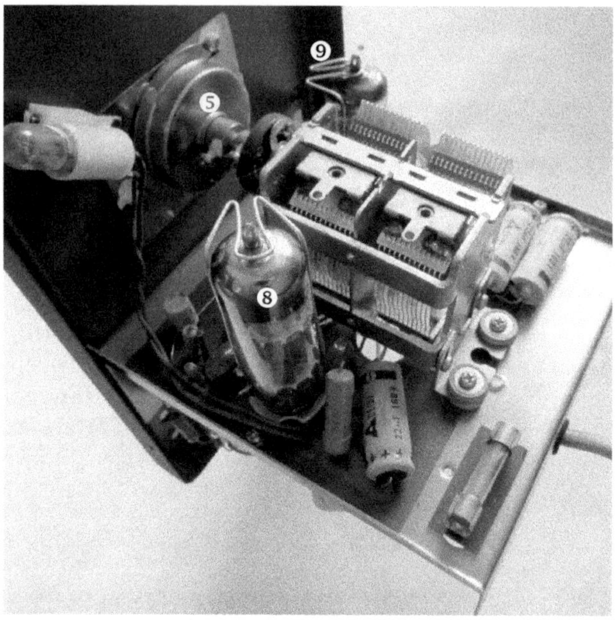

ABOVE: Tech TE-22D internal view, variable capacitor (made by ALPS Japan) on rubber mounts (frequency control), 6BM8 triode-pentode (8) and 6AQ8 duo-triode (9)

LOW DISTORTION OSCILLATORS

Philips PM5109

Philips PM5109 is a 1980s "RC generator", a very capable and valuable test instrument. Its 10Hz-100kHz range is adjustable in four steps (1) and continuously (2). The user can choose between fast settlement time (after frequency changes) or low distortion (3).

The grounded (asymmetrical or "single-ended") output through the BNC connector (4) can be configured for 600Ω or 50Ω output impedance (5). Its amplitude can be adjusted in 10dB steps (6) up to the maximum of -60dB attenuation (all three buttons pressed). As the attenuation is selected, the LEDs automatically indicate the analog meter's and output's full scale voltage (7), from 10V to 0.01V (10mV).

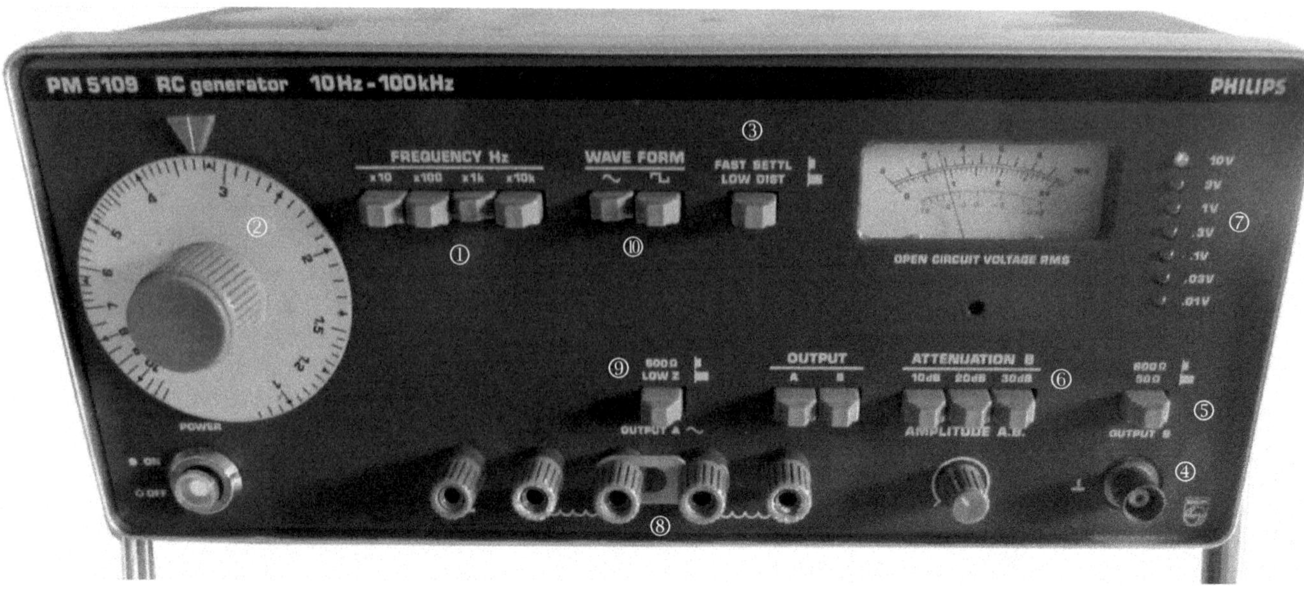

The balanced floating output through an isolation output transformer (8) also has its output impedance selectable, 600Ω or "Low Z" (9). The two secondary windings are separate and can be connected in series (as per the photo) or out-of-phase, meaning they can provide push-pull signals.

Sine and square waveforms are available (10). The maximum output is 30V_{PP} asymmetrical or 10V_{PP} symmetrical. Both output voltages are continuously adjustable. The output amplitude is displayed in both volts and decibels on an analog voltmeter. The distortion between 300Hz and 20kHz is only 0.03% at 1kHz, rising to 0.7% outside that range.

There's even a DIN loudspeaker output (at the rear, not shown), enabling PM5109 to be used for direct testing of loudspeakers (without an additional amplifier), a handy feature for speaker designers & builders in our midst.

Gould J3B

The functionality of Gould J3B is almost identical to that of PM5109, as is its frequency range (10Hz-100kHz). Gould has a Wien-bridge oscillator coupled to a transistor power amplifier driving a split-primary output transformer and the output attenuator.

There are four outputs. 30V_{RMS} (15V-0-15V) is available from the balanced floating main output of 600Ω (300Ω - 0 - 300Ω) impedance (1).

The low (1Ω) impedance output (2) can supply 3V_{RMS} into a 5Ω load. This is similar to Philips' loudspeaker output. The distortion is less than 0.1% above 100Hz, rising to 0.5% at 10kHz. Two attenuator pushbuttons (3) are provided 1/10 (-20dB) and 1/100 (-40dB), with both pressed -60dB is achieved.

The 10 Hz-100 kHz frequency range is switchable through four-decade ranges (6), and the 6:1 reduction drive with capacitor tuning (not shown in the photo) gives high-frequency resolution with minimum bounce.

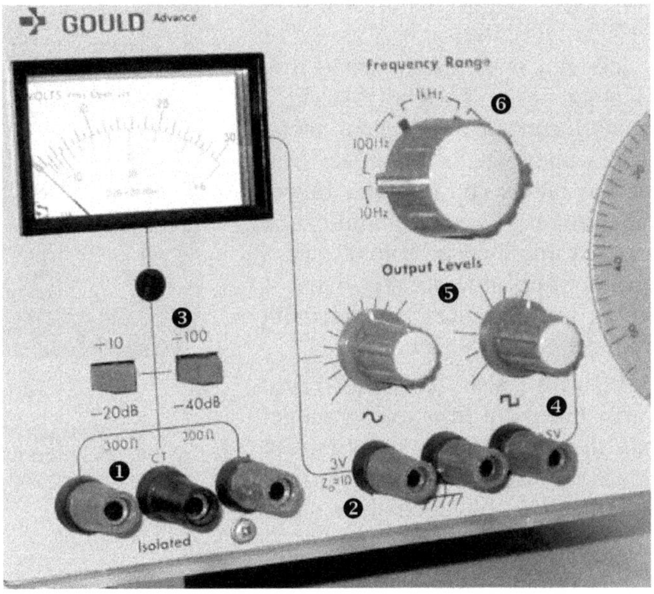

The 0-5V square wave output (1kΩ source impedance) has its own separate binding post (4) while the GND is shared. Each has its own output level control (5), a handy feature so once you adjust the square wave amplitude for amplifier testing, and then move on to sine wave tests, you don't have to spend time adjusting it again. The output amplitude is displayed on an analog voltmeter in both volts and dB. The outputs are protected against short circuits.

The low distortion output is located at the rear of the instrument, typically 2.5V_{RMS} from 5kΩ output impedance. The distortion is lower than 0.02% above 200Hz, rising to 0.2% at 10kHz, lower than PM5109.

Some other useful vintage low distortion oscillators are Krohn-Hite 4100A, Stanford Research SRS DS360, and J E Sugden SI 453, which even includes attenuation position marked "RIAA", meaning it has an inbuilt inverse RIAA filter for testing phono preamplifiers.

RIGHT: Partial block diagram of GOULD J3B (power supply, metering circuit and squarer not shown)

BELOW: Total harmonic distortion of a typical low distortion oscillator versus output voltage level (LEFT) and THD versus signal frequency (RIGHT)

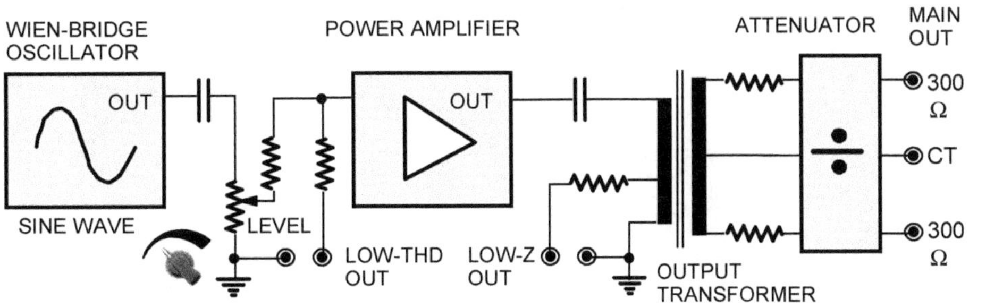

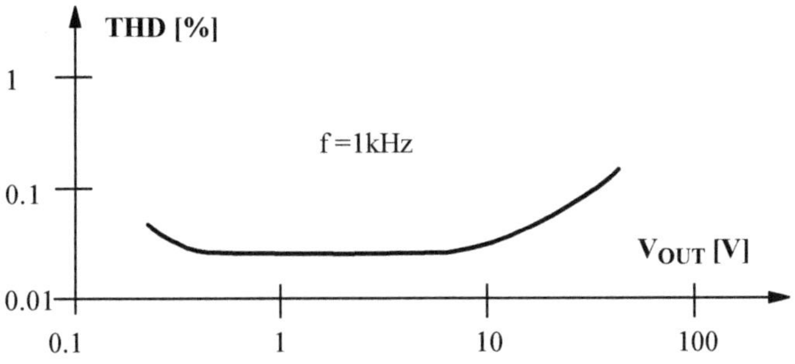

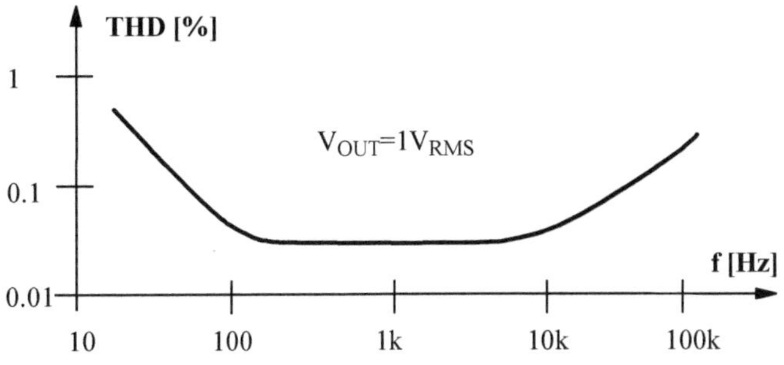

SIGNAL TRACERS

Signal tracers as learning platforms

In the 1950s & 60s, oscilloscopes were expensive, so signal tracers found their place in amateur workshops. They weren't that popular even then and are all but forgotten now as troubleshooting tools. However, tube signal tracers have an educational value as learning platforms.

It does not matter what brand it is (Heathkit, Eico, Knight, PACO); a signal tracer is a complete mono single-ended audio amplifier. You get the mains (power) and output transformers, two amplification stages, a single-ended output stage, a tube rectifier, and a speaker. There is even a visual indicator circuit built around a magic eye tube.

There are also transistor-based signal tracers, a few years younger than these tube-based beauties, but they aren't nearly as cool or good sounding (if converted to a small guitar amplifier, for instance).

ABOVE: Most signal tracers by Eico, Heathkit, PACO, KnightKit, and other similar T&M makers of the 1960s are functionally identical, meaning their front controls and terminals were also conceptualized and laid out, often even marked, in a similar way. Most can operate as a high gain (RF) tracer, low gain (audio) tracer, noise locator, and a primitive Wattmeter.

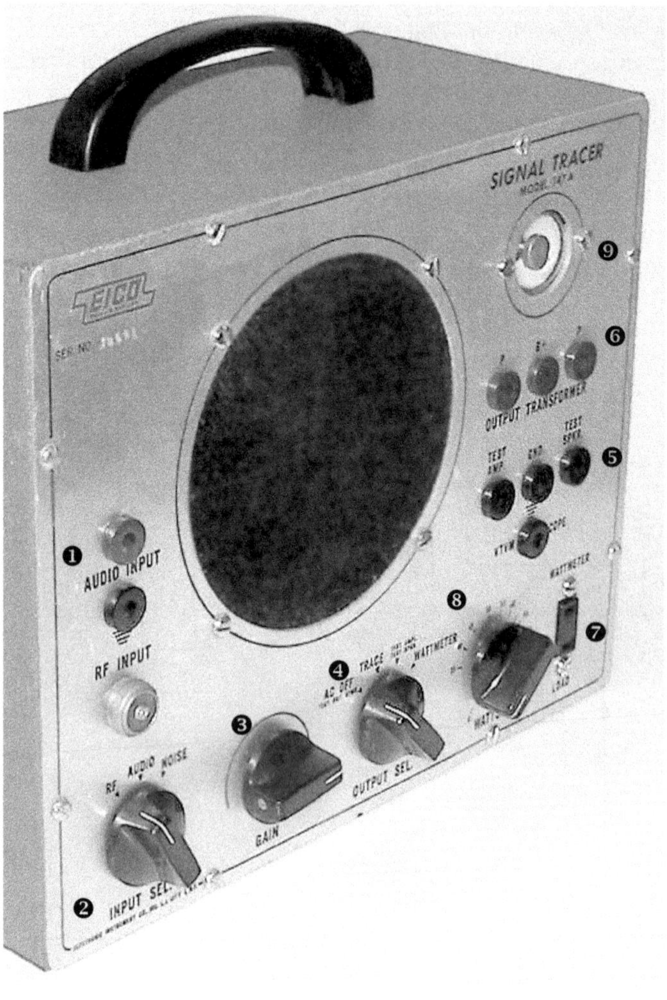

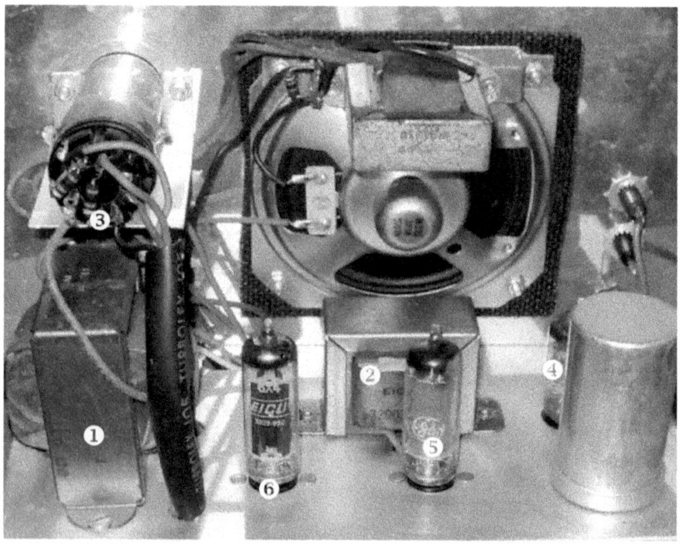

ABOVE: Internal view of Eico 147A signal tracer. Mains transformer (1), output transformer (2), magic eye tube (3), 12AX7 preamp tube (4), 6AQ5 output tube (5), 6X4 rectifier tube (6).

LEFT: Eico 147A's control panel

1) Audio (J1 & J2) and RF Inputs (J3)
2) "RF - Audio - Noise" input selector switch
3) "Gain" potentiometer
4) "AC off (test output transformer) - Trace - Test amp/Test speaker - Wattmeter" switch
5) Test amp (J10), Test speaker (J7), VTVM/Scope and GND terminals
6) Output transformer terminals
7) Wattmeter Load receptacle (2-pin USA type)
8) Wattmeter calibrated rheostat
9) Magic eye tube (operational only in Wattmeter mode)

Audio signal tracing

When used as a troubleshooting tool for tracing the signal through audio amplifiers, signal tracers have one significant advantage over oscilloscopes and multimeters (voltmeters) - they make checks much faster and easier. You don't have to take your eyes of the test point & probe and look at the screen or display to read the value. There is no need to adjust scope controls (its vertical scale or time base) or keep changing voltage ranges on a non-auto-ranging meter. You simply listen to the quality and loudness of the tone through the tracer's amplifier and speaker.

So, for the readers brought up in a digital world (and that will mean most of you), this is not a laughable test gear of yesteryear that belongs in a technical museum, but a quick & helpful troubleshooting tool.

Multimeters aren't very useful in picking up intermittent shorts or breaks in the circuit or identifying amplifying stages that distort the signal. One never knows if the flickering needle or digital display is due to the range change internally, bad contact with the probes, or one of the faults just mentioned. A signal tracer will pick these faults immediately, you will hear the breaking tone due to poor contacts, and you will identify distorted signals with ease!

Tracers help identify the absence of audio signals in critical points (for instance, grids and anodes of preamp and power tubes) and can also detect the presence of AC signal or filtering ripple (hum) in points where there should be none.

For instance, if the cathode resistor of the output stage is bypassed by a capacitor (as in most tube amps), there should be no audio signal present at that point (the capacitor is supposed to bypass it to the ground). If signal voltage is picked up at the cathode, the electrolytic bypass capacitor is either open or has lost its capacitance (dried out).

Operation as a wattmeter and noise locator

The noise locator test will identify noisy and intermittent passive components such as fixed resistors, capacitors, volume and tone control pots, and cold solder joints. The suspect component should be jiggled or prodded with an insulated wand (wooden or plastic chopsticks are great for that purpose).

Once the device-under-test (load) such as an amplifier is plugged into the receptacle (US 2-pin ungrounded type), the Wattmeter function is used to ascertain if its power consumption is within normal limits or if an overload exists due to an internal fault. That could be a short in the power transformer, rectifier, or filtering circuit, or a problem with output (power) tubes drawing too much current.

Once the "Wattmeter" control rheostat is adjusted, so the shadow angle on the eye tube closes with no overlapping of the bright sector's edges, the approximate power consumption of the load in Watts can be read from the calibrated dial.

Since the wattmeter circuit is responsive only to the current drawn and the dial calibration is based on the average US mains voltage of $117V_{AC}$, the accuracy of the reading does not take into account the power factor of the load, so it is actually a VA (apparent power in VoltAmperes) rather than Watts (real power) readout, and as such its accuracy is pretty poor, no greater than 10%! With cheap digital plug-in type wattmeters now available (see page 93), this function is now superfluous.

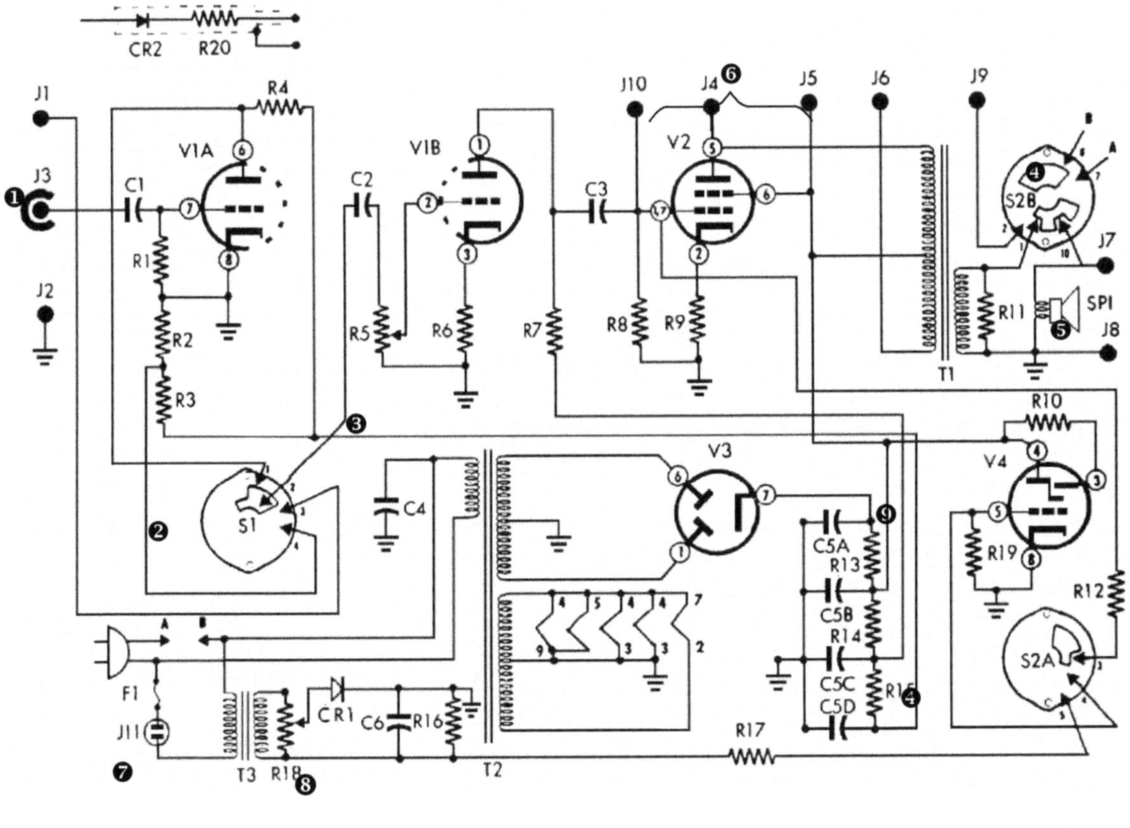

ABOVE: Circuit diagram of Eico 147A signal tracer. Points 1 to 9 correspond to the front panel features on the previous page.

Forward and backward signal tracing methods

There are two ways to signal trace an amp or preamp during troubleshooting. In forward signal tracing, the audio oscillator is set to the required input voltage and left connected to the input. The input to the test instrument (voltmeter, oscilloscope, or signal tracer) is moved from the output of the first stage down the signal path.

In backward signal tracing, the test instrument is connected to the amp or preamp output, set for a required (expected) voltage, and not touched again. Instead, the function generator's output is constantly adjusted for the required voltage and moved from the input of the final stage back towards the input.

In most cases, audio oscillators cannot provide high enough voltage to drive the output stage of amplifiers, so for that reason alone, forward tracing is more practical. Backward tracing is a viable method for troubleshooting preamplifiers, though.

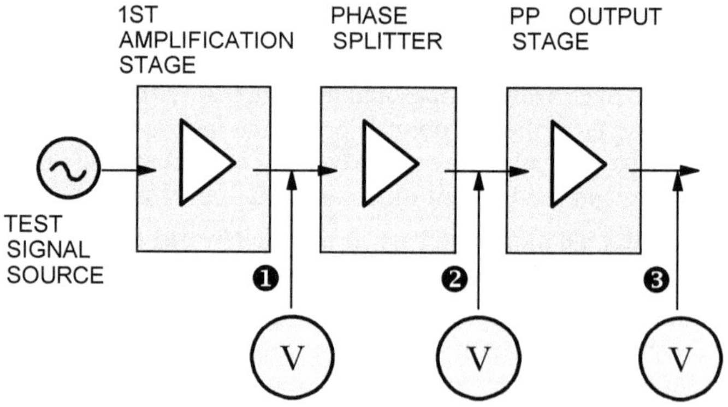

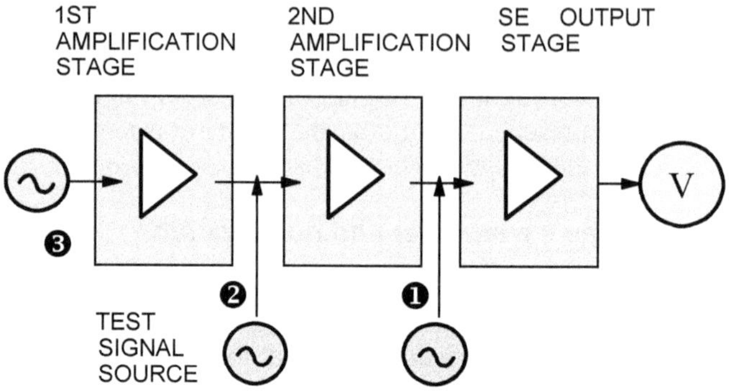

ABOVE RIGHT: Forward signal tracing.

RIGHT: Backward signal tracing.

Safe powering up of amplifier or preamplifiers

Once you have completed the visual and mechanical inspection of an amplifier (or any other electronic device you are troubleshooting or repairing) and performed the cold checks of all the major components (transformers, chokes, tubes, coupling, and filtering capacitors) with an ohmmeter and LCR-meter, you are ready to power the amplifier up.

Variac® power-up. Bring up the voltages gradually up with a variable autotransformer. This would give the power supply electrolytic capacitors some time to reform the electrolyte. Most variable autotransformers can exceed the rated mains voltage; you may use that capability to see if an amplifier will cope with wide power fluctuations and if the power supply capacitors have enough voltage margin. Beware, in some amps, the caps are rated close to the operating voltage, say $400V_{DC}$, while the anode voltage is $390V_{DC}$, so exceeding the mains voltage by 10% (say 130V instead of the nominal 117V) will bring that voltage up to $440V_{DC}$, way above the capacitors' rating. As a result, some marginal caps may start smoking and even explode violently, so keep your face away from the amp while performing this test, or you may end up with hot toxic gunk in your eyes and face.

ABOVE: Variable autotransformers of 500W-1kW rating can be used to gradually increase the supply voltage to a device.

Light-bulb power-up. Wire a 40-60W light bulb in series with the amp's AC cord. It must be an incandescent globe (remember those, in many countries, you cannot even buy them anymore)! Fluorescent, LED or halogen types are not suitable for this purpose.

Turn the amp on. If there are no shorts in the power supply, you should see the globe flash briefly to full brightness and then go down to dimly lit. If the lamp stays at full brightness, there is a short circuit within the amplifier. By wiring the lamp in series with the amp's AC supply, you limit the current that can flow through the circuit to safe levels and prevent any damage if a short is present.

Staged power-up. Unplug all tubes from the amplifier, except the tube rectifier, if used. Disconnect high voltage after the CRC or CLC filtering. Power the amplifier up. If the fuse does not blow, measure power supply DC voltages. They should be higher than nominal due to the lack of anode loads.

Plug in preamp tubes. Turn the amp back on. If the fuse does not blow, measure DC voltages on grids, plates, and cathodes of preamp tubes. If within +/- 10% of nominal, turn the amp off again and plug in the power tubes. Turn the amp back on. If the fuse does not blow and the power tubes' anodes do not glow red, measure DC voltages at main points again.

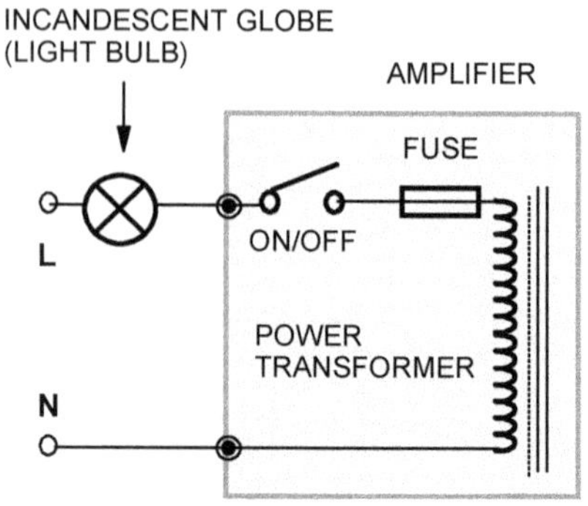

ABOVE: The globe acts as a load in series with the mains transformer's primary winding.

ELECTRONIC FILTERS

Ithaco 4213 electronic filter

A relatively modern USA-made instrument (the late 1970s, early 1980s), 4213 is a typical example of a general-purpose variable analog filter, in this case with two filters, a high pass (HP) and low-pass (LP). Both cutoff frequencies are adjustable in steps (1) for the HP filter and (3) for the LP filter and multiplied by the multiplier settings (2) and (4), respectively. The Mode switch (5) selects one of three modes, "Normal," "Pulse," or "Reject."

An optional DC-coupled amplifier (6) provides 0-40dB of gain in 10dB steps. Its upper cutoff frequency is 2.5 MHz. The block diagram of the filter in the "Normal" mode (below) illustrates where in the signal chain that amplifier sits (in the LP filter, between the 1st and 2nd quadratic module). The "OUT" position of the multiplier (as in the photo) simply bypasses the high pass filter and feeds the signal from the output buffer straight into the LP filter.

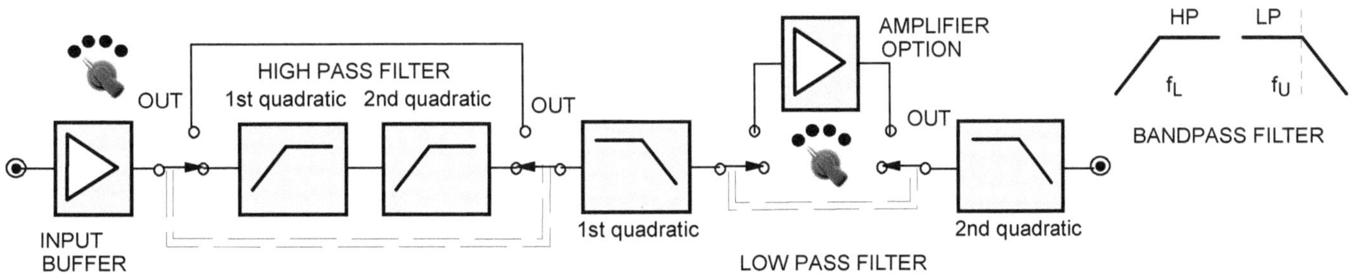

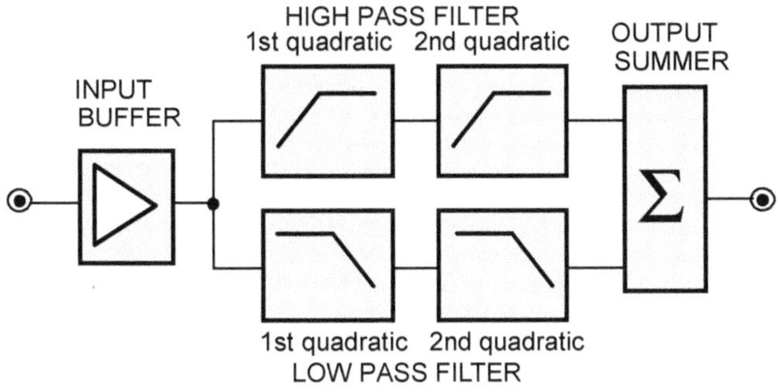

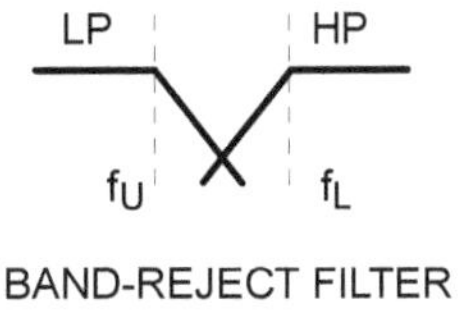

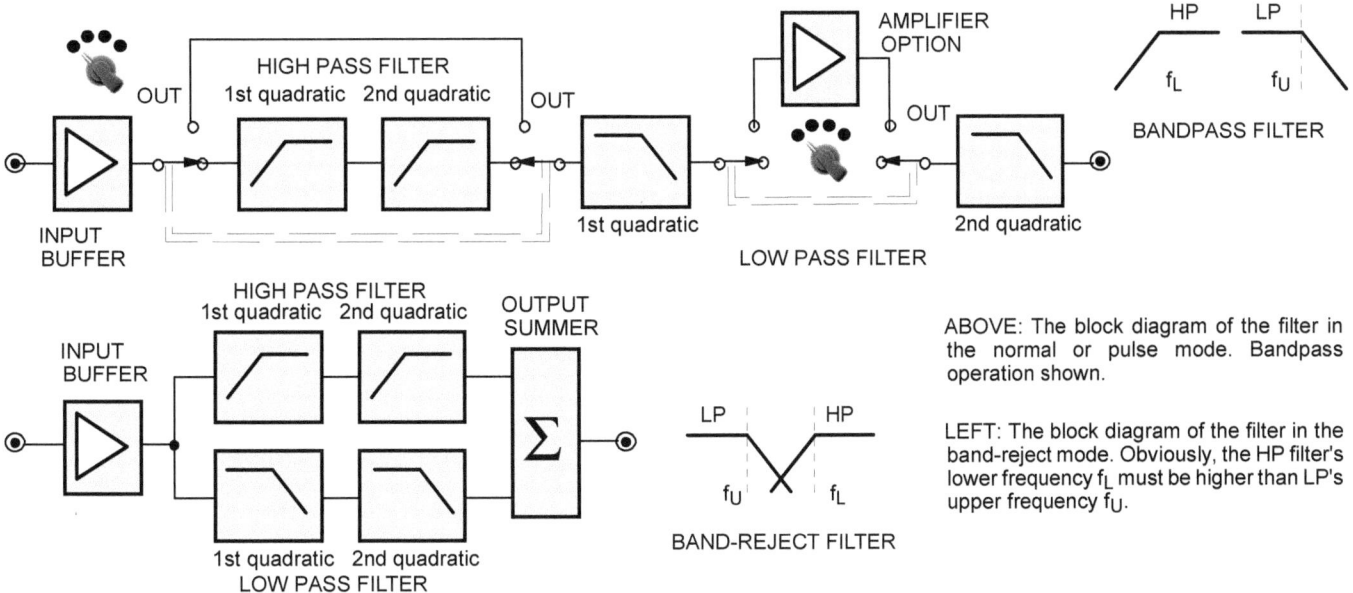

ABOVE: The block diagram of the filter in the normal or pulse mode. Bandpass operation shown.

LEFT: The block diagram of the filter in the band-reject mode. Obviously, the HP filter's lower frequency f_L must be higher than LP's upper frequency f_U.

The HP (high pass) filter's lower frequency f_L in the bandpass mode is much lower than LP's upper-frequency f_U. In the band-reject mode, the situation is reversed; the HP filter's lower frequency f_L is higher than LP's upper-frequency f_U. The filter rejects the band of frequencies between f_U and f_L!

In the band-reject mode, the two filters are connected in parallel, and their output signals are added (block diagram on the previous page).4-pole Butterworth filters are used, with 24dB/octave roll-off (slope). Remember, an octave is a 2:1 ratio of frequencies, for instance, 2kHz and 1 kHz. A decade is a 10:1 ratio (1kHz - 10kHz, or 1kHz - 100Hz).

A Bessel filter is used in the pulse mode to achieve a maximally flat time delay (linear phase response).

The graph shows 4213's normalized response in the band-reject mode (as a sharp notch filter). If used to measure Total Harmonic Distortion (THD), the 2nd harmonic at fN=2 would be attenuated around -7dbV, so 43dBV above the fundamental at fN=1. This would introduce a small error (the fundamental is not "completely" attenuated, say 70 or 80dBV), but the error is so tiny that it can be neglected for practical purposes.

THD is the sum of all harmonics V_H divided by the RMS value of the fundamental voltage signal V_F or

THD= V_H/V_F*100 [%].

V_H is the RMS voltage measured after the notch filter. The filter passes all harmonics, f_N=2, f_N=3, f_N=4, etc. but attenuates and all but removes the fundamental at f_N=1.

See chapter 7 for THD measurements.

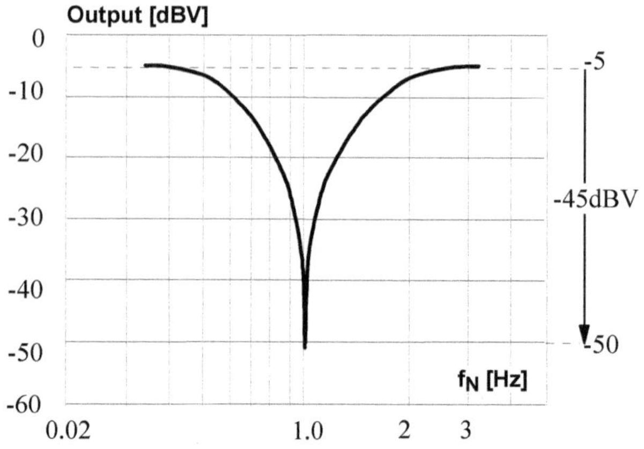

ABOVE: The normalized attenuation curve of Ithaco 4213 electronic filter in the band-reject mode, working as a notch filter, the type that would be used in measuring the relative amplitudes of harmonic distortion components (wave filter)

3 MULTIMETERS - TYPES, OPERATING PRINCIPLES AND FUNCTIONS

- MULTIMETERS
- ANALOG (MOVING COIL) METERS
- DC VOLTMETERS
- AC VOLTMETERS
- OHMMETERS
- VACUUM TUBE VOLTMETERS (VTVM)
- SOLID STATE ELECTRONIC MULTIMETERS
- DIGITAL MULTIMETERS

> "It is really just as bad technique to make a measurement more accurately than is necessary as it is to make it not accurately enough."
>
> Arthur David Ritchie

MULTIMETERS

A multimeter is the most-often used piece of test gear, so buy the best one you can afford. Modern ones are digital and auto-ranging and thus very easy and quick to use. However, many amplifier builders prefer analog meters, not for their romantic and old-fashioned "feel," but for the smooth behavior of the analog indicator.

For quick checks, it isn't necessary to read the exact figures on the scale; one glance at the needle is enough to ascertain that the voltage is "in the ballpark," usually 1-5V_{DC} on cathodes and 100-200V_{DC} on the anodes of vacuum tubes.

Also, when testing potentiometers for tracking and smooth resistance change (to detect any sudden jumps or breaks), analog ohmmeters are the only way to do it. Likewise, with unstable AC or DC voltages, the displayed values of digital meters bounce around and never settle on one value, thus making readings impossible. Due to the inertia of the moving coil mechanism, the analog ones will settle at some average value.

Types of multimeters

A passive analog multimeter is an elaborate switching arrangement of shunt and multiplier resistors, enabling a permanent magnet moving coil (PMMC) meter to measure DC voltage, DC current, and resistance. Most also have a rectifier to measure the average value of rectified sinewave voltage and current. There are no active or amplifying elements or components.

Electronic or "active" multimeters have an amplifier; two amplifying elements in a differential arrangement are used (tubes, FETs, or bipolar transistors).

- Passive analog non-RMS meters, also called VOM (Volt-Ohm meters): very basic and cheap, manually-selectable ranges, low input resistance, so they affect the measured circuit due to the voltmeter loading effect (more on that soon). No active components (tubes, transistors or integrated circuits). Cannot measure frequencies above 1-3 kHz, thus not suitable for DIY audio.
- Active analog non-RMS meters, VTVM (Vacuum Tube Volt Meters) or FET meters: high input impedance, can measure frequencies way above 20 kHz, manually-selectable ranges. A good old-fashioned choice. Examples: Philips PM2505 Electronic VAΩ Meter (1986) and Micronta FET Analog Multitester (1988)
- Active digital non-RMS meters: a wide range of quality and capability levels, from cheap and cheerful to so-so. There is no point in buying them when quality True-RMS meters are widely available and more affordable than ever before!
- Active digital True-RMS meters: the best choice for the digital generation.

Starting on the next page, we will analyze the operation of various types of meters in that order, from the simplest passive analog meters (VOMs), progressing through active analog meters to modern digital multimeters.

Get to know your multimeter

You should decide what multimeter suits your needs (both current and future) and your budget, and then research various brands and models, their capabilities, and limitations. Once you narrow your choice down to a few models, if possible, find them in your local electronic store and try them out. Different instruments have a different "feel," and you may or may not like some of them, although they look good "on paper" or online.

Instrument symbols

On their meter or elsewhere on the control panel, analog test instruments usually feature one or more important symbols that indicate critical aspects of its performance or operation. Digital instruments, on the other hand, don't use these at all, so reading their user or operation manuals is mandatory. Here are some of the most common symbols and markings:

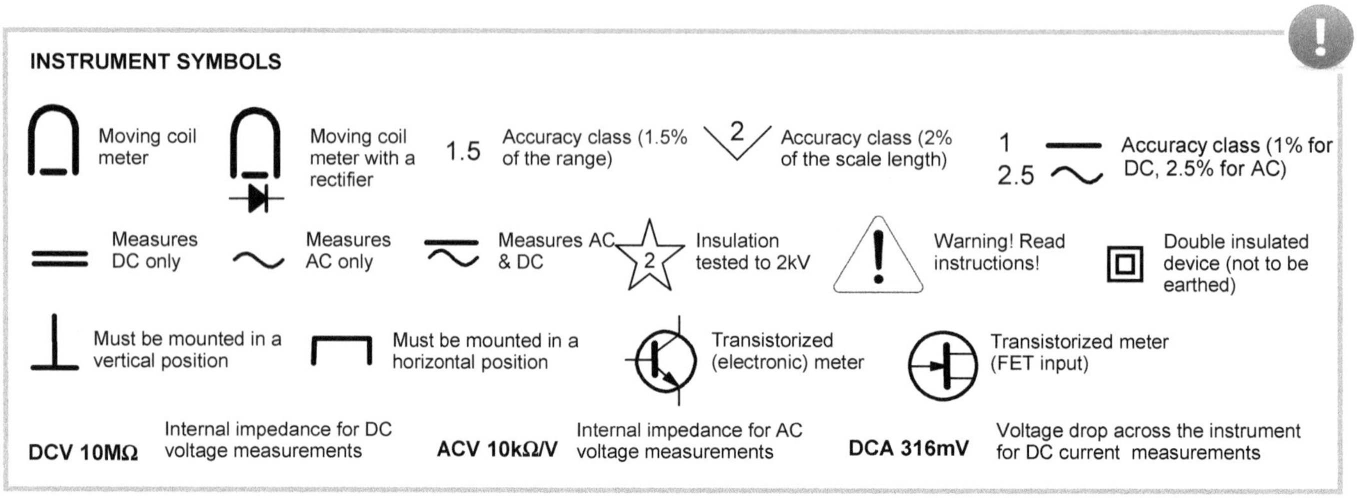

INSTRUMENT SYMBOLS
Moving coil meter
Moving coil meter with a rectifier
1.5
Accuracy class (1.5% of the range)
2
Accuracy class (2% of the scale length)
1
2.5
Accuracy class (1% for DC, 2.5% for AC)
Measures DC only
Measures AC only
Measures AC & DC
2
Insulation tested to 2kV
Warning! Read instructions!
Double insulated device (not to be earthed)
Must be mounted in a vertical position
Must be mounted in a horizontal position
Transistorized (electronic) meter
Transistorized meter (FET input)
DCV 10MΩ
Internal impedance for DC voltage measurements
ACV 10kΩ/V
Internal impedance for AC voltage measurements
DCA 316mV
Voltage drop across the instrument for DC current measurements

ANALOG (MOVING COIL) METERS

The operating principle and internal construction

In 1881 d'Arsonval patented a permanent magnet moving coil meter movement that remains in widespread use to this day. The horseshoe-shaped permanent magnet (1) creates a DC magnetic field that is made uniform across the airgap by inserting a soft iron cylinder (2). Inside the airgap is a lightweight spool (3) on which a coil with N windings of a fine magnet wire is wound. The spool is suspended on a shaft, freely rotating, pivoting on two jewel bearings (5). A fine lightweight needle-shaped "pointer" or "indicator" is attached to the spool.

The resting position (no current through the coil) of the spool/pointer can be manually adjusted by inserting a small screwdriver into a slot on the meter's face. This rotates a pin and the lever (6) attached to the spool/pointer assembly and mechanically zeroes the meter.

When an electric current flows through a conductor in a homogenous magnetic field, a force acts on such a current, $F=N*B*I*l$, and, as a result, a torque develops, $T=F*r$, where r = radial distance, N=number of turns in the coil, I = current through the coil, l= the coil's length perpendicular to the magnetic field, and B=magnetic field density. This electromagnetic torque is opposed by the mechanical torque of two conductive springs (4), whose strength is precisely calibrated so that a known DC current causes a rotation of the pointer to a certain angle. The two springs also provide an electrical connection between the moving coil and the outside meter terminals (7 & 8).

The torque determines the sensitivity of the meter; the greater the torque, the lower the DC currents that can be detected and indicated. Factors B, l, and r are limited by the permanent magnet material properties and by the miniature size of the movement, so the meter sensitivity can generally only be increased by a higher number of turns. However, there is no free lunch in life or electronics, for that matter. Increasing N increases the weight of the coil and, more importantly, the total length of the winding and internal resistance R_I of the coil. Ideally, such resistance should be zero, so the higher R_I, the more the meter will depart from the ideal case and the higher the error if such a meter is used to measure current.

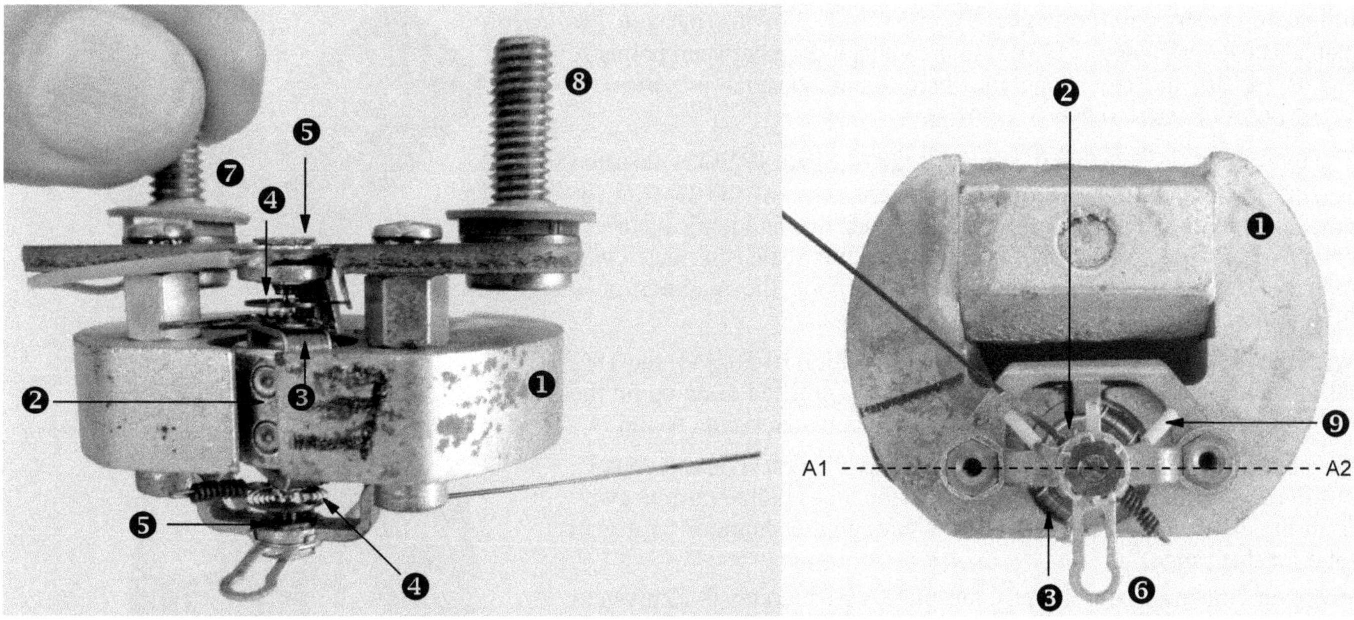

ABOVE: The construction details of a typical moving-coil meter mechanism 1) "Horseshoe" permanent magnet 2) Cylindrical soft iron core 3) Moving coil on a spool 4) Restraining springs 5) Shaft and bearings 6) Mechanical zero adjustment lever 7) "+" threaded external meter terminal, connected by wire to one end of the moving coil through the upper spring 8) "-" threaded terminal, connected mechanically to the permanent magnet, lower spring and the metal assembly 9) End stops (limiters) for the pointer

RIGHT: Simplified view of the cross-section A1-A2, with dimensions "r" and "*l*" marked

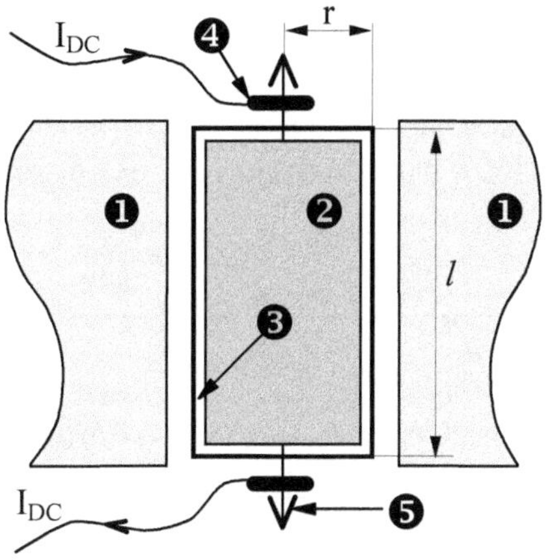

If all the constants are known, and the meter's face (scale) is suitably marked and calibrated, a moving coil meter measures only DC current. All other electrical quantities, such as AC current, AC & DC voltage, and resistance, will have to be converted to or expressed in terms of DC current.

Likewise, to measure current above the meter's natural full-scale deflection (FSD) current, some sort of switch-selectable shunting arrangement must be included to divert the higher currents away from the meter's fragile coil, which can quickly burn out if overloaded.

Most meters are unidirectional, having zero at the left end of the scale, but there are also bidirectional meters, primarily used in bridges and balanced instruments, where zero is in the center of the scale.

REVERSE ENGINEERING EXERCISE: Shunts and scales for DC current measurements

This large moving coil meter was salvaged from a vintage Japanese-made TEC transistor tester, where it was used to display collector current. The meter was made in the USA by Precision Meter Co., Inc. The front markings of interest are "accuracy 2% FS" (1), where FS stands for "Full Scale" and "F.S. 5mA" (2).

The mounting board at the meter's back carries three shunt resistors, starting from the meter's positive terminal (3); their values are 0.2Ω (4), 1.8Ω (5), and 18.0Ω (6). A hookup wire was soldered to each of the junctions, four in total, meaning the transistor tester had three current ranges. How did this circuit work, and what were the ranges?

To perform this analysis, we are only missing one figure: the meter's internal resistance, which we measured as 20.5Ω. I hooked up the digital multimeter to the analog meter with short test leads terminated with crocodile clips to take the photo.

Before doing this kind of low resistance measurement, make sure your multimeter or ohmmeter displays 0 ohms when you short its test probes. If it doesn't (many cheaper meters read 0.2-0.6 Ω instead of zero), you'll have to subtract that figure from all test results. However, one of the most important lessons learned in T&M field is never to use cheap extension leads for low resistance measurements. When I measured the meter directly, the result was 19.9Ω, meaning the 300mm long test leads had incredibly high resistance, 0.6Ω!

Let's start with all three resistors in the circuit. In all cases, we look at full-scale currents and voltages. There is 5mA flowing through the meter whose resistance is 19.9Ω, so the voltage drop between points A and B (across the meter and the three shunt resistors in series) is V_M=0.005A*19.9Ω= 0.0995V or 99.5mV_{DC}.

The total shunt resistance is 0.2+1.8+18=20W. Since 19.9Ω is the same as 20Ω (within the margin of measurement error of the DMM) for practical intents and purposes, we can conclude that an identical current of 5mA will flow through the shunt. So, at full scale, the meter will indicate 10mA, and the user will have to divide the readings on a 0-100mA scale by 10.

With the "Range" switch in the following position (0-50mA), the 18Ω resistor will be short-circuited, and we get a simplified drawing on the right. Now we need to figure out the current between points A and D.

We have a 20Ω meter and a 2Ω shunt. Intuitively, you should feel that a 10-time lower shunt resistance will divert ten times higher current away from the meter so that 50mA will flow through the shunt, or 55 mA in total for that range.

However, the transistor tester's faceplate was marked 0-10mA and 0-50mA, not 0-55mA! The shunt current needs to be 45mA, and since the voltage across the meter and the shunt is the same, 99.5mV, the total shunt resistance needs to be 99.5/45 = 2.21Ω instead of 2.0Ω. One explanation is that the instrument's designers allowed for the additional 0.2Ω contribution of the test probes and internal connections.

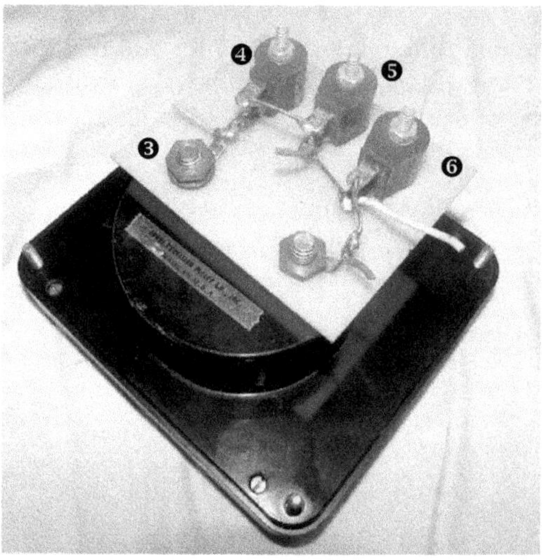

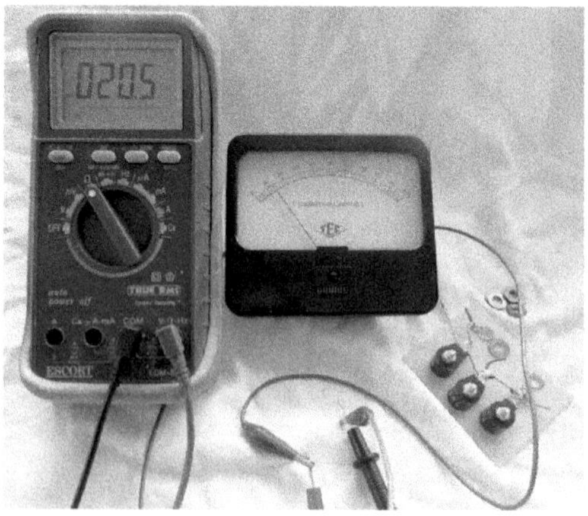

ABOVE: Using "universal" leads and crocodile clips the internal resistance of the meter was measured as 20.5 ohms. The correct value, without extension leans and clips was 19.9 ohms!

BELOW LEFT: 0-10mA range on a 0-5mA meter

BELOW MIDDLE: 0-50mA range on a 0-5mA meter

BELOW RIGHT: 0-500mA range on a 0-5mA meter

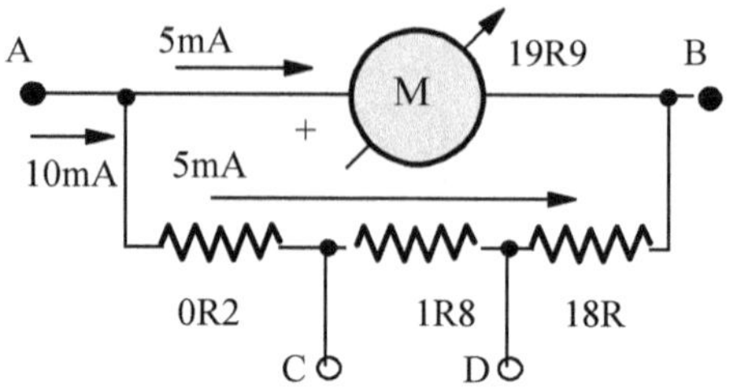

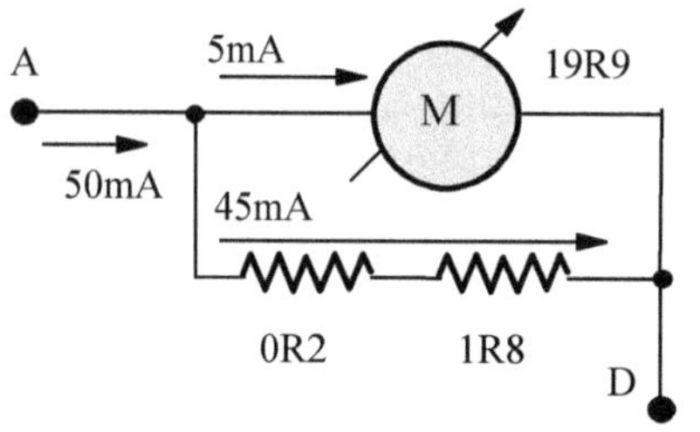

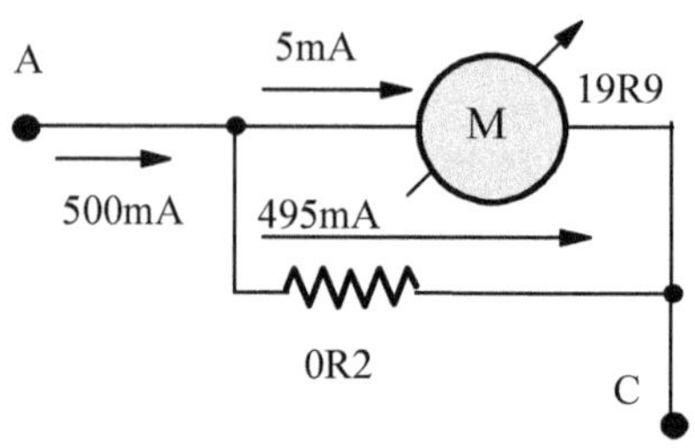

Finally, with the "Range" switch into the third position, both 18R and 1R8 resistor will be short-circuited and we get a simplified drawing so we can determine the current between points A and C.

0.2R ("R" is often used for "ohms") is a 100X lower resistance than 20R, so shunt current will be 100 times higher than meter current, or 500mA. The tester's range was 0-500mA, so the shunt current needed to be 500-5=495mA, meaning the shunt resistance of 99.5/495=0.201 Ω was needed. Indeed, we measured 0.2Ω (a multimeter cannot measure 1mΩ)!

Moving coil meters (MCM): measuring internal resistance and FSD (full scale deflection) current

Apart from using a precise ohmmeter, such as the one inside a digital multimeter, other methods can be used to determine an analog meter's internal resistance. In the test circuit below, turn the 200 or 250 kΩ potentiometer R_S (linear taper) to maximum resistance and connect the battery. Adjust R_S until the unknown meter M_X shows Full-Scale Deflection. Connect 5 kΩ R_P pot (linear taper) in parallel with the meter and adjust the wiper until the meter under test shows 1/2 FSD. Disconnect R_P and measure its resistance between the wiper and the upper end, equal to R_I (internal meter resistance).

To measure the analog meter's FSD current, turn the 200-250 kΩ pot R_S to maximum resistance (slider or wiper to the far right) and the 5 kΩ pot to the minimum (wiper in the uppermost position). Connect the 1.5V battery or 1.5V DC power supply. Adjust the two pots until the tested meter M_X shows FSD. Read the FSD current on the reference ammeter MR. This can be another calibrated moving coil meter or a multimeter set on the mA range.

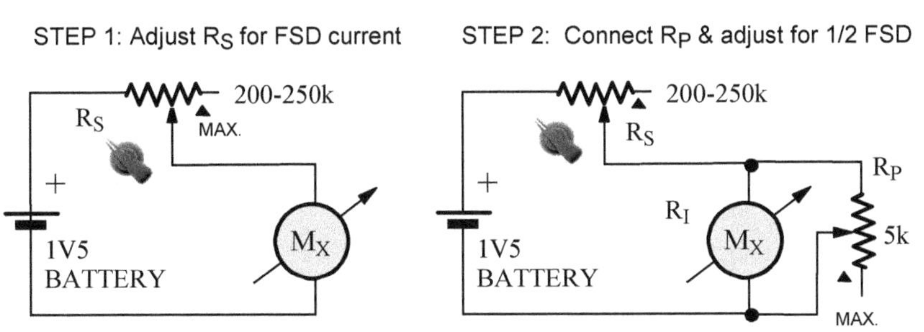

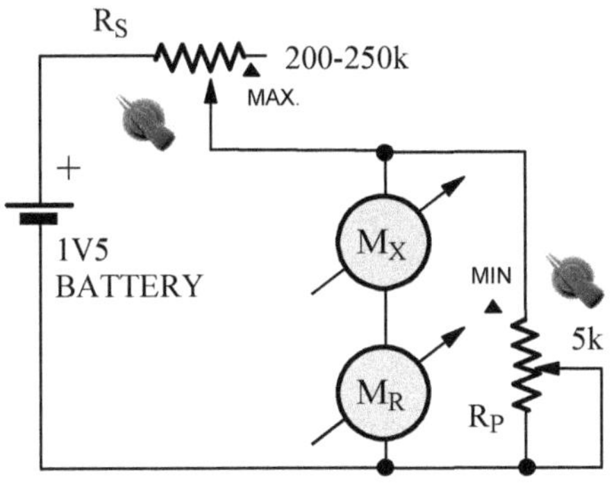

ABOVE LEFT: measuring MCM's internal resistance R_I

ABOVE RIGHT: measuring MCM's FSD

Replacing analog meters

If you have an instrument whose meter is damaged, sticking, or has a burned-out moving coil, you can replace it not just with an identical meter (often impossible to find) but also with a meter of different (higher) sensitivity and lower internal resistance. As a rule, the replacement meter must be of equal or lower FSD (Full-Scale Deflection) and of equal or lower internal resistance! It cannot be of higher FSD or higher internal resistance.

Say you need to replace the analog meter on the Hickok KS15560 tube tester. The meter has an FSD of 200mA and internal resistance of 2,365Ω. You have a 100mA FSD meter from a dead Hickok 600A tester. That meter has an internal resistance of 1,165Ω. Calculate the required values of the shunt RS and series resistor RA to be added.

Step #1: Determine the current to be diverted through the shunt resistor. In this case, I_2=100mA will go through the new meter (M2) at full-scale deflection, the old meter (M1) had I_1=200mA through it, so the shunt has to divert $I_S=I_1-I_2$=100mA.

Step #2: determine R_S. Since half the old current has to go through the shunt, R_S has to have the same value as the new meter's resistance, 1,165Ω. Each will take half of the 200μA! Generally, you will have to calculate the R_S from the formula for the voltage drop, $I_2 \times R_{M2} = I_S \times R_S$. You know all three, except R_S, so you can calculate it easily.

Step #3: Determine R_A. Calculate the total resistance of the new meter and R_S in parallel ($R_{M2}||R_S$). Then, subtract that figure from the original meter's resistance and that is the value of the required R_A.

We have 1,165Ω and 1,165Ω in parallel, so the total resistance is half of 1,165 or 582Ω.

The old meter had R_{M1}=2,365Ω, so we need R_A= 2,365 - 582 = 1,783 Ω in series. A standard 1k8 (1,800Ω) value should be sufficiently close.

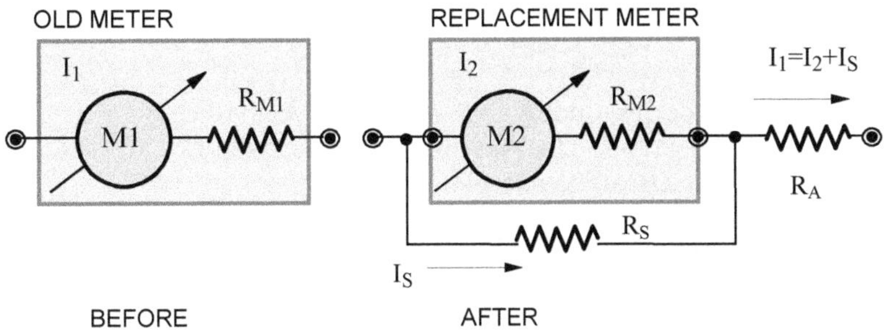

DC VOLTMETERS

The evolution of a multirange voltmeter

We know by now that each "analog" or moving coil meter can be fully specified by two of its three electrical parameters, either by its FSD (Full-Scale Deflection) current and the voltage drop across its coil, or, more commonly, by its FSD current and internal resistance. The third parameter can be easily calculated using Ohm's Law in both cases.

Say we have a 50μA meter movement (FSD) with R_M=2,000Ω coil resistance. At full scale, the voltage across its terminals will be V=FSD*R_M =50*10^{-6}*2,000 = 0.1V or 100mV, which means we already have a simple, single-range milliVolt meter. If the scale is calibrated 0-100 with 100 fine graduations, that instrument could measure down to 1mV_{DC}, which is fairly sensitive.

But how do we measure higher voltages with it? Simple! Just add a resistor in series, and voila, to reach the same FSD, a much higher DC voltage will now be needed. That is why such a resistor R_S is called a multiplier resistor since it multiplies the voltage range of the instrument. Say we add a 98,000 Ω resistor in series (for the total resistance of 100 kΩ), the measured DC voltage that will produce a full-scale indication will now be V_X= FSD*(R_M+R_S) = 50mA*100kΩ = 5V! We get a multirange DC voltmeter by switching various resistors in series with the meter. All we need is a few resistors and a selector ("RANGE") switch, as illustrated below.

There is nothing wrong with such a parallel arrangement of resistors, except that various nonstandard resistance values are required, and those are expensive to manufacture (must be custom ordered in large quantities), so a series arrangement is almost universally used (above right). In that case, only the first resistor (R1) is of nonstandard value; all others have standard resistance values and can be obtained commercially in precision tolerances (1% or less).

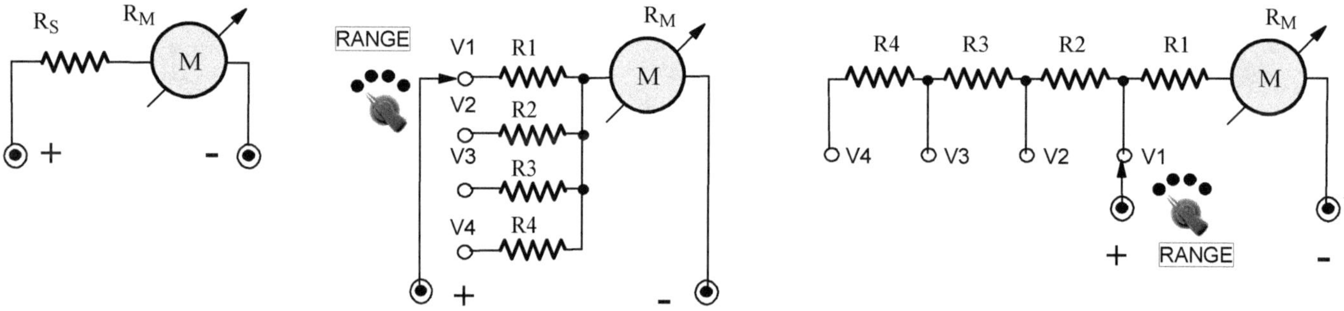

ABOVE LEFT: The basic DC voltmeter (single range) is simply the multiplier resistor R_S in series with the meter ABOVE MIDDLE: Multirange DC voltmeter (parallel multiplier resistors) ABOVE RIGHT: The more practical (series) arrangement of multiplier resistors

Practical (commercial) circuit

Using the vintage Simpson 260 VOM as an example, with six DCV ranges, only the first resistor (48k) is of nonstandard value, although it could easily be made up of the standard 1k and 47k values in series.

On its lowest range (2.5V), the total resistance is 48+2=50kΩ, so the FSD will be 2.5/50,000 = 50μA, the exact spec of the meter used! If you are really keen or don't trust me, prove to yourself that the FSD on all ranges will be the same, 50μA.

This also means that the voltmeter's sensitivity will also be the same on all ranges.

SENSITIVITY [Ω/V] = 1/FSD [μA] so 1/50*10-6 = 20,000 Ω/V. The importance of this figure will be discussed shortly when we take a deeper look into how the connection of such a voltmeter affects the measured circuit, also called the voltmeter loading effect.

To use the highest voltage range (5,000V), the range switch (1) must be in its highest position, marked 1,000/5,000V, and instead of the (+) terminal, the positive test probe must be plugged into a separate socket (2).

There is also a corresponding 5kV socket for AC voltage measurements next to it. This was necessary due to the 1kV limit on the contacts of the rotary switch used.

On the 5kV range, the additional 80MΩ multiplier resistor in series drops the voltage down to the 1kV level at the (+) terminal and 1kV terminals, which are switched together.

THE SENSITIVITY ON DC VOLTAGE RANGES DEPENDS ON THE METER'S FULL SCALE DEFLECTION CURRENT

SENSITIVITY [Ω/V] = 1/FSD [μA]

BELOW: Multiplier resistors in a commercial DC voltmeter with 6 ranges (Simpson 260 VOM)

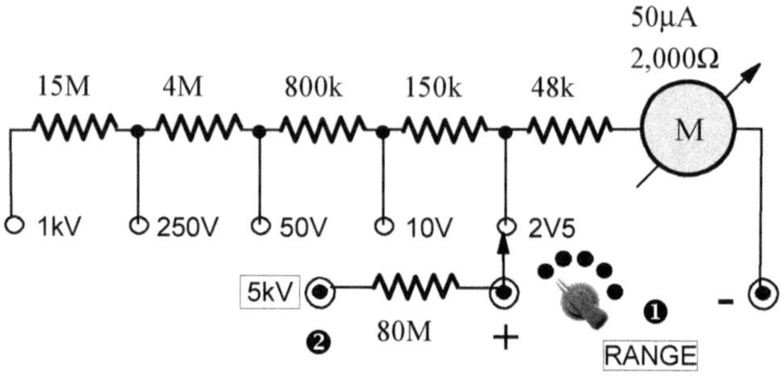

The voltmeter loading effect

When it comes to tests & measurements, the first common mistake is measuring the wrong thing, and the second is measuring the right thing the wrong way. This case is about the later error. Say you want to measure a DC voltage between points A and B of this simple circuit (below left). Since R1 and R2 are equal, the voltage across each resistor is half the battery voltage of 12V or 6V. The whole circuit is a simple resistive voltage divider. You have a VOM such as Simpson 260 just discussed, and on its scale, it says "DC: 20 kΩ/V" but you pay no attention to that detail. You turn the selector knob to the $10V_{DC}$ range". What will be the reading of the multimeter?

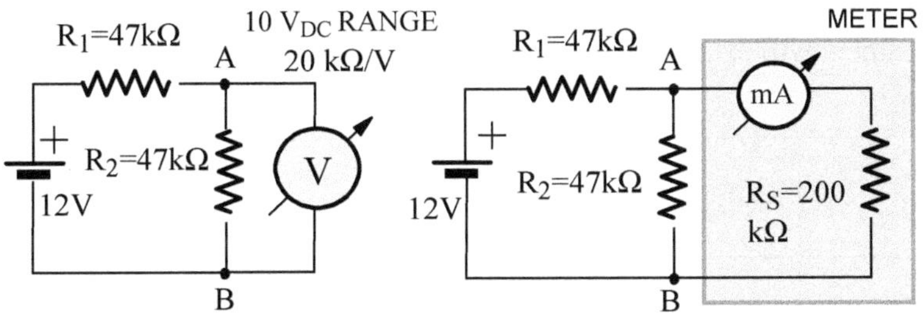

LEFT: How a voltmeter with a low input resistance (20 kΩ/V) loads the measured circuit and produces inaccurate results!

We have seen that the "voltmeter" in a simple VOM is just a milliammeter with a series resistor added. On a 10V range, that resistor is 200 kΩ; that is why its input resistance is specified as 20kΩ/V or 200kΩ in this case. The voltmeter's presence changes the conditions in the circuit. That "upset" is called the voltmeter loading effect since the voltmeter's finite internal resistance is "loading" the circuit.

R_{EQ}=47*200/(47+200) =38kΩ. The equivalent resistance is the 47kΩ resistor in parallel with a 200kΩ internal resistance of the voltmeter, around 38kΩ. The voltage measured between points A and B is now $V_M=12*R_{EQ}/(R_1+R_{EQ})$ = 12*38/(47+38) = 12*0.447 =5.36V.

The error introduced by the voltmeter's internal resistance is $\varepsilon = (V_M-V)/V*100\% = (5.36-6)/6*100\% = -10.67\%$! V_M is the measured voltage, V is the true voltage. The error is negative since the measured value is lower than the true value. Of course, for troubleshooting purposes, a +/-10% error may be acceptable, but in some high impedance tube circuits, the error could be 50% or even larger. The voltmeter loading effect is only one reason why cheap "passive" multimeters and voltmeters are not suitable for any serious audio work. Invest in a good quality, high input impedance (10MΩ+) true-RMS multimeter!

AC VOLTMETERS

To measure AC voltages and currents, VOMs use a simple half-wave rectifier (D1) and the same MC analog meter used for DC measurements. The additional diode D2 (often packed together with the main rectifier diode D1 into one package called "instrument rectifier") is reverse-biased on the positive half cycle of the sine input voltage. It does not affect the rest of the circuit. When forward-biased during the negative half-cycles, D2 provides an alternative path for the leakage current of D1, which is now reverse-biased. That reverse current would otherwise flow through D1 and the meter.

As with DC voltmeter circuits, R_S is the multiplier resistor; only one is shown here, but five or six of them would be switch-selectable, depending on the particular VOM. The purpose of the shunt resistor R_{SH} is to increase the current flow through the rectifier D1 so that it operates in a more linear portion of its characteristic curve. Its resistance is usually in the 200-400 Ω range.

Rectifiers tend to be nonlinear at low current levels, which shows at the bottom (left) end of the scale, so shunting the meter solves this problem, if not entirely, then at least cheaply and easily.

That shunting action decreases the current through the meter, and further reduces the sensitivity of the VOM on AC ranges. The principal reduction in sensitivity is due to half-wave rectification, where the average of the "DC" voltage (or rather pulsating peaks) is only 45% of the RMS or effective value of the measured AC sine voltage. So, the AC voltmeter's sensitivity is practically reduced by the same percentage. With the added impact of R_{SH}, the sensitivity of a typical VOM such as Simpson 260 goes down from 20kΩ/V on DC ranges to only 5kΩ/V for AC (25%)!

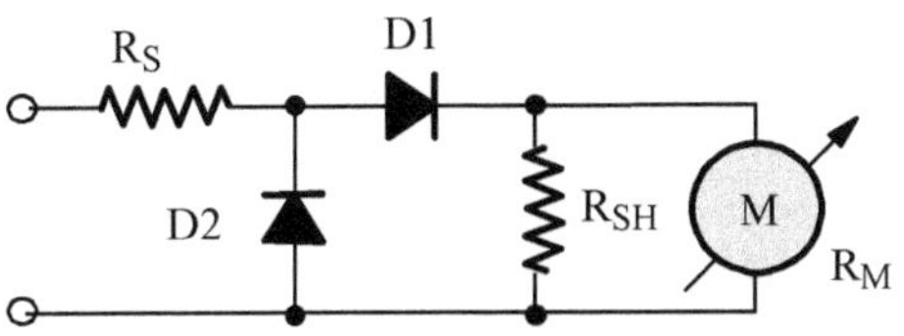

ABOVE: Half-wave rectification is commonly used in VOMs to measure AC voltages

True RMS multimeters versus average-responding RMS-calibrated (ARRC) multimeters

The Root-Mean-Square (RMS) value of a signal such as an AC voltage or current is that value that produces the same heating effect (dissipates the same power) on a resistor as the equivalent DC voltage. For that reason, it is also called an *effective* value (same heating or work-producing *effect*). It is mathematically defined as

$$V_{RMS}=\sqrt{\frac{1}{T}\int_{t_0}^{t_0+T} v^2(t)dt}$$

The waveform is first squared, then the average (or "mean") value is calculated through the integration over one period ("T") of the signal. Finally, the square root is taken of the mean value. For sine wave $v(t)=V_P\sin(\omega t)$, where $\omega=2\pi f$ and frequency $f=1/T$ (T=period). After integrating and taking the square root, $V_{RMS}= V_P/\sqrt{2}=0.71V_P$!

The three values of an AC signal, the PEAK, the AVERAGE and the RMS (Root-Mean-Square) or effective value, are related through two factors. The CREST-factor is the ratio of its peak (crest) to its RMS value, or 1.41 for a sine wave. The FORM- factor of an AC signal is the ratio of its RMS value to its average value, or 1.11 for sine wave.

Most non-RMS instruments such as VOMs and cheaper digital multimeters measure average value, but their scale is then calibrated in RMS value by multiplying everything by 1.11. They are sometimes called average- responding- RMS-calibrated meters (ARRC). Therefore, they are *only* accurate for a pure (undistorted) sine wave! For a square wave the RMS value is the same as its average, so an ARRC instrument will display 11.1% more! The half-wave rectified pulses (1) are very common in tube testers and cheaper instruments.

Waveform	Crest Factor CF Form Factor FF	ARRC meter's indication	Relative error of ARRC instruments $\varepsilon=(V-V_{RMS})/V_{RMS}*100\%$
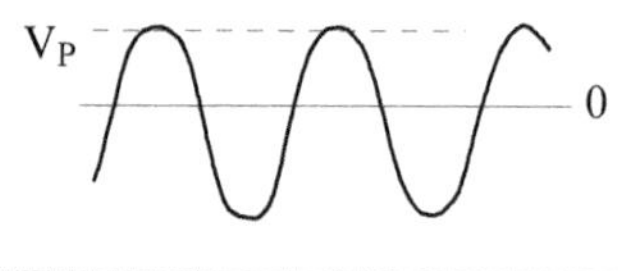	$V_{RMS}= V_P/\sqrt{2}=0.71V_P$ $V_{AV}=2V_P/\pi = 0.637V_P$ $CF=V_P/V_{RMS}=1.41$ $FF=V_{RMS}/V_{AV}=1.11$	$V=V_{AV}*1.11 =V_{RMS}$	0
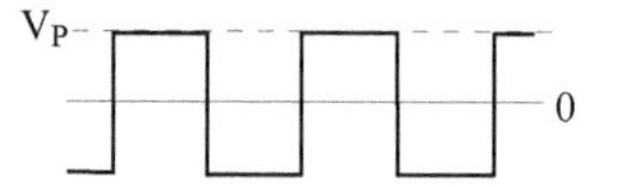	$V_{RMS}= V_P$ $V_{AV}=V_P$ $CF=V_P/V_{RMS}=1$ $FF=V_{RMS}/V_{AV}=1$	$V=V_{AV}*1.11 =1.11*V_{RMS}$	+11%
	$V_{RMS}= V_P/2=0.5V_P$ $V_{AV}=V_P/\pi = 0.318V_P$ $CF=V_P/V_{RMS}=2$ $FF=V_{RMS}/V_{AV}=\pi/2=1.57$	$V=V_{AV}*1.11 =0.71*V_{RMS}$	-29%
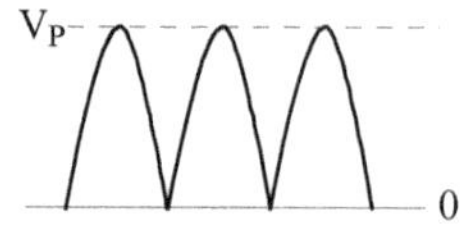	$V_{RMS}= V_P/\sqrt{2}=0.71V_P$ $V_{AV}=2V_P/\pi = 0.637V_P$ $CF=V_P/V_{RMS}=1.41$ $FF=V_{RMS}/V_{AV}=1.11$	$V=V_{AV}*1.11 =0.71*V_{RMS}$	0

Case study: Same tubes, same model tube tester, different anode currents

You sell a matched pair of NOS Telefunken EL34 tubes on eBay for a small fortune to a guy who believes that vintage German tubes are the best of the best. In a week, the buyer lodges a complaint with PayPal claiming that goods aren't "as described," and PayPal removes the money from your account and refunds the buyer, no questions asked!

Using the precision resistor method (a 10Ω resistor in series with TUT's anode), you measured the plate currents on your Triplett 3423 tester. You got the readings of 486mV (corresponding to the anode current of 48.6 mA) and 482mV (48.2 mA), as close a match as one could hope for.

The buyer had the same current test circuit installed in his 3423 tester (both learned that trick from my book "How to Use, Calibrate, Repair and Upgrade Vacuum Tube Testers"), re-tested the tubes, but his plate currents were only 34 mA. He claims that it is way too low, that the tubes are not new, and their emission is way down. What would you do?

In this case, the clue is not in the calibration difference between your two testers but in the multimeters used to measure plate currents and how this type of tester operates. The TUT (tube-under-test) has AC voltages on its anode and grid, so the tested tube works as a rectifier (it's connected as a simple diode), providing its own anode and bias "DC" voltages (right).

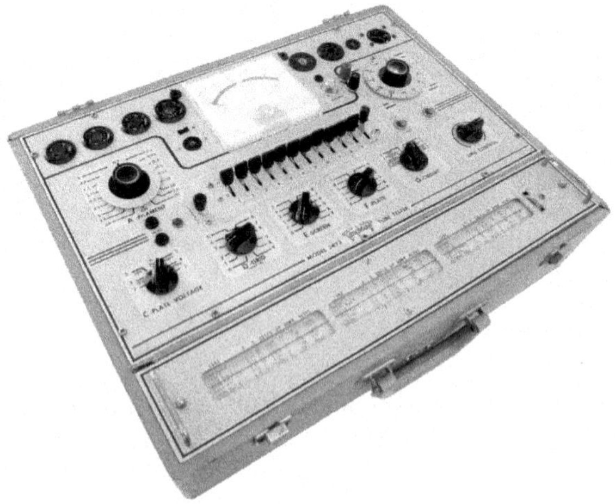

ABOVE: Tube testers such as Triplett 3423 use AC voltages for anode and grid bias, so TUTs' anode currents are self-rectified half-wave sinusoidal pulses (1)

BELOW: The simplified equivalent test circuit

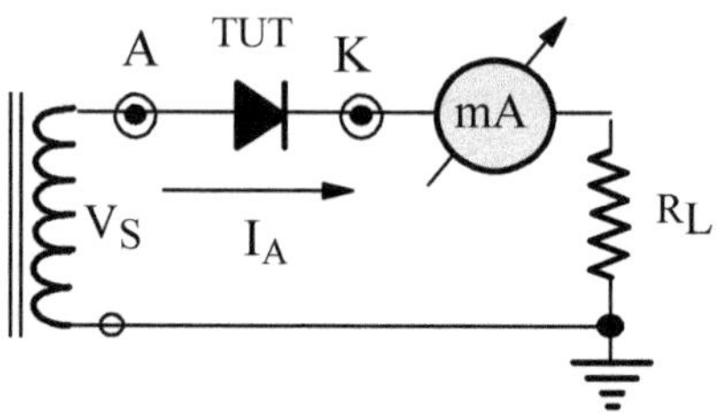

However, the "DC" is not the "proper" or smooth DC, but a series of rectified sine pulses as illustrated in the table above (2). A true-RMS meter will measure any waveform accurately. That is the meter you've used. Your buyer used a cheap analog ARRC multimeter, which measures an average value of the signal, but is calibrated in the RMS values for a sine wave, which is an average multiplied by 1.11 (11% more). Your buyer had no clue about that issue!

As we have seen (table on the previous page), for this type of pulsating waveform, the RMS value is 1/2 of the peak value V_{MAX}, and the average value is 31.8% of peak. The same applies to the current. A true-RMS meter will measure $0.5I_{MAX}$. The average-responding meter calibrated in RMS values will measure I_{AV}=$0.318I_{MAX}$, but will show 0.318*1.11*I_{MAX}= $0.353I_{MAX}$, so the ratio will be $0.353I_{MAX}/0.5I_{MAX}$ = 0.71!

A non-RMS meter will indicate 71% of the true value (an error of -29%). So, instead of the correct figures of around 48mA, the buyer got 0.71*48 = 34 mA!

OHMMETERS

Ohmmeter function on VOMs

When analog VOMs (Volt-Ohm-Milliammeters) work in ohmmeter mode, they still essentially measure current, which, according to Ohm's law, is $I=V_B/R_X$, proportional to the internal battery voltage VB but inversely proportional to the resistance of the unknown resistor R_X. This results in a highly nonlinear ohms scale, in contrast to voltage and current scales, which are always linear.

The second point of differentiation is that the zero on the ohms scale(s) is to the right when the FSD (Full-scale deflection) current flows through the meter. Again, with current and voltage measurements, the opposite is the case.Finally, the accuracy and resolution of the measurement depend not just on the range used, but due to the nonlinearity of the scale, on the actual value of the resistor measured.

For instance, looking at the vintage VOM by Simpson (model 260), with 220 Ω resistor on X1 range, the reading would be ambiguous and ultimately imprecise since the difference between 200 Ω and 300 Ω is only one minor division (1). Changing to the X100 range, the reading at (2) is fine since one minor division is now 20 Ω. A 22 kΩ resistor can also be relatively precisely measured on the X10,000 range at (2), although now one minor division is, just as the range is, 100X bigger, or 2,000 Ω!

What about higher resistance values? Say you need a 2M2 resistor (2.2 MΩ) for a tube circuit (which are of high impedance variety compared to low impedance bipolar transistor circuits). You have no choice but to use the 10,000 scale, so the reading will again be at (1). The 150 marking on the scale is now 1M5 (1,500,000 Ω), followed by 2M and 3M, so now the pointer will be somewhere between 2 and 3M, meaning the resolution of this test will be very low.

The annoying drawbacks of such an instrument are apparent. Each time the range is changed, the "Zero ohms" control must be adjusted, slowing down the use of analog ohmmeters. The accuracy is reduced as the battery ages since the "zero adjust" must eventually be set to a subnormal value (too low), upsetting the proportions of the ohmmeter circuit.

Without going into (unnecessary) detailed circuit calculations, the circuit (next page) can be analyzed from a functional perspective. On the X1 range, the 1,138 Ω and 21,850 Ω resistors are in series with the 10,000 Ω "Zero adjust" potentiometer and the meter itself, whose internal resistance is 2,000Ω. Thus, depending on the position of the potentiometer, we have up to 23k+10k=2k= 45 kΩ of resistance in parallel with the 11.5 Ω resistor so that such high parallel resistance can be neglected.

The 1.5V battery has an internal resistance, and so do test leads, so the actual input resistance of the X1 ohmmeter is around 12Ω. We have a battery in series with a 12Ω resistance, so if test leads are connected across a resistor with the same resistance (12Ω), the meter indicator will deflect to half scale.

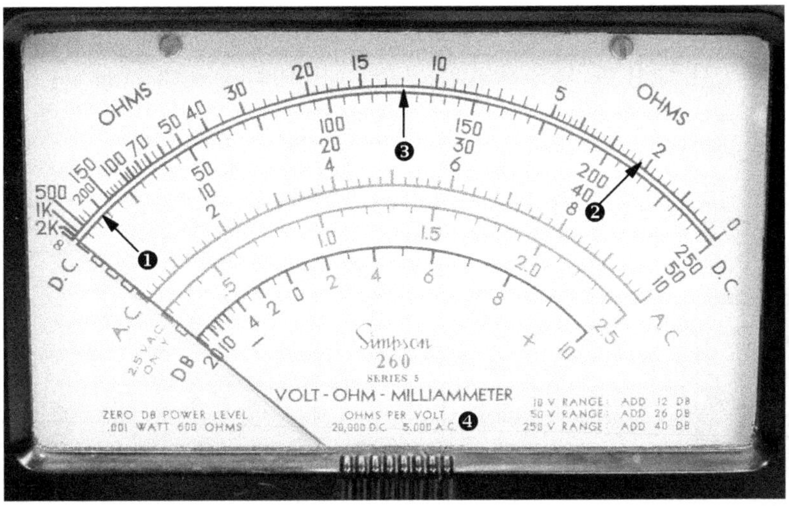

The meter faceplate (LEFT) and controls (BELOW) of a typical vintage analog VOM, Simpson 260

Notice its sensitivity of 20kΩ/V for DC and only 5kΩ/V for AC measurements (4). 20kΩ/V for DC means the meter is 1/20,000 or 50μA full scale.

Looking again at the meter's "OHMS" scale, we see that 12Ω is indeed at its center (3).

Since the same scale is used on all ohmmeter ranges, the test circuit proportions must be kept constant on all ranges. On the X100 range, 11R5 and 1,138R are in series (a total of 1149.5Ω), paralleled by the 2k meter, 10k pot and 21k85 resistor in series, or up to approx. 34kΩ. Again, 34kΩ is a much larger (30 times) shunt resistance compared to 1.1495kΩ, so it can be ignored.

However, to preserve the ratios between the ranges, a 110Ω resistor had to be added in series with one of the test leads. The input resistance is now 1,149.5+110=1,259.5Ω, not quite 1,200Ω (12Ω from X1 range multiplied by 100) but close enough. That is why the instrument must now be zeroed on the new range using the "Zero ohms" control (5), so the pointer will deflect to half-scale (3) when measuring a 1,200Ω resistor (1k2), one of the standard values.

The same scaling principle applies to the X10,000 range, 100 times the X100 range. However, we see an additional battery (6V) in the circuit. Higher resistances require a higher voltage to push the same current through, so one 1.5V battery would not be sufficient.

Electronic ohmmeters

Electronic ohmmeters (such as those inside digital multimeters) feature a CCS (Constant Current Source), which pushes a known and stable DC current through the unknown resistance R_X. The voltage drop V_X across the measured resistor is proportionate to its resistance (Ohm's Law, $V_X=I_0*R_X$), and that DC voltage is amplified by a DC amplifier and indicated on an analog or digital meter.

The accuracy depends on the stability of the regulated DC power supply, the CC Source, the DC amplifier and the meter.

Get to know your ohmmeter!

In some multimeters the positive lead (red) is internally connected to the - (negative) pole of the battery, as illustrated below. You need to mentally reverse the connections when testing with such ohmmeters since the black (negative) lead has the positive DC voltage on it. In that case it would seem that everything is opposite.

For instance, when testing semiconductor diodes, a diode would conduct when *seemingly* reversely polarized, which would not make sense until you understand the ohmmeter's internal battery connection. To figure it out on your ohmmeter simply connect it to a DC voltmeter, its reading will tell you if the red or the black probe is positive.

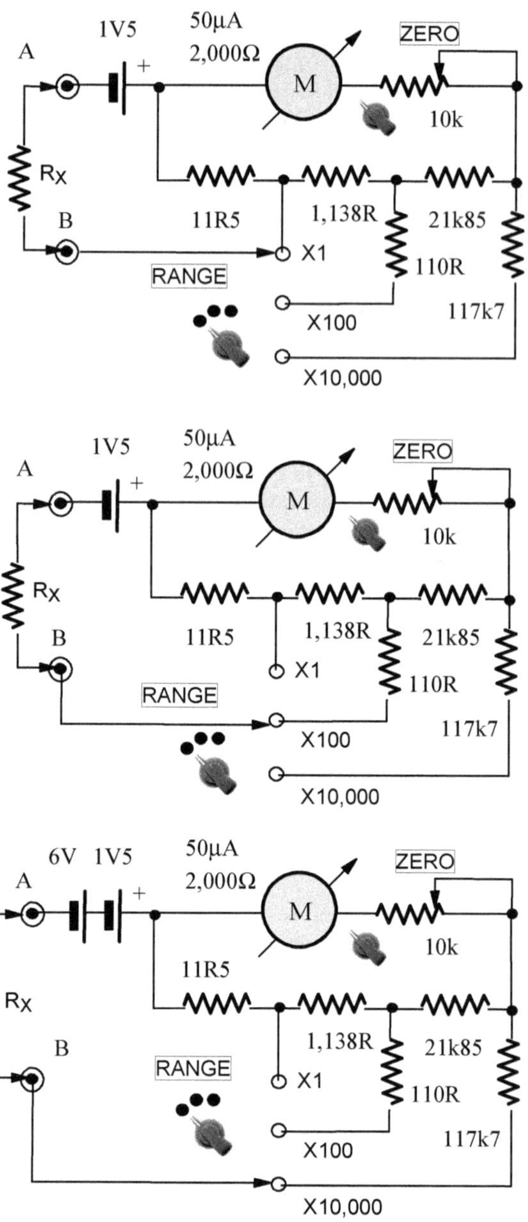

ABOVE: Resistance measurement on Simpson 260 on each of its three ranges

BELOW RIGHT: Electronic ohmmeters are quite simple

BELOW LEFT: Two types of ohmmeters, with +DC voltage on the red test lead and with the negative (-) DC voltage on the red lead

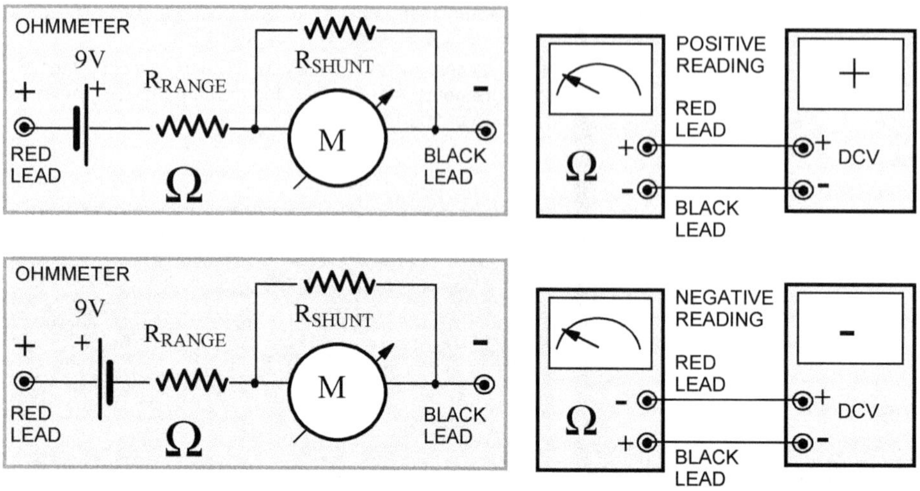

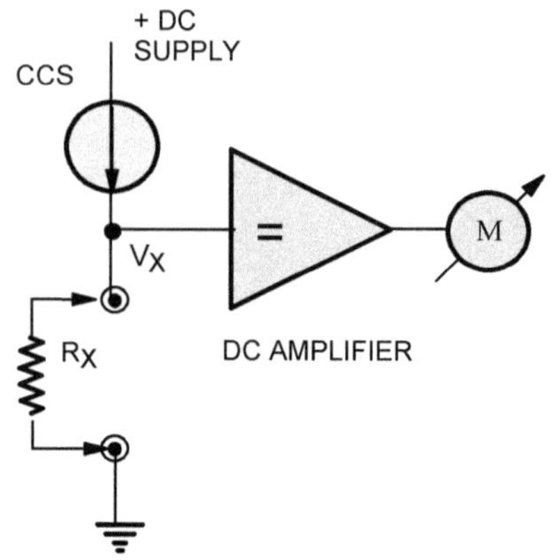

VACUUM TUBE VOLTMETERS (VTVM)

Principle of operation

The earliest VTVMs used a single triode as a linear amplifier. While the circuit had distinct advantages over passive VOMs (such as Simpson 260 used in our discussions), the chief of which was its much higher input impedance, its accuracy was affected by various factors such as the aging and weakening of the tube. Since the cost of duo-triodes wasn't any higher than that of single ones, in their wisdom, somebody finally realized that using two triodes in a differential amplifier configuration would solve all those problems and provide a much more stable and accurate test platform.

Both triodes inside one glass envelope would age at the same rate; the circuit would be much more immune to drift, interference, and other internal and external disturbances, which would affect both tubes equally and have no effect on test results due to the inherent balanced nature of the differential amplifier.

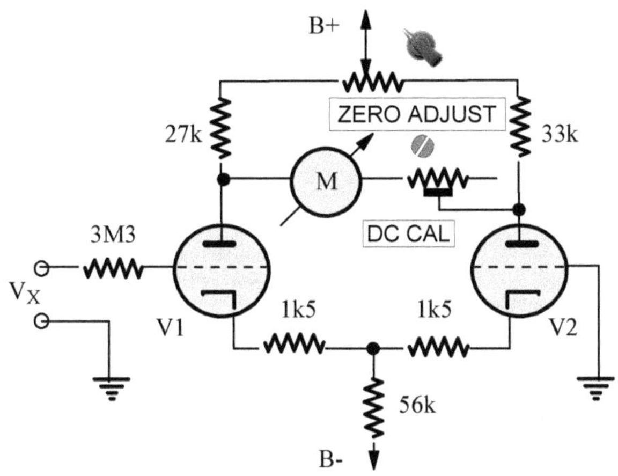

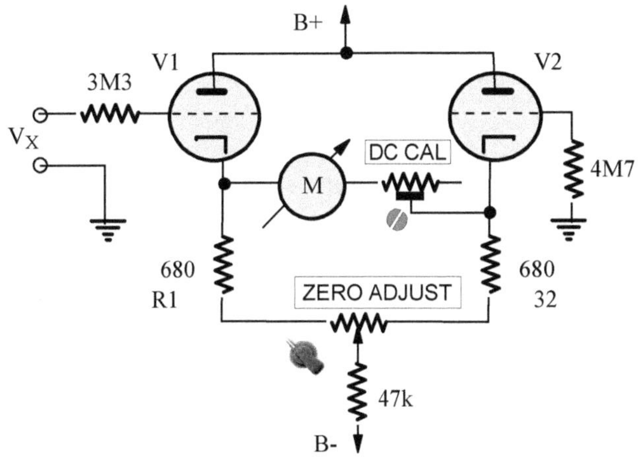

ABOVE LEFT: The simplified VTVM circuit with MC meter across the two anodes, as used in Eico models 221 and 235, RCA WV-98C, B&K 177-V95, Leader LV-76B, and Sencore 177. Resistor values and tube types may vary.

ABOVE RIGHT: The simplified VTVM circuit used in Eico models 222, 249, and 232, and Triplett 850 with MC meter across the two cathodes

RIGHT: While the earlier Heathkit VTVMs had only six ranges, later models and those of other T&M equipment makers, such as this PACO V-70 VTVM, had seven voltage and resistance ranges

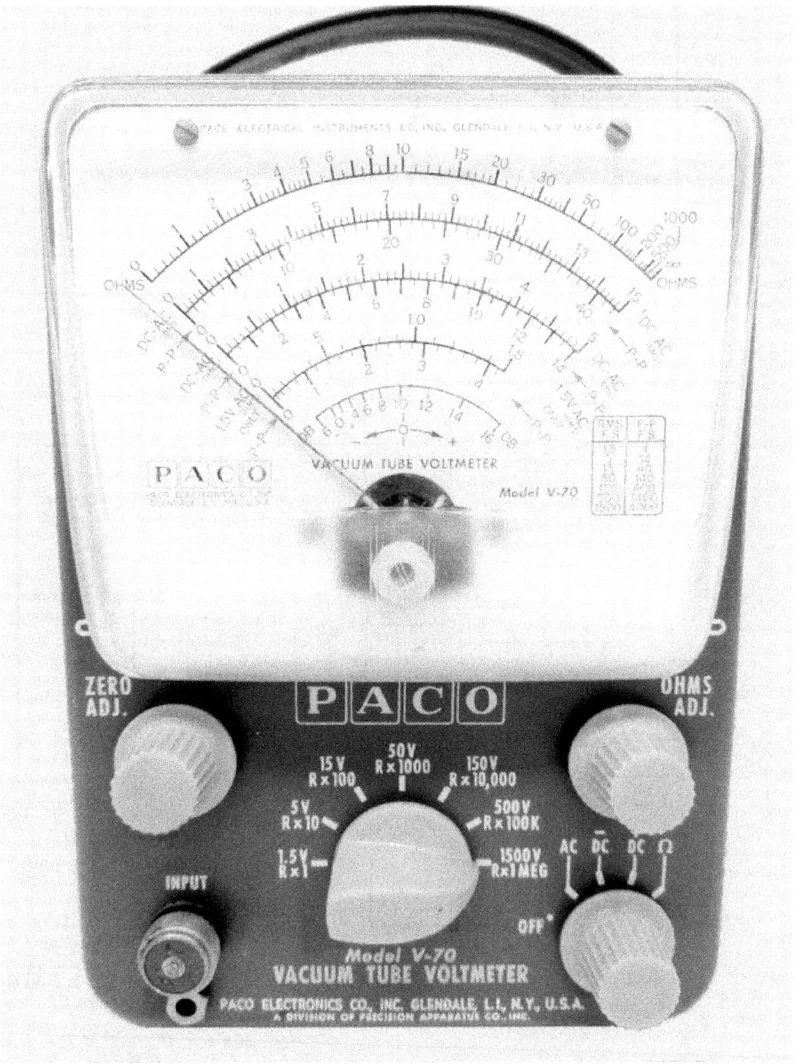

The two "halves" of duo-triodes had to be reasonably matched, though, but that was not difficult or expensive to achieve. Earlier VTVMs used Octal tubes such as 6SN7, the later ones Noval triodes such as 12AU7. Both of these are low m tubes (voltage gain); the high gain of the differential amplifier isn't needed in this application.

My book "Audiophile Vacuum Tube Amplifiers" (Volume 1) provides a detailed analysis of differential amplifiers, so we will not go into such detail here. The meter is connected in a bridge arrangement, either between the two anodes or the two cathodes. Tubes V1 and V2 comprise two of the bridge arms and resistors R1 and R2 the other two. Anode (plate) resistances of V1 and V2 change when an input (measured) voltage VX is applied to the grid of V1. The plate resistance of V1 will decrease, while due to higher bias on its cathode, caused by the higher current through V1, the resistance of V2 will increase.

Thus, the bridge is unbalanced, and a current flows through the meter. The mathematical analysis will show that such current is linearly proportional to the measured input DC voltage VX! Before such signal was applied, the anodes and cathodes of both tubes were on the same potential, and there was no current through the meter. The "Zero Adjust" control compensated for minor differences between the two bridge halves.

The "Ohms" zero still needs to be adjusted, as in cheaper VOMs, and there are typically three internal trimmer calibration potentiometers, "AC cal.", "DC cal." and "AC Balance." In an early Heathkit V-5A model VTVM circuit below, notice a separate "AC & ohms" input, later models such as PACO V-70 had only one input.

Some VTVM models (Eico 232, Heathkit V-7, RCA VoltOhmyst WV-77E, and WV-98C) are peak responding on AC ranges, using a peak-to-peak rectifier circuit, also called a "voltage doubler" or "diode pump," just like the voltage doubler inside the power supplies of some audiophile amplifiers.

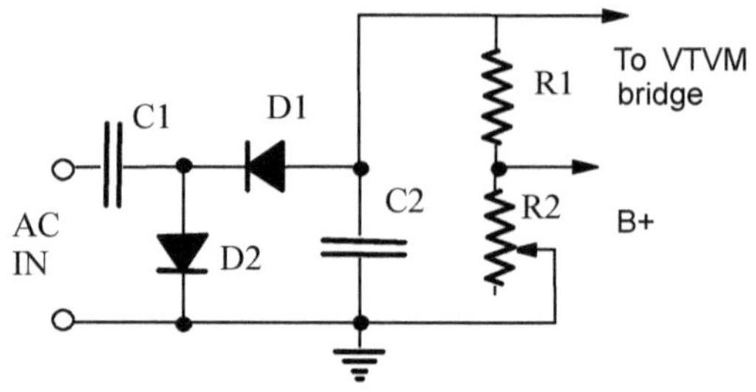

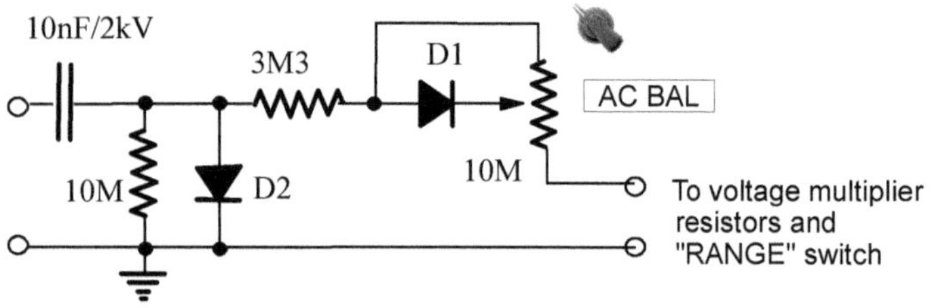

ABOVE: Voltage doubler in peak-to-peak AC VTVM

BELOW: Circuit diagram of Heathkit V-5A VTVM with major functional blocks and controls identified and marked © Heathkit

LEFT: Heathkit V-5A tube rectifier circuit redrawn to help understand its operation

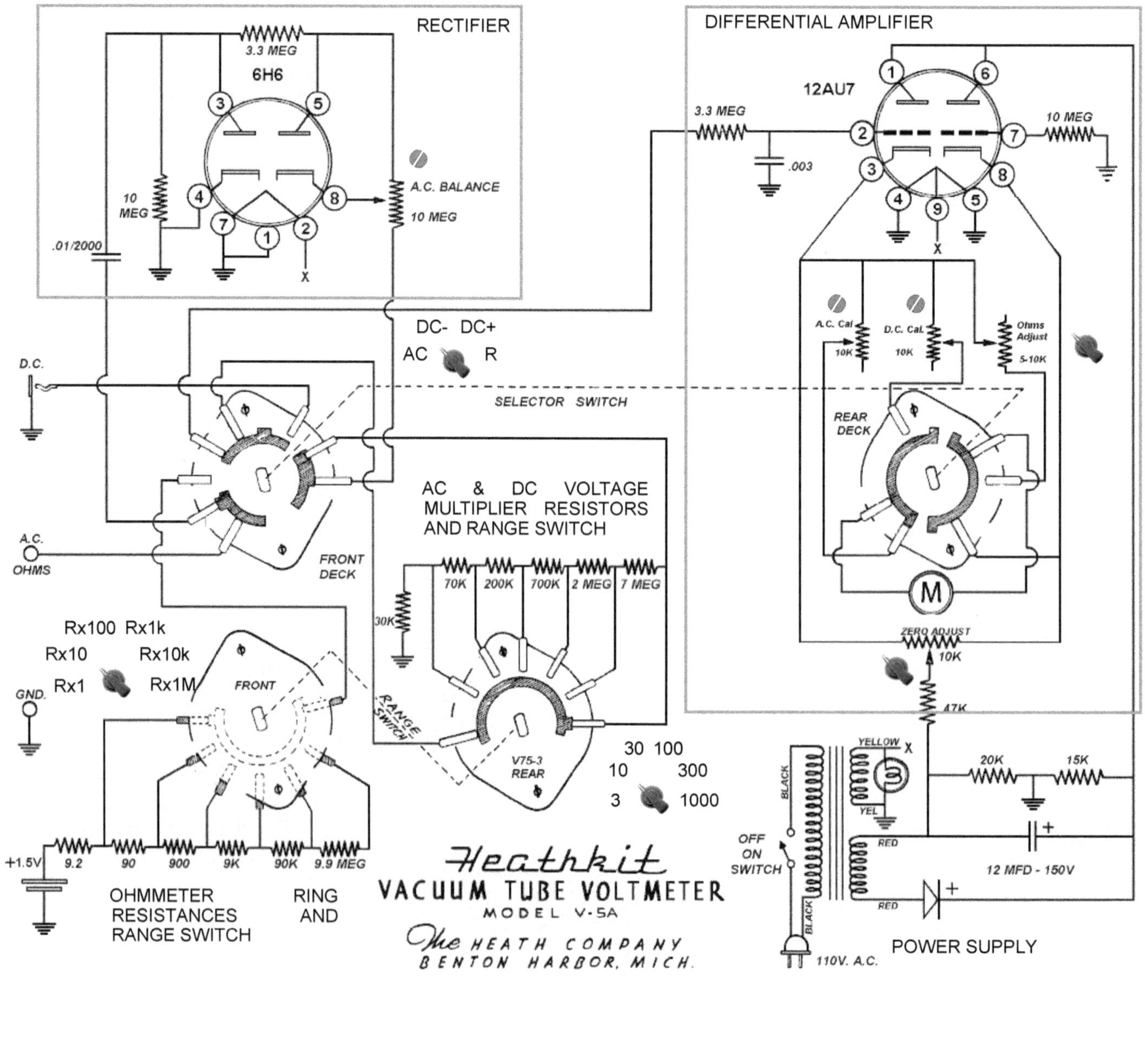

Audio VTVM

While general-purpose vacuum tube and solid-state voltmeters measure both AC and DC voltage plus resistance, why would Heathkit release a VTVM that only measured AC voltage?

While the lowest AC voltage range on VTVMs is usually 1.5 V_{RMS} full-scale, Heathkit AV-3 has the lowest range of 0.01 V_{RMS} (10 mV), allowing very low audio signals to be measured (1), such as those encountered in microphone preamplifiers and phono stages. Heathkit Audio Analyzer AA-1 also features the same circuit and 10mV full-scale sensitivity. The frequency response is quite wide, specified as 10Hz-400kHz (-1dB), and the input impedance is 1MΩ at 1kHz.

On higher input ranges (10-300V), the input is fed directly into its voltage divider (2) through the coupling capacitor and a precision series resistor. On lower 0.01-3 V ranges, the input AC signal first goes through a cathode follower (3) with a 6C4 triode and then into the voltage divider. The cathode follower is an impedance converter with high Z_{IN} and low Z_{OUT}.

After the voltage dividers, the measured signal is amplified by a cascode amplifier (12AT7 dual triode). Its output is directly coupled (4) to another cathode follower stage (5) with one half of the second 12AT7 tube, AC-coupled (6) to the other half, a common cathode amplifier stage. The un-bypassed cathode resistor (7) introduces a mild local negative feedback.

This stage drives the full-wave rectifier, a full bridge with four crystal diodes (8), and the moving coil meter. The other AC node (return) of the diode bridge is fed back to the cathode of the cascode amplifier's lower tube through an internal trimmer potentiometer (9), which changes the strength of the negative feedback and so sets the gain of the whole meter (calibration). Aside from the input CF, the whole instrument is a 3-stage audio amplifier with strong negative feedback to improve linearity and reduce noise and distortion.

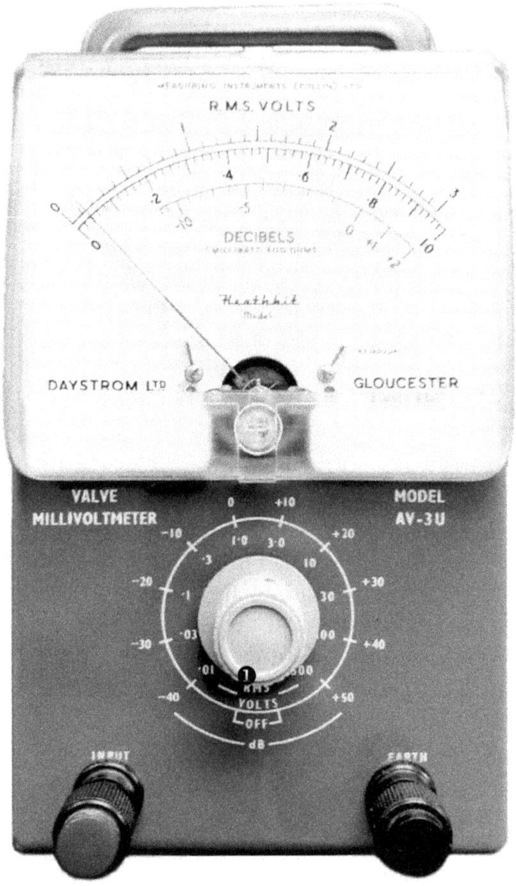

ABOVE: Some Heathkit equipment, such as this AV-3U audio VTVM, was also manufactured in the United Kingdom

Circuit diagram of Heathkit AV-3 valve (tube) millivoltmeter © Heathkit

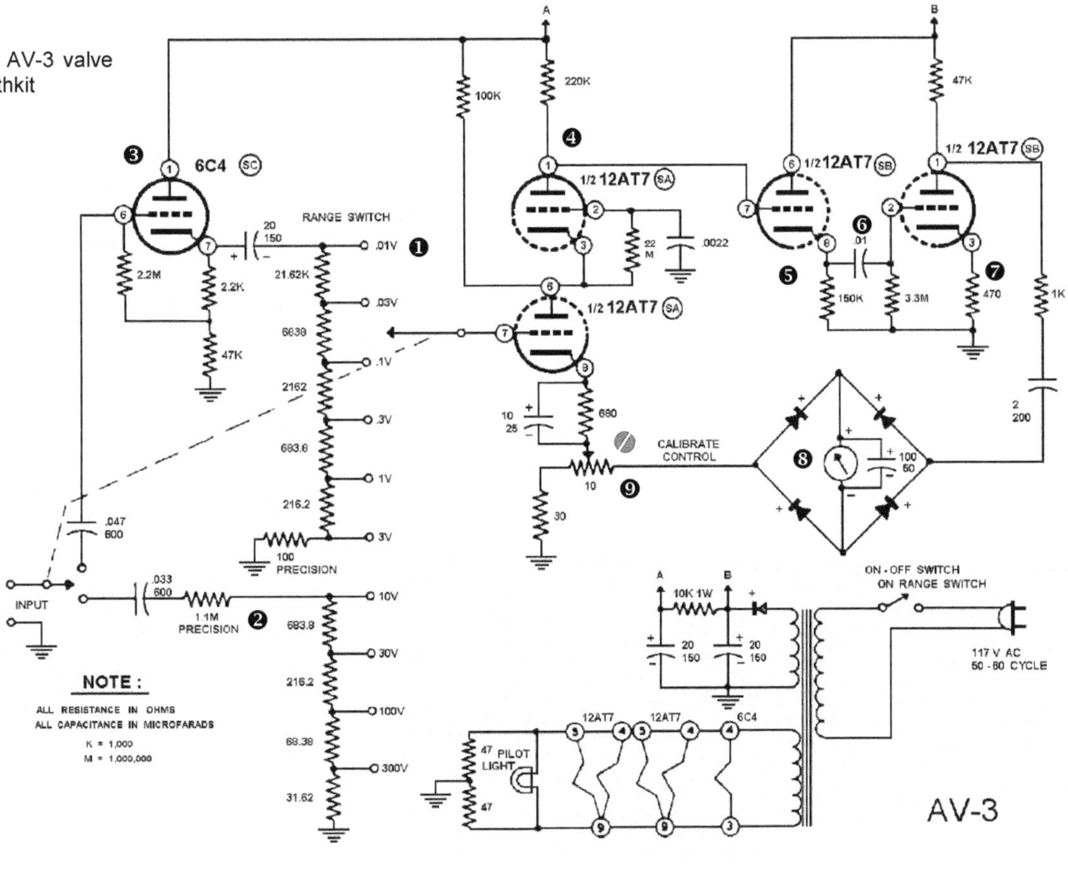

SOLID STATE ELECTRONIC MULTIMETERS

FET analog multimeter: Micronta FET Analog Multitester Cat No. 22-220

FET multimeters replaced their VTVM predecessors in the early 1970s. The black case of this Micronta model (sold by Radio Shack and Tandy chains but made in Japan and later in South Korea) makes it look dated, and it is dated; it hails from the late eighties, although its case design goes way back to late 1960s! The case is identical to its predecessor, model 22-208, of 1968 vintage. Both are quality test instruments, with a large and precise mirrored scale, model 22-220 being much more complex and technically improved. Priced at US$49.95 on its release in 1988, this multitester wasn't cheap by any standard.

Although it says that it's "fuse protected," the fuse did not save it when we left it on "ohms" and connected it to a high DC voltage inside an amplifier. The resistance measurement stopped working altogether. A circuit diagram is available on the web, so it can be easily fixed.

The "dual FET" symbol (1) indicates a FET differential amplifier. A 10MΩ input impedance for DCV is respectable (2). When the instrument is used as a DC ammeter, the voltage drop is 316mV.The additional sockets complement the standard COM (3) and V-Ω-A (4) sockets for $10A_{DC}$ (5) and 1kV AC and DC connections to bypass the selector switch (8). The standard "Ohms adj." and "Zero adj." potentiometer controls are provided (9), just as with VTVMs.

Since some of you may be interested in parallels between audio amplifiers and test instruments (and there are many!), here is the circuit diagram of the differential amplifier inside this Micronta FET VOM.

The input attenuator is not shown for clarity. Notice the 6.2V Zener diode at the DC input, which means that this amplifier works below that voltage level.

Micronta FET Analog Multitester

- DC voltage ranges: 1-3-10-30-100-300V up
- Accuracy +/-3% of full scale value
- Resistance ranges: x1, x10, x1k, x10k, x100k (up to 20MΩ)
- DC Volts input impedance 10MΩ (32MΩ on 1kV range!)
- Frequency response 30Hz-200 kHz at 3V and 10 V ranges
- Sensitivity: 10kΩ/Volt for AC voltage, 10 MΩ for DCV
- 315 mV voltage drop for DC current
- 5" meter (25μA FSD) with 3-colour mirrored scale

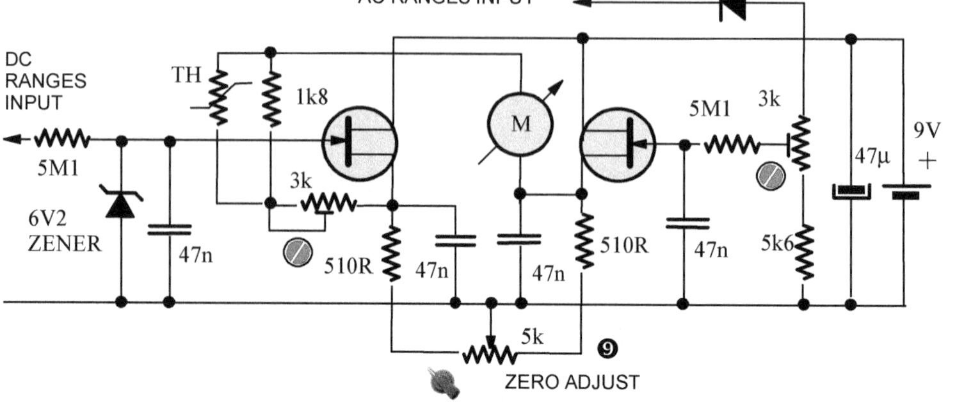

All measured voltages are attenuated to below $5V_{DC}$.

The moving coil meter is connected between the sources of the two JFETs, so the differential FET amplifier works as a cathode follower. Thermistor (temperature-dependent resistor) TH in series with the meter provides temperature compensation.

The JFET differential amplifier (LEFT) is the heart of Micronta Dual FET multimeter (ABOVE)

Notice that the 9V battery voltage is not regulated at all; as the battery ages, the voltage on FET's drains will fall, but that does not affect the accuracy of the measurements as long as the FETs stay in their linear operating range. That is one of the beauties of the differential arrangement, where both transistors (in this case) or tubes are equally affected by the gradual drop in supply voltage, so there is no overall impact on test results.

Philips PM2505 Electronic VAΩ Meter

Philips PM2505 Electronic VAΩ Meter also hails from late 1980s. Just above the On-Off switch is the main DC-AC-Ω selector (1). All test probe connections are in one line on the left side, the COM (2), the triple function V-Ω-continuity terminal (3), the μA-mA connection (4), and the A (higher current) terminal on top (5).

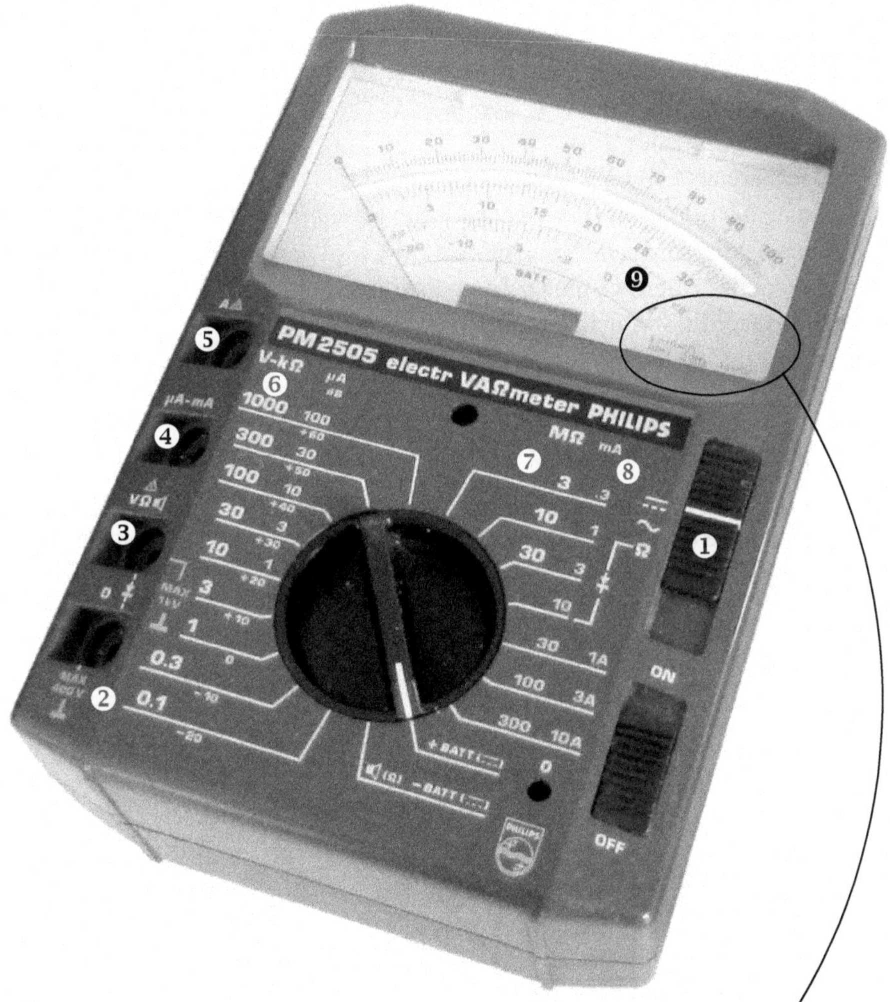

The range markings seem confusing at first glance, but one soon gets used to their logic. The "V-kΩ" markings are identical, from 0.1 to 1,000 in nine ranges on one side (6), the resistance ranges continue on the other side with three MΩ ranges (7), next to which are the seven mA and three ampere ranges (1-3-10A).

Notice the bottom log scale (9) in dB (-20 to +2dB) and two linear scales, 0-30 and 0-100, both with a few minor divisions "overruns," which is always a good idea.

There is no nonlinear resistance scale like on most other analog multimeters, which means that the instrument uses a constant current source to convert the resistance value into a DC voltage (see page 54). A linear relationship exists between the DC voltage across a resistor and its resistance (V=I*R).

This enlarged detail (right) illustrates a few important specifications of its meter deduced from the printed symbols.

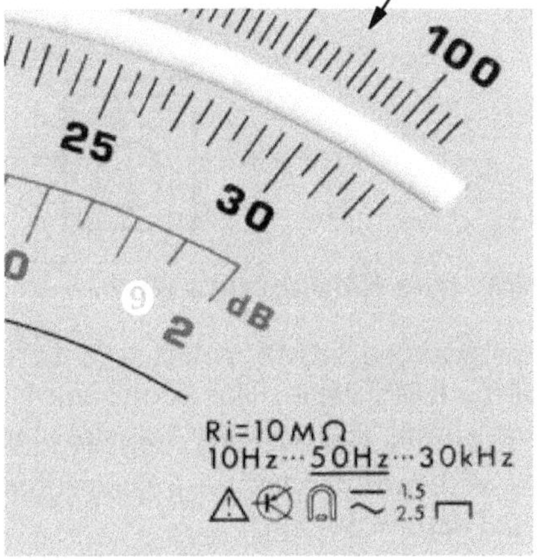

The internal resistance of the VAΩ meter is 10 MΩ, and the frequency range of the signals it can measure (within the specified accuracy) is 10 Hz to 30 kHz. The range extends to 70 kHz, but the accuracy is not guaranteed above 30 kHz. The underlined 50Hz frequency means that its accuracy is higher for 50Hz signals (in this case, 2.5%, compared to 5% on other frequencies).

As for the symbols in the last row, the time has come to explain their meanings since these are international symbols and thus widely used.

The triangle with the exclamation mark warns the user to read the manual first before operating the meter. The transistor symbol indicates an active instrument with bipolar transistors (usually in a differential amplifier circuit). The third symbol indicates a permanent magnet moving coil meter.

The DC accuracy class of the instrument is 1.5% of the range used. Notice a higher DC accuracy (1.5%) than 2.5% for AC signals due to the non-linearity of the solid-state rectifier (AC/DC converter).

Finally, the last symbol indicates that the instrument must be mounted in a horizontal position on the bench, usually due to its analog moving coil meter. It will operate at any angle (there is an adjustable angled bracket provided at the back so it can be tilted at approx. 30 degrees as in the photo) and even horizontally, but then the specified accuracy is not guaranteed.

DIGITAL MULTIMETERS

True-RMS digital multimeters

Escort's model EDM-89S is a typical example of the mid-late 1990s handheld digital multimeters. We have been using it continuously until the present day without any trouble. Truly a quality, reliable and dependable instrument.

- 5,000 count digital display + 53 segment analog bargraph
- 0.1% basic accuracy, True RMS, Auto/Manual ranging
- Sampling rate: Digital 3.3 times per second, Analog 20 times per second
- Selector: DC V, DC mV, AC V, µA, mA, A, Ω, diode check, audible continuity, frequency, capacitance
- dBm measurement with 20 selectable impedances
- Frequency measurement 1Hz-10 MHz, with resolution of up to 10mHz
- Capacitance measurement from 5nF to 50µF, with up to 1pF resolution
- Relative mode, Tolerance mode, Zoom mode
- Dynamic recording with time stamp
- "Auto Power Off", "Sleep Mode", mA & A ranges overload fuse protection

How digital multimeters work

Conceptually, digital multimeters are easy to understand. The analog side comprises the attenuators for DCV and ACV, the current-to-voltage converter on current measuring ranges, and the constant current source for the ohmmeter function. The ACV attenuator is capacitively coupled to remove the DC component of the voltage, followed by the rectifier (AC/DC converter). The DC scaled voltage thus obtained is converted into a digital signal in the A/D (Analog-to-Digital) converter. Early chips needed a decade counter to interface them with a digital readout, but modern A/D converters drive display (digital readout) chips directly.

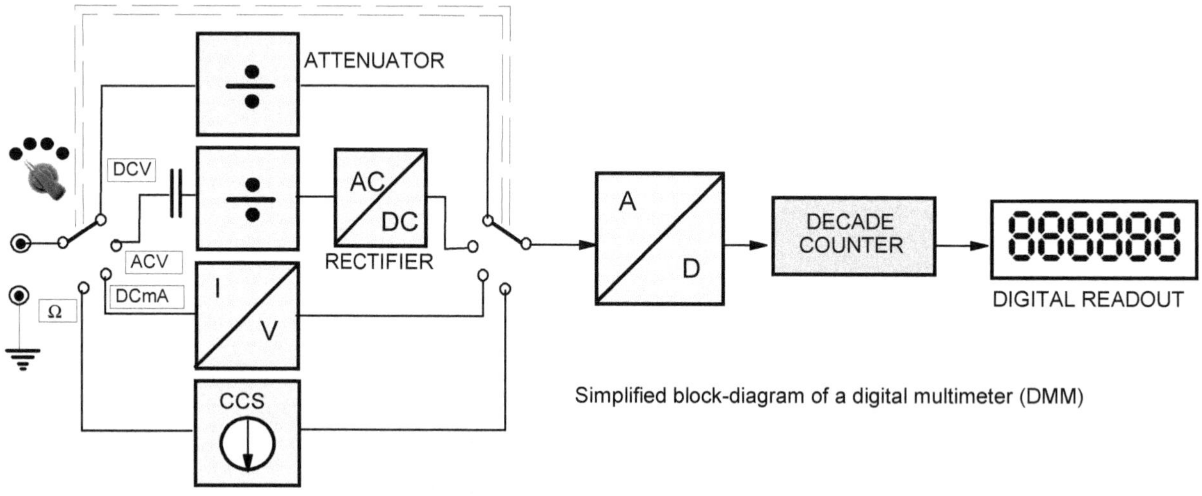

Simplified block-diagram of a digital multimeter (DMM)

How true RMS meters work

Early analog TRMS meters used thermocouples to convert measured AC voltages into heat, which is the actual definition of the RMS value. This equivalent DC value would produce the same heat (energy) as the measured AC voltage of any waveform. The response was slow, and these instruments were expensive and cumbersome.

As with most modern instruments, there are two ways to obtain the true RMS value of an AC signal: the software and the hardware way.

The heart of a digital True-RMS voltmeter is an A/D (Analog-to-Digital) converter chip that samples and digitizes the waveform of the measured signal, after which the true root-mean-square value is calculated by the algorithm in the meter's firmware (software), usually residing in a microprocessor or microcontroller chip. Various other numerical and statistical operations may also be performed, depending on the functionality of the multimeter, for instance.

Some digital oscilloscopes have such functions and the capability to calculate the RMS value of the AC signal displayed.

The measurement's accuracy and the instrument's frequency range (bandwidth) heavily depend on the precision and resolution of the analog to digital conversion.

The hardware way requires special-purpose ICs (integrated circuits) to perform the required functions. AD636 is one such chip, a low-power monolithic IC that uses a logarithmic circuit to perform true RMS to DC conversion on low-level signals. It even includes an auxiliary dB output, whose level is the logarithm of the true RMS output. An external current can set the 0 dB reference level to any input level from 0 dBm (774.6 mV) to -20 dBm (77.46 mV).

An averaging capacitor is the only external component required, its value guided by the tradeoff between low-frequency accuracy, ripple, and settling time. An optional on-chip amplifier can be used as a buffer for the input or the output signals. When used in the input circuit, it interfaces the converter to standard 10 MΩ input attenuators.

As with any log circuit, the converter's bandwidth is proportional to the signal level. For instance, a $1V_{RMS}$ signal produces less than 1% (of the instrument's indication) additional error up to 220 kHz. The upper frequency for the same 1% reading accuracy for a 10 mV signal drops to 14 kHz.

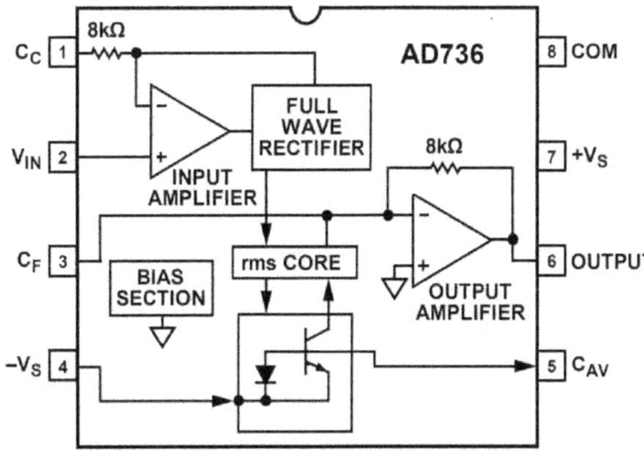

ABOVE: Functional block diagram of AD736, another low cost, low power, True RMS-to-DC Converter © Analog Devices

"Counts" and "digits" - what's the difference and how it impacts multimeters' readouts and resolution

If a DMM is specified as having 4.5 digits (also sometimes written using fractions as "4½ digits"), the first number is the number of full digits that can be displayed, meaning 0, 1, 2, all the way up to 9. The 1/2 digit always refers to the first position or the first digit, which in this case can only be either "0" or '1".

Say the reading of a 4.5 digit DMM is 1.999V. The resolution on that range is 0.001V. The 3.5 digit meter will also show 1.999V with 0.001V resolution. If the measured DC voltage changes to say 2.001V, the 4.5 digit DMM will display it as 2.001V with the same resolution as before.

However, since it cannot have 2 as its first digit, a 3.5 digit meter will move the decimal point to the right and display 02.00V, so the resolution will reduce 10-fold to 0.01V! Thus, a 4.5 digit DMM's display can show 00000 to 19999 (the first digit can show 0 or 1, and the proceeding digits can show 0-9). If we count from 00000 to 19999, we will count 20,000 times, so a 4.5 digit DMM could also be specified as a 20,000 count meter.

The reading of a 3.5-digit multimeter is in the 0.001-1999 range or 2000 counts, while a full 4-digit multimeter would have 10,000 counts. Multimeters with 4000, 6000, and 20000 counts are also common.

For some reason (probably because only a decimal point features on an LCD display, there is no comma), the counts are written without the comma (6000 instead of 6,000).

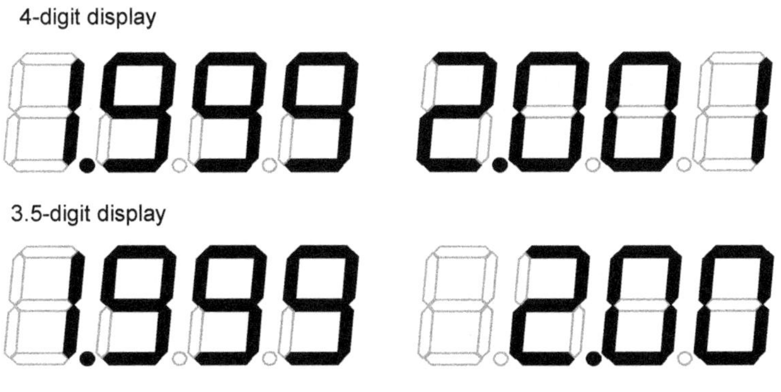

Thus, on a $2V_{DC}$ range, a "2,000 count" or 3.5 digit DMM will count up to 2,000 pulses of the master clock, while the 4.5 digit DMM will have to count up to 20,000 counts to measure the same voltage! Assuming the same clock frequency, that will take ten times (!) longer, or, alternatively, a 10-times faster sampling rate will have to be used to achieve the same measurement speed.

LEFT: The limitation of a 3.5 digit display means a loss of resolution with signal changing from 1.999V to 2.001V or any higher value. The 4.0 or 4.5 digit DMM's resolution is not affected.

Three budget True-RMS digital multimeters side-by-side

The range of currently (2019/20) available digital multimeters seems to be ever-growing. What do you get for your money? Three very affordable True-RMS DMMs are pictured on the next page. Compared to expensive alternatives such as Fluke DMMs, these represent an incredible value.

While both Aneng An882B and Zotek VC17B+ sell for AU$27-30, HoldPeak HP-770HC is priced slightly higher, around AU$37-40. All are auto-ranging true-RMS multimeters that can also measure temperature (using a supplied temperature probe) and capacitance. Also, based on the number of "counts" (6,000), all are in the same class.

The high current test probe input is separate (1), 10A on Aneng, and 20A on Zotek and HoldPeak. HoldPeak unused test probe sockets are automatically blanked off (covered), an added safety feature (2) that isn't necessary. It is impossible to push one's finger into those recessed sockets; their internal diameter is just too small.

The "Select" push button (4) changes through alternate measurement modes (depending on the main rotary switch setting), such as DCmV/ACmV/Frequency/Duty Cycle, Resistance/Continuity/Diode in the ohmmeter mode (5), DCA/ACA or DCmA/ACmA on the ammeter settings.

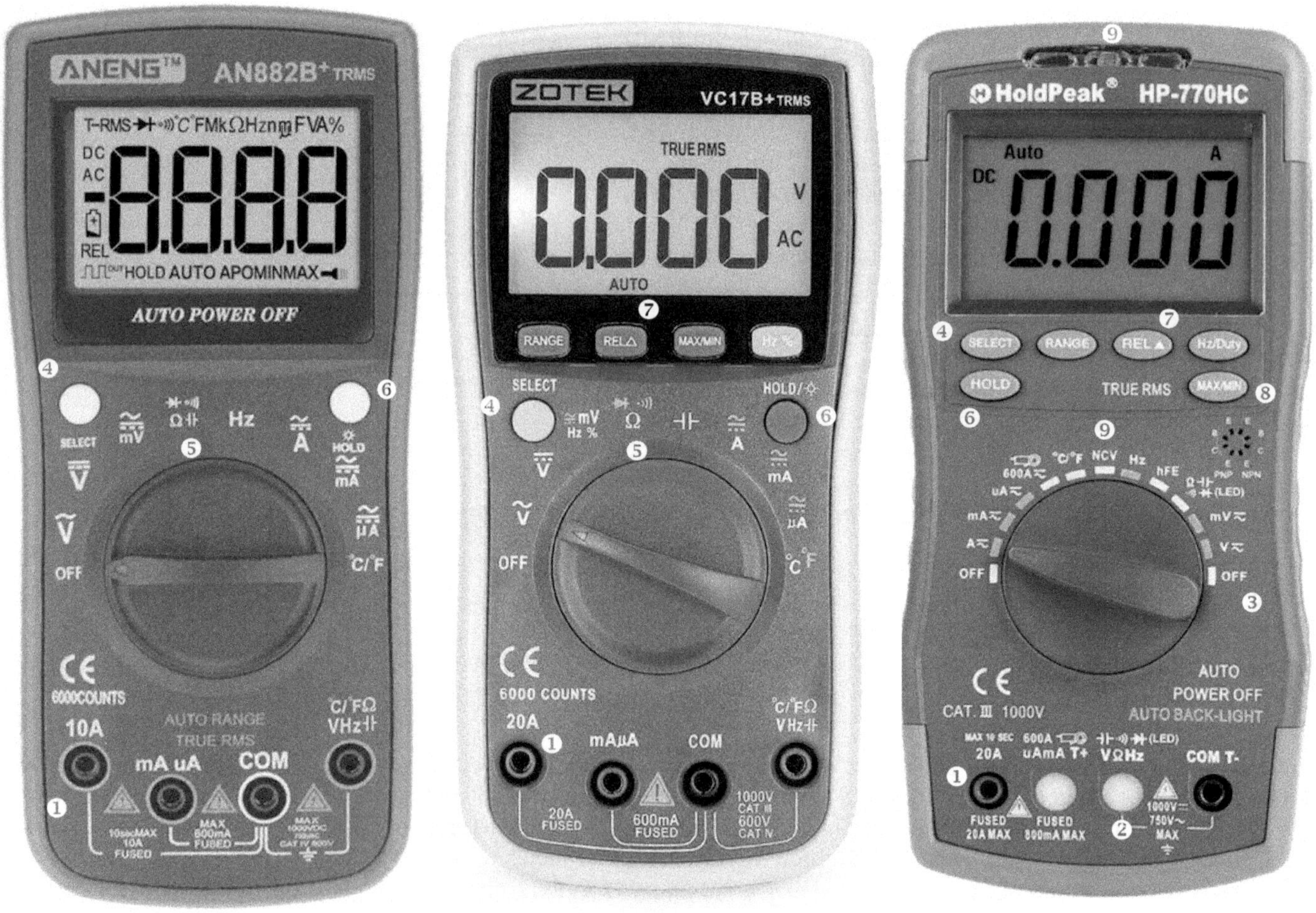

All have the "Hold" button to hold the last measured value (6), while Zotek and HoldPeak also feature the "REL" function (7) and "Max/Min" display (8). The NCV (Non-Contact Voltage detection) is an added feature on HoldPeak (9).

Apart from temperature, all can measure frequency ("Hz"), Zotek and HoldPeak have the capacitance testing capability (although they cannot measure low capacitance values in the pF range), and HoldPeak also added a bipolar transistor DC beta (h_{FE}) test function, making it my personal choice out of the three.

HoldPeak HP-770HC

- DCV ranges: 60mV, 600mV, 6V, 60V, 600V, 1000V
- ACV ranges: 60mV, 600mV, 6V, 60V, 600V, 750V
- DCC ranges: 600μA, 6,000μA, 60mA, 600mA, 6A, 20A
- ACC ranges: 600μA, 6,000μA, 60mA, 600mA, 6A, 20A
- Capacitance: 9.999nF - 99.99mF
- Frequency ranges: 9.99Hz, 99.99Hz, 999.9Hz, 9.999 kHz, 99.99 kHz, 999.9 kHz, 9.999MHz
- Resistance ranges: 600R, 6k, 60k, 600k, 6M, 60M
- Transistor h_{FE} range: 0 - 1,000

CALCULATION

A 3½ digit DMM, when used on its 60V range (the maximum reading is 59.99V), has a specified accuracy of +/-0.5% of the reading + 2 counts. What is the absolute error if the true value of the measured voltage is 12.6V?

Solution: One count is the smallest change that the display can show (its resolution), which in this case is 0.01V (10mV). The error is $\varepsilon = +/-(0.005*12.6) + 2*0.01 = 0.083$V or 83mV. The DMM's reading could be anywhere from 12.6-0.083 = 12.517 V to 12.6+0.083 = 12.683 V!

4 OSCILLOSCOPES - HOW THEY WORK & HOW TO USE THEM

- CATHODE-RAY OSCILLOSCOPES (ANALOG OSCILLOSCOPES)
- A CRASH COURSE IN OSCILLOSCOPE CONTROLS
- COMPENSATED OSCILLOSCOPE PROBES
- DIGITAL OSCILLOSCOPES

> In our lust for measurement, we frequently measure that which we can rather than that which we wish to measure... and forget that there is a difference.
>
> Udny Yule

CATHODE-RAY OSCILLOSCOPES (ANALOG OSCILLOSCOPES)

An oscilloscope ("scope" or CRO - Cathode Ray Oscilloscope) is a test instrument of such importance that whole books have been written about its operation and use. It can be considered a visual amplifier since its main task is to display signals in the time-domain, i.e., their waveforms. Instead of driving a loudspeaker and converting the electrical energy into sound, as audio amplifiers do, the oscilloscope amplifiers (vertical & horizontal) drive the visual display ("screen"), which is either a Cathode Ray Tube (CRT) or, in newer digital oscilloscopes, a color LCD display.

The main requirements on oscilloscopes are identical to those demanded from hi-fi amplifiers: high input impedance, so they don't load the test circuit, minimal distortion, and wide bandwidth.

There are various types of oscilloscopes, with memory, sampling, etc., but here we will focus on the basic type used in audio work. Some models have a fixed configuration or functionality; others use various plug-in modules to enhance their capabilities for specific tasks. Most notable are Tektronix modular units, still highly regarded amongst the users. Older vintage models use vacuum tubes, then there are hybrid units from the late 1960s, and finally, totally solid-state units from the 1970s onwards.

Cheaper or low-spec oscilloscopes are one-channel units with a low upper frequency, typically between 100 kHz and 1 MHz. Better ones go up to 20, 40, 100 MHz, and even higher. Obviously, for audio work, even a 1 MHz scope is acceptable. The only warning is that vintage scopes such ac Eico, Heathkit, Hickok, and similar brands (many of them were kits) were not triggered scopes as all modern scopes are, so they don't even make the minimum grade!

The recommended top vintage brands are Tektronix, LeCroy, Hameg, Fluke-Philips, Iwatsu, Beckman, HP, Meguro, Philips, Gould, Trio-Panasonic, Hitachi, Kenwood, Kikusui, Leader, Telequipment, Labtech, Goldstar, Nicolet, GW, BWD, B&K Precision, and Sencore.

The functionality and external controls of an analog oscilloscope

I have selected this low-priced scope by Taiwanese firm Labtech firstly because it has been serving us well for over 20 years and secondly because it is typical of scopes in its class. It has well-laid out commands and thus can serve as an educational platform. There are four functional groupings framed in light gray.

<u>1. CRT or display controls</u>

Intensity changes the brightness of the beam trace on the screen, while focus makes it sharper and thinner if it is too thick or smudged. *Trace rotation* adjusts the horizontal line (without any signal at the input) a few degrees each way to ensure it is perfectly horizontal.

Some scopes have an additional control called *Astigmatism*, which is used to correct focusing in X and Y directions. If uncorrected, a dot on the screen that is supposed to be a perfect circle has a shape of an ellipse, i.e., it is distorted.

Likewise, *Scale illumination* control changes the brightness of the lamps or LEDs used to illuminate the screen's perimeter, similar to regulating the illumination on your car's instrument panel. Many oscilloscopes have a *Beam finder* or *Trace finder* button, which helps locate the beam if the controls are maladjusted and a trace cannot be seen.

CAL (short for "calibration") does not belong to this family of cathode-ray controls but was most likely placed there for convenience. It is an output from an internal square wave generator, a signal of fixed amplitude and frequency (usually 1kHz) that can be used to calibrate the vertical and horizontal controls and to compensate the probes.

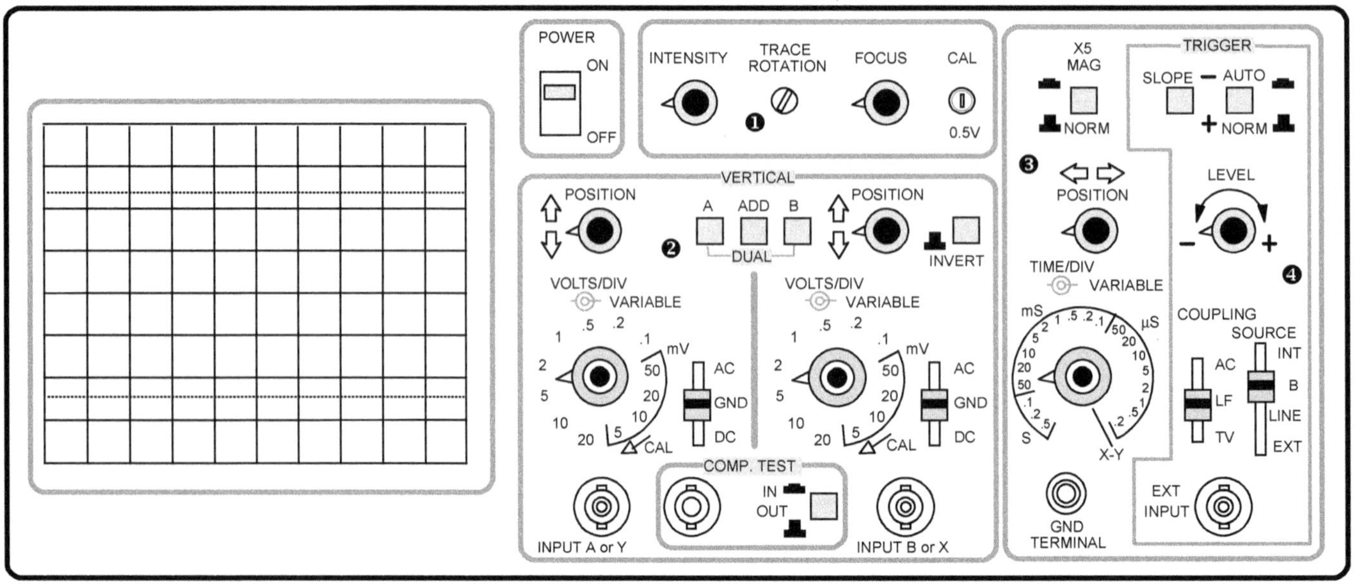

ABOVE: Controls of a typical CRO (Cathode Ray Oscilloscope), based on Labtech 3502c, a mid-1990's 20Mhz Dual Channel Oscilloscope

2. Vertical controls

The VOLTS/DIV attenuator changes the scale of either channel, in this case from 5mV/division to 20V/division. The potentiometer mounted on the same shaft as the switch (variable control) enables continuous "stretching" or "shrinking" of the signal, but in that case, the calibration of the display is lost, and the amplitudes cannot be read from the screen. Only when the continuous control is turned fully anticlockwise in the CAL position does the display scale match the figures indicated by the VOLTS/DIV selector!

The signal coupling switch has three positions. In GND, there is no vertical or horizontal deflection (the input is grounded), and a horizontal line will be displayed. The *Position* control will move that line or the signal waveform up and down (for vertical inputs) or left-right for the horizontal or time-base channel. In the DC position, both the AC and DC components of the signal are displayed.

In contrast, in AC position, a capacitor is connected in series with the input, removing a DC component from the signal (if any). This is useful for observing a small AC voltage (ripple) of say $1V_{PP}$ on top of a high DC voltage, say 300V from a power supply, which in DC position would be impossible.

The *Invert* button inverts the phase of one channel (channel B in this case).

This scope has a Component test function, a simple low voltage low current I-V curve tracer for diodes, Zeners, and passive components (a slanted line for resistors, an ellipse for inductors and capacitors), purely qualitative so next to useless, probably included as a marketing gimmick to attract beginners.

The three pushbuttons select the "mode." "A" displays only channel A or Y signal, pushing B displays only channel B or X waveform, pushing both displays both channels in "dual-trace" mode.

Both signals are chopped at a high frequency in CHOPPER mode, so a section of ch. A signal is displayed, then a segment of ch. B, end so on. During each horizontal sweep, the shopped segments fall into a different position, but due to the inertia of the human eye, we get an impression of a continuous waveform. The phase relationship of the two signals is preserved. In the ALTERNATE mode, one sweep displays ch. A signal, the next sweep displays ch. B signal, and so on, but the phase relationship between two signals is lost and cannot be measured.

Some scopes have manual control of CHOP or ALT mode. This particular scope automatically decides for you; time-base settings 0.5sec/div to 1msec/div use CHOP mode at around 200kHz. The other ranges automatically use ALT mode. The user has no control over this choice.

ADD sums the two signals, while pressing ADD and INVERT displays the difference of the two signals (SUB for subtraction mode).

3. Horizontal controls

The TIME/DIV selector switch for the horizontal channel is akin to the V/DIV selectors for vertical channels. The number selected is a time period corresponding to one major division in all positions. The X5 MAG (magnifying) button "zooms in" by multiplying the sweep time five times, so if five full periods of a signal are displayed, pushing this button will display only one period (akin to "magnifying" the signal under a magnifying glass).In the last (fully clockwise) position, called X-Y, the internal time-base (sawtooth voltage driving the beam left-to-right) is disconnected, and the signal from the external horizontal input is brought in.

4. Time-base and trigger controls

AUTO-NORMAL switch in AUTO position selects a free running time base, with horizontal deflection of the beam regardless of whether a measured signal is present or not. In NORMAL (often called TRIGG. for "triggered" mode), the generator of the sawtooth deflection voltage is inactive until the measured (displayed) signal triggers it. Some scopes also have a third option, called SINGLE, where only a single sweep is generated.

The SLOPE button selects if the positive (rising) signal edge or the falling (negative) signal will trigger the time base. LEVEL is a potentiometer that selects at what level of the displayed signal the time base signal will be triggered.

The COUPLING selector switch selects one of three modes. The trigger signal is brought in through a coupling capacitor in AC trigger mode, so any DC component is removed. In DC or HF REJ (for high-frequency reject) mode, the trigger signal is run through a low pass filter, removing high-frequency components. This is useful if the signal is "dirty" with HF spuriae or in TV servicing to remove the RF composite of the sync signal.

Finally, the SOURCE switch selects the signal to trigger the time base. INT means internal signals; channels A and B are added and used as trigger signals. In "B", only channel B triggers the time base, LINE selects a power line voltage as a trigger signal, and EXT uses the external signal on the EXT INPUT BNC connector to trigger the time base.

The internal workings of an oscilloscope

This functional block diagram and controls of a typical triggered-sweep scope are similar to those behind the front panel of the "typical" scope illustrated on the previous page. The functionality partially depends on the actual circuit design and the level of integration used in solid-state instruments.

In some budget vintage tube scopes, the horizontal amplifier is simpler and inferior in performance to the vertical amplifier. Better quality instruments and modern solid-state models have identical vertical and horizontal amplifiers. The only additional stage in the vertical (or "Y") amplifier is the delay line, as illustrated on the next page.

The input signals are attenuated first (to get them to the standard range) and then amplified before passing through a phase splitter and the final push-pull power amplifier driving the deflection plates. The trigger signal is taken after the preamp stage, goes through a cathode follower (high input impedance) to separate it from the vertical circuit, so it doesn't load it, passes two switchable controls, "trigger source" (in "internal" position") and trigger coupling, which has options of eliminating high-frequency components or radio-frequency composite signal for TV servicing.

The sync signal is then amplified and converted into narrow pulses, triggering the sawtooth generator. In the "INT" mode of the horizontal channel, that linear signal passes through the horizontal amplification chain. It drives the horizontal deflection plates, so linear horizontal deflection is achieved (like the horizontal sweep on TV screens). The vertical position of the beam is thus proportional to the time, so a time-domain visual display of the signal is achieved.

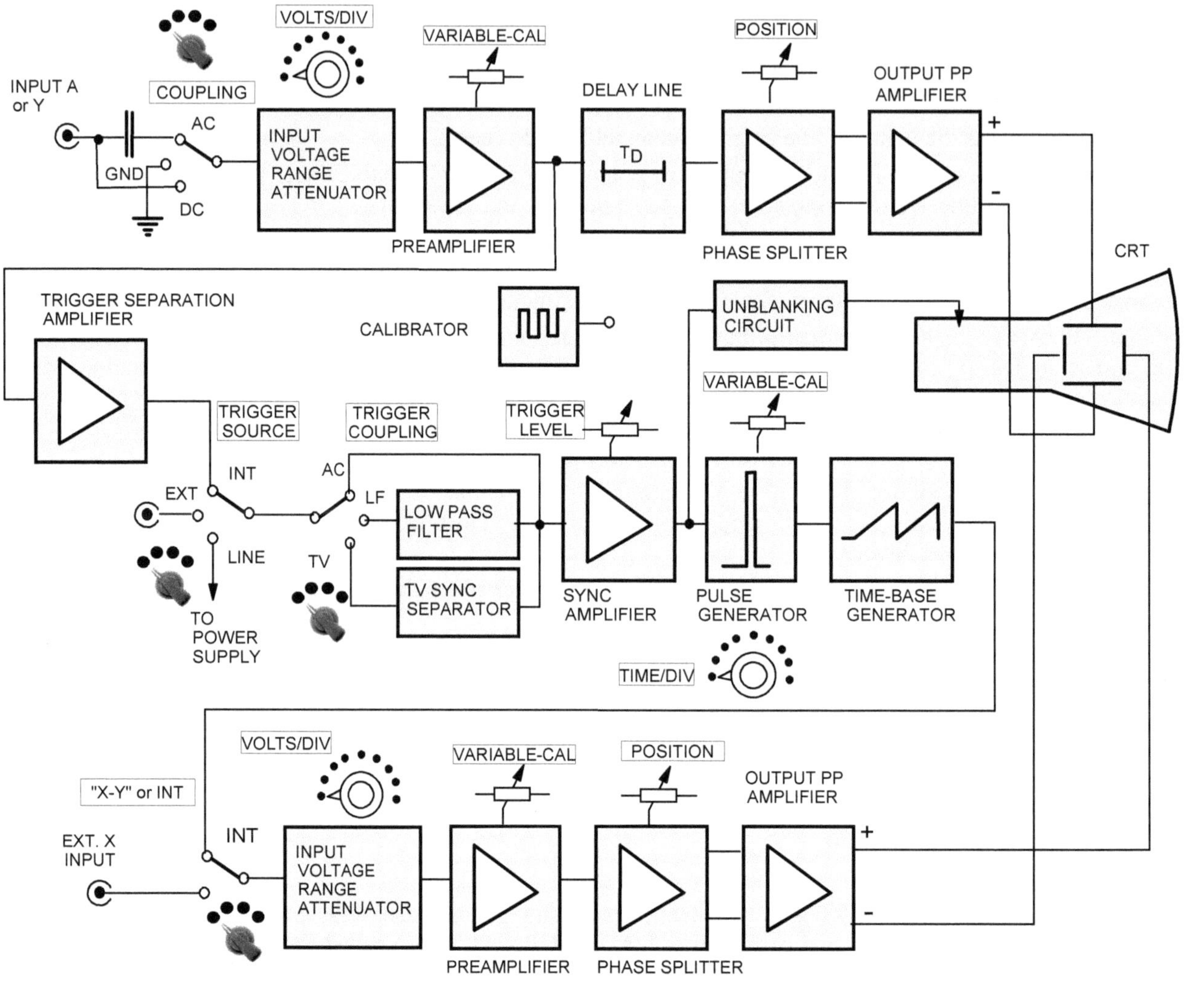

ABOVE: Functional block-diagram of a typical analog oscilloscope
BELOW: Typical CRT (Cathode Ray Tube) connections and controls

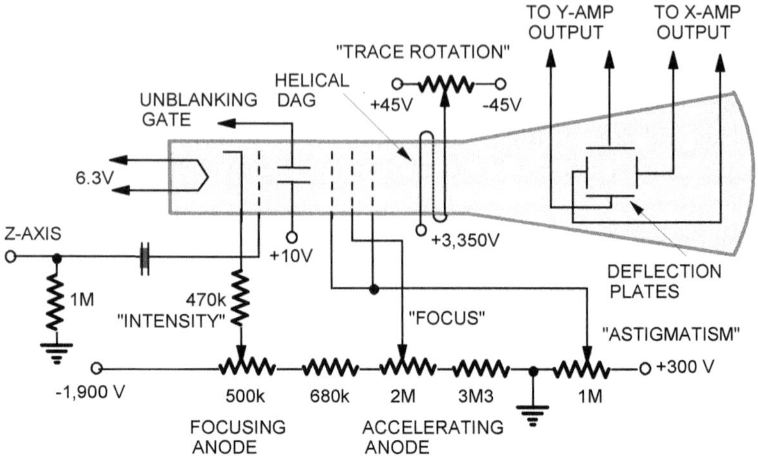

Cathode ray tube

Fundamentally, the cathode-ray tube functions just like any other vacuum tube but is more complex since its aim is not just to control the flow of electrons but to control the deflection of the electron beam and thus the point at which it hits the fluorescent screen surface.

The intensity control adjusts the cathode potential while the z-axis electrode is akin to the control grid. Since it is the closest to the cathode, it controls the number of electrons that flow through to deflection plates.

The unblanking gate deflects the beam of the screen, so there is no trace during the beam's return ("retrace") and when the sweep is not operating.

Focus and astigmatism are more refined controls further down the path for sharpening the beam's trace.

A CRASH COURSE IN OSCILLOSCOPE CONTROLS

As with most things in engineering, it is easier and faster to get your hands on a real oscilloscope and learn by trial & error than learning it from a book. For some reason it looks much more complicated when described here than in reality. Anyway, to get you started, let's look at a few basic oscilloscope controls, what they do and how to use two or more of them together to display the signal waveform in a way that will give you the desired information.

Input coupling

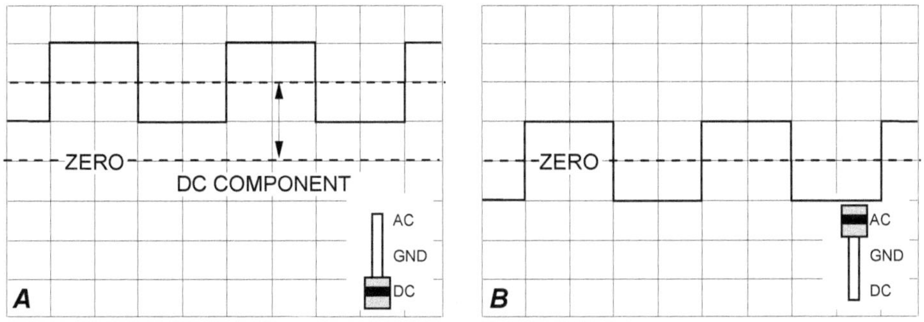

The left oscillogram (A) shows an output of a function generator with a DC offset. Since the input coupling switch is in the DC position, this DC offset is displayed on the screen. The zero is in the middle of the screen.

Once the coupling switch is moved to the AC position (as in B), the input capacitor does not allow the DC component of the signal to pass through, so now the waveform is symmetrical around the zero level.

Vertical scale (VOLTS/DIV) and vertical position

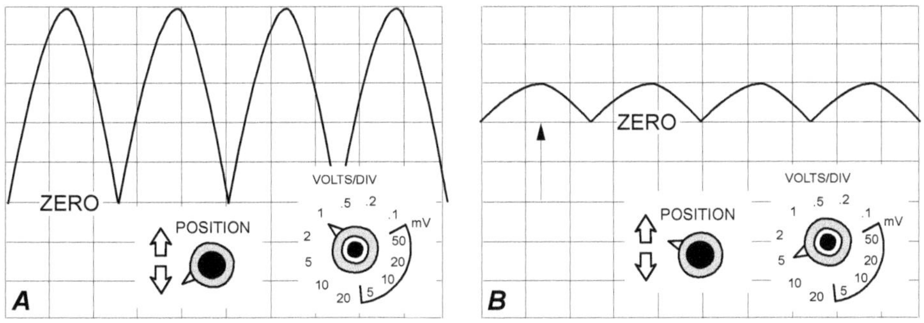

The output of a full-wave rectifier with a peak of 5 divisions or 5V is displayed in A.

Changing Volts/div from 1 to 5 shrinks the amplitude from 5 to 1 division.

The zero level was also changed (by the "POSITION" control), as indicated by the arrow in B, moved two divisions upward.

Since both cases display four pulses, the horizontal sensitivity (time base) was not changed.

Horizontal scale (TIME/DIV)

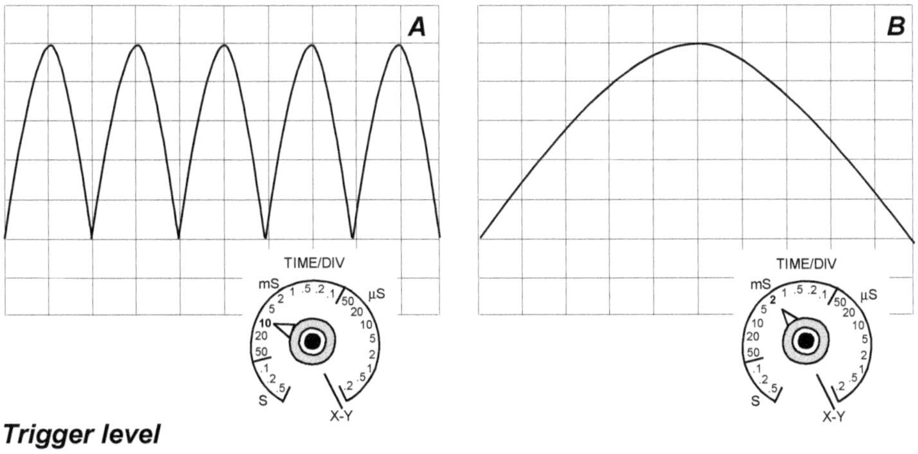

The left oscillogram (A) shows five pulses; the period is 2 divisions times 10ms/div or 20 ms. To get only one pulse as in B, the horizontal sensitivity (time base) must be changed, so 10 horizontal divisions equal 20ms, or 20/10=2ms/div!

The vertical level and vertical sensitivity (V/div) were not changed.

Trigger level

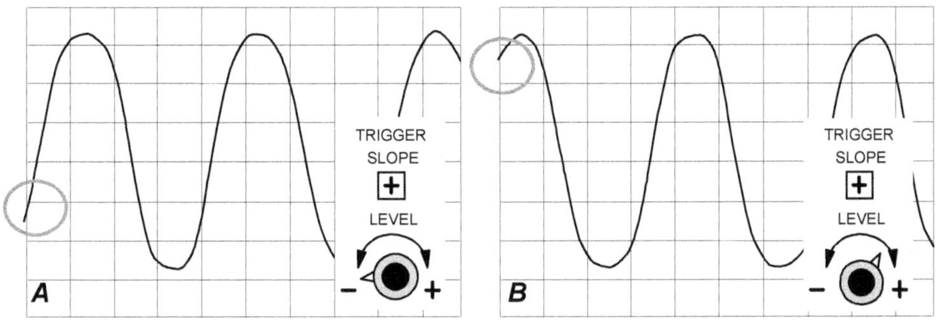

The trigger slope in both cases is positive (the rising signal edge), but the trigger level in oscillogram (A) is much lower than in the right one (B). The starting point of the display in B (circled for emphasis) is thus much higher on the waveform (closer to the peak of the sinewave) than in A.

Trigger slope

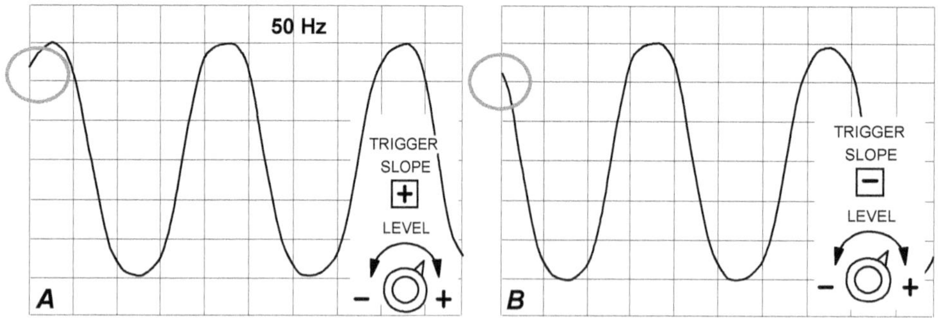

The trigger level in both cases is the same (close to the peak of the sinewave). The trigger slope in the left oscillogram (A) is positive (the rising signal, while the slope in B is negative (the falling signal).

Problem #1:

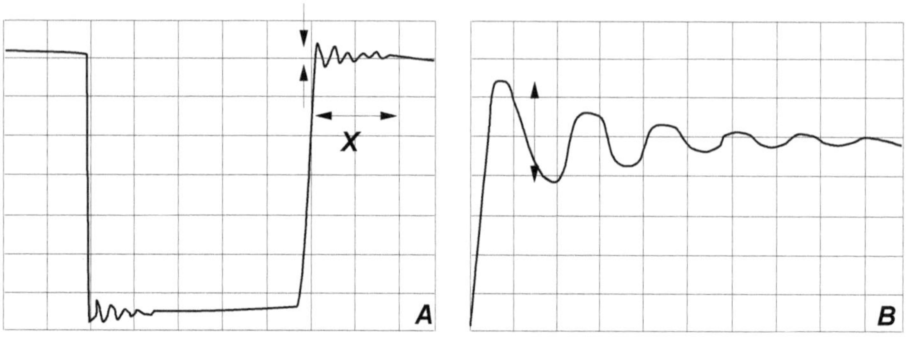

A tube power amplifier is tested using a square wave signal, and the output waveform shows signs of oscillation, as in A. What controls do we need to adjust to enlarge the section of interest (indicated by "X" span), as in B, to study the nature of this potential instability?

ANSWER: The part of the waveform in A we are interested in(marked "X") stretches across two horizontal divisions. In B), this part stretches the whole width of the screen or 10 divisions. Thus, the horizontal sensitivity must be increased 10/2= 5 times, for instance, from 10ms/div to 2ms/div! Notice that the signal in B) has also been vertically enlarged ("zoomed in"), meaning that vertical sensitivity was also increased. In A), the first peak is approx. 1/2 divisions tall, while in B) it is approx. 2.5 divisions tall, so it seems that the vertical sensitivity was increased 5 times, from, say, 1V/div in A) to 0.2V/div in B).

Problem #2:

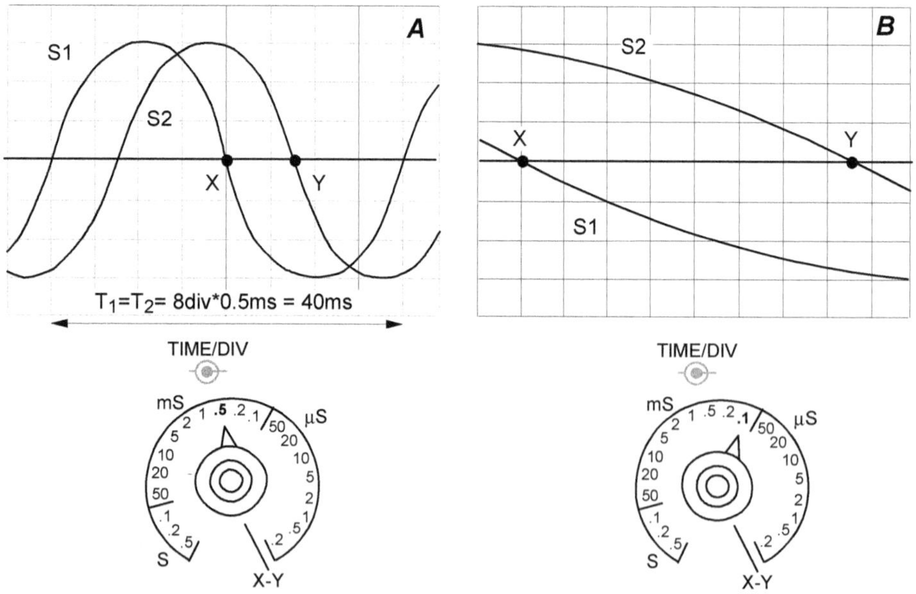

The input and output signals of a preamplifier are made to have the same amplitude (by adjusting the volume control). Their phase shift is studied on the CRO screen. The time base in the left oscillogram (A) is set at 0.5ms/div.

1. What is the frequency of the test signal?
2. What is the phase shift in degrees and radians?
3. What oscilloscope controls need to be adjusted so that the oscillogram looks like the illustration on the right (B)?

ANSWER: 1. The period of both signals is the same (an amplifier does not and can not change the frequency of the test signal. The period is T=8div*0.5ms = 4ms, so frequency is f = 1/T = 1/0.004 = 250Hz.

2. The phase shift is the distance between points X and Y, which is 1.5div in A) . Since the period is 8div, the phase shift is 1.5/8 = 0.1875 or 18.75% of the period, and since the period is 360°, the phase shift is Φ=0.1875*360° = 67.5^{o}!

3. To help you find this answer, we have marked on B) the same two points as before, X and Y. Now their distance is 7.5 divisions, so XY=7.5div=0.75ms so 1div=0.1ms! Since the amplitude is still 3 divisions, the only control that has changed is the time base, from 0.5ms/div in A) to 0.1ms/div in B), or a horizontal "zoom" of 5 times!

Problem #3:

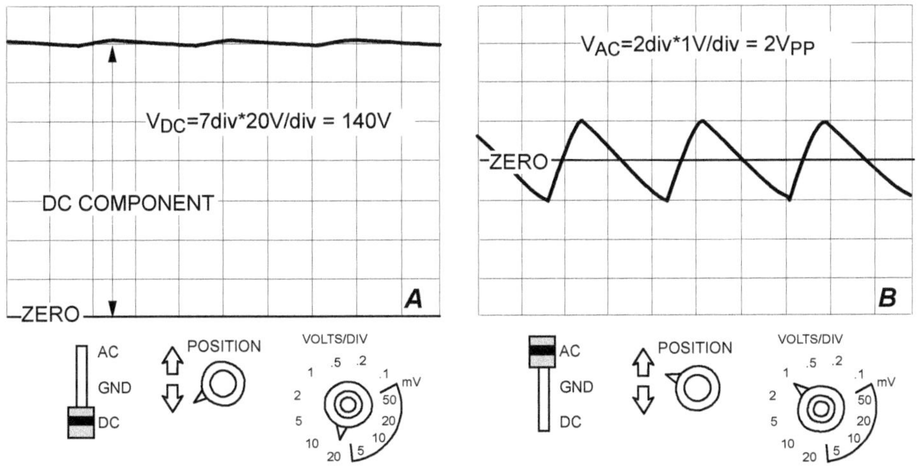

A ripple of a DC power supply is observed, and an oscillogram is obtained as in illustration A).

1. What oscilloscope controls must be adjusted to get the oscillogram as illustrated in B)?
2. How many volts is the DC component of the power supply?
3. What is the peak-to-peak value of the ripple (AC component)?

ANSWER: 1. The "Coupling" switch must be moved from DC into AC position to remove the DC component. The ripple waveform would then be at the bottom of the screen because that's where the zero level was set at. To bring it up to the middle of the screen as illustrated in B), the vertical position control needs to be adjusted in the clockwise direction (up). Finally, the vertical sensitivity needs to be increased to "stretch" the signal and increase its amplitude, so the V/div needs to be changed from 20V/div in A) to 1V/div in B), a vertical zoom of 20 times!

2. In A), the DC component is 7 vertical divisions, and since each division is 20V, V_{DC}=7div*20V/div = 140V!

3. In B), the peak-to-peak value of the ripple is two divisions, and since 1div=1V, V_{AC}=2div*1V/div = $2V_{PP}$!

COMPENSATED OSCILLOSCOPE PROBES

The frequency compensation principle applies to attenuators, oscilloscope probes, and many other systems requiring a uniform response over a wide range of frequencies. The best-known application of the frequency compensation principle is the compensation of test probes for oscilloscopes and vacuum tube voltmeters.

The capacitance of the coaxial cable used to connect the tip of the probe to the oscilloscope and the input capacitance of the oscilloscope itself, together with the signal source impedance RS, form a low-pass filter. At higher frequencies, the reactance of the two capacitances in parallel becomes smaller and smaller, and thus high frequencies are shorted to ground. This severely limits the upper-frequency limit f_U of oscilloscopes and other test equipment.

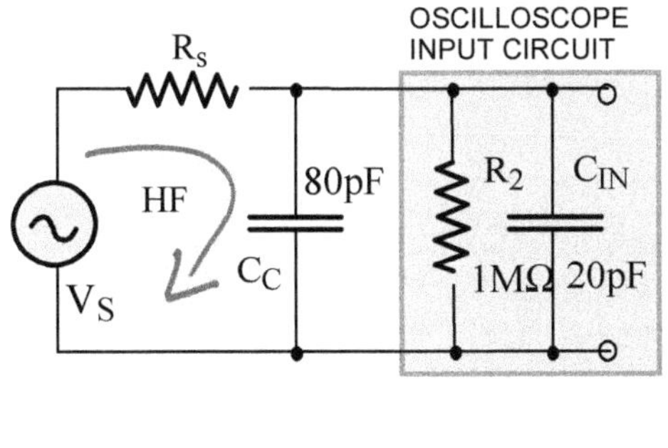

LEFT: The capacitance C_C of the shielded test cable, together with the input capacitance C_{IN} of the oscilloscope, attenuates high frequencies

RIGHT: The solution? A frequency-compensated probe!

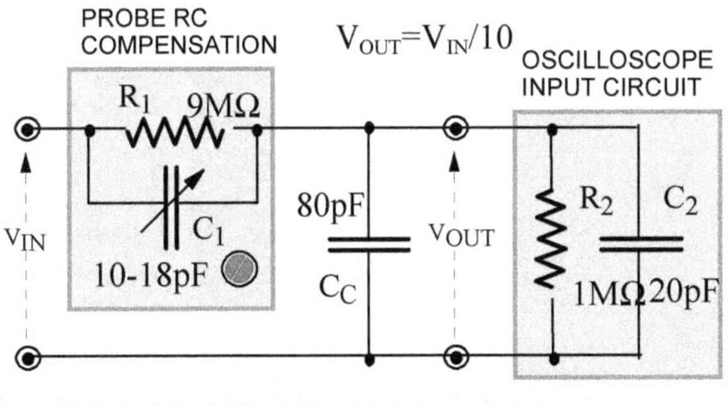

For the compensated voltage divider $R_1C_1=R_2C_2$, so its transfer function becomes independent of the frequency: $A = R_2/(R_1+R_2)$.

Compensation made the whole input circuit behave as a simple resistive voltage divider and increased the oscilloscope's input impedance. However, this has been achieved at the expense of the signal voltage reduction in the $R_2/(R_1+R_2)$ ratio, which is usually the attenuation of 10:1! In the typical case (as illustrated), R_1=9MΩ and C_1=100/9 = 11.1 pF.

Frequency compensation is a typical example of the gain-bandwidth compromise. Just as for vacuum tube amplifiers, the gain-bandwidth product is constant. We widen the useful bandwidth of a test instrument, but we pay the price in a tenfold reduction in input signals!

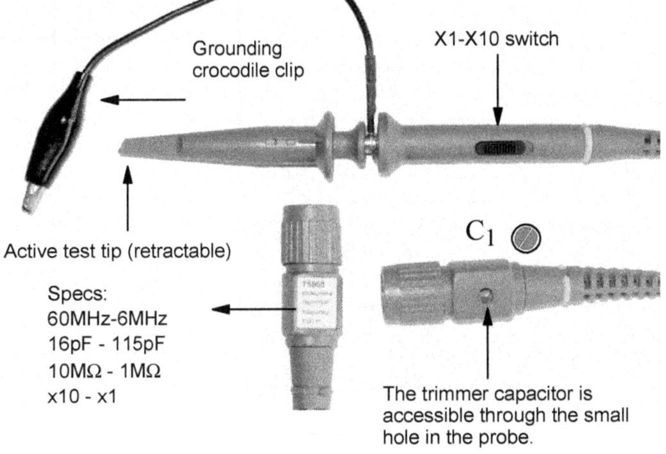

ABOVE: A typical oscilloscope probe

DIGITAL OSCILLOSCOPES

You will either like or hate digital oscilloscopes. I prefer the analog type. Owon PDS 5022S is a 25Mhz 2-channel digital oscilloscope with a color LCD screen. Pressing four different buttons to go through various menus on a low-resolution low-brightness LCD screen to get to a specific control is annoying. The main menus (CH1, CH2, MATH, and TRIG) have their own buttons (1), which then assign different meanings to the column of F1-F5 menu buttons (2). The function of those buttons is permanently displayed on the right side of the LCD screen. If there are more options (page 2 and 3), as in this case, page 1 of 2 is displayed (3), the second screen is accessible by pressing the F5 button.

The settings are displayed at the bottom of the screen in barely visible font size (4). The main flaw is that the lowest vertical sensitivity is only 5V/div (5) instead of 20V/div or even more on most analog scopes.

That means that the 1:10 probe must be used if large amplitude signals are to be displayed! The two cursors, which tell you the signal's amplitude in the chosen point, are a handy feature.

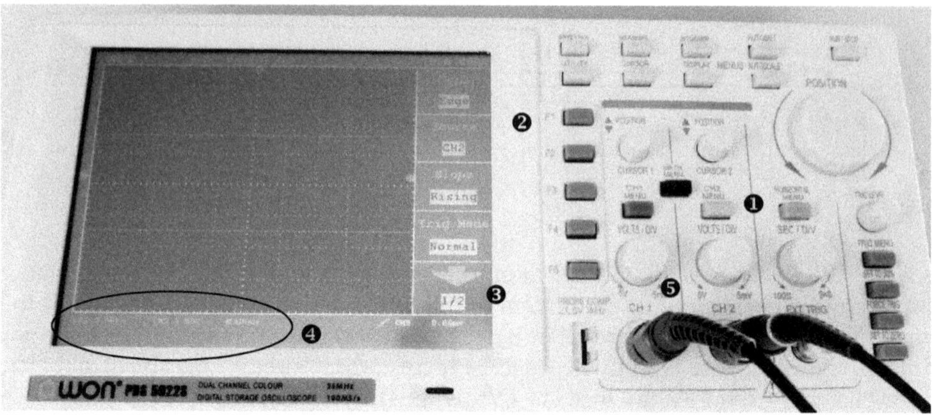

Combo 2-in-1 and 3-in-1 test instruments

Hantek 2D72 and 2D42 are 3-in-1 instruments comprising of an oscilloscope, digital multimeter, and function generator (2CH+DMM+AWG). Hantek 2C72 and 2C42 are the 2-in-1 versions of the "D" series, missing the function generator (2CH+DMM). "7" in all models number denotes 70MHz bandwidth in oscilloscope mode, and "4" indicates a 40MHz version of the instrument. Obviously, the "C" models are cheaper than their "D" counterparts, and the 40MHz versions are cheaper than the 70 MHz versions. For audio work, 40Mhz is fine.

The 4,000 counts DMM (voltage, current, resistance, capacitance) is of standard capabilities, with a maximum input voltage of 600V_{AC} or 800V_{DC} and 10MΩ input impedance.

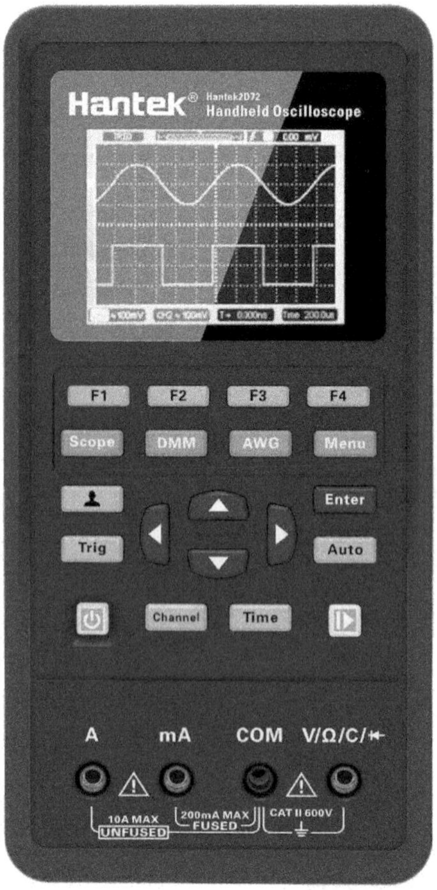

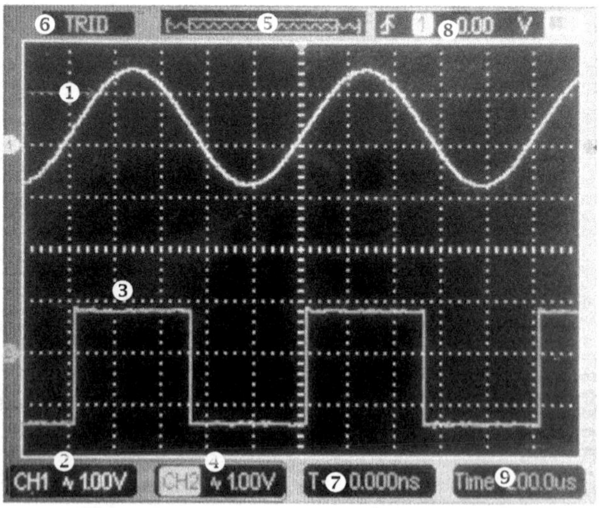

1) CH1 trace
2) CH1 vertical sensitivity
3) CH2 trace
4) CH2 vertical sensitivity
5) Main time base window showing free running mode
6) Trigger status
7) Trigger time
8) Trigger level (rising edge)
9) Time base setting (200μs/div)

The 200 x 100 x40 mm (approx.) instrument weights just over 600 grams.

The arbitrary waveform generator produces sine (1Hz-25MHz), square, sweep, and exponential waveforms of up to 5V amplitude (with high impedance loads such as a 47kΩ or 100kΩ input impedance tube amplifier). With a low impedance 50Ω load, amplitude drops to half (2.5V), which means its internal or output impedance is also 50Ω.

5 TESTING PASSIVE ELECTRONIC COMPONENTS (RESISTORS, CAPACITORS & INDUCTORS)

- TESTING FIXED AND VARIABLE RESISTORS
- TESTING CAPACITORS
- THE BRIDGE METHODS OF TESTING CAPACITORS
- INDUCTANCE MEASUREMENT METHODS AND INDUCTANCE BRIDGES
- DIGITAL LCR METERS

> "The same applies to the concept of force as does to any other physical concept: Verbal definitions are meaningless; real definitions are given through a measuring process.
>
> Arnold Sommerfeld"

TESTING FIXED AND VARIABLE RESISTORS

Testing fixed resistors

In this experiment, we measured three different resistors, ranging from a low resistance/high power wire wound resistor used as a dummy load for amplifier testing to a high resistance metal film resistor. The measurement of the 4k7 resistor was straightforward on all three instruments. Philips AMM's readout was limited due to the inherent imprecision of an analog scale, so only accurate to one decimal place.

With an 8Ω power resistor, the decimal point on the 5,000 count DMM shifted once more to the right, so we only had one decimal place, 8.1Ω. Again, the 10,000 count LCR meter was superior, with a readout to 3 decimal places. However, due to its AC signal testing regime, its reading changed with the test frequency, from 8.159Ω @ 1kHz to 8.112Ω @ 120Hz! The Philips AMM was way off the mark, indicating 11Ω.

Moving towards high resistance values, the 60+ years old carbon composition 2M2 resistor was measured as 2.458M by the DMM and 2.5M by Philips AMM. Interestingly, using a 1kHz test signal, the LCR meter measured the exact figure as its DMM brother, 2.458M, rising to 2.468M at 120Hz.

Tested component	Escort ELC-131D LCR meter (LCRM)	Escort EDM-89S true-RMS digital multimeter (DMM)	Philips PM2505 analog multimeter (AMM)
4k7 resistor	4.745 kΩ	04.74kΩ	4.8kΩ
8Ω 100 Watt power resistor	8.159Ω @ 1kHz 8.112Ω @ 120Hz	008.1Ω	11Ω
2M2 resistor	2.458MΩ @ 1kHz	2.458M	2.5M

Measuring high resistances: PACO tube testers 650 & T-62 as MΩ-meters

Precision Apparatus Company (PACO) models 650 and T-62 tube testers are functionally identical. Under the hood, it is the same grid circuit analyzer and megohmmeter. The photo shows a true RMS digital multimeter measuring the test voltage on "Megohmmeter" test terminals (76.4V_{DC}). Even a quality instrument such as this Escort meter is not an ideal voltmeter (infinite input impedance), but has a declared Z_{IN}=10MΩ. Indeed, notice the MEGAOHMS scale (1) indicating around 11MΩ! In this instance, the two test instruments are "testing" each other, measuring each other's internal parameters they are supposed to measure as chosen by the selector switches (2) on MΩ and (3) on V_{DC}.

The MΩ meter is very sensitive, its scale extending up to "1k", meaning 1,000 MΩ or 1GΩ (6)!

A small trimmer pot (4) accessible from the top (just under the meter) is marked "VTVM ADJ." Does this mean there is a vacuum tube voltmeter inside? It certainly does.

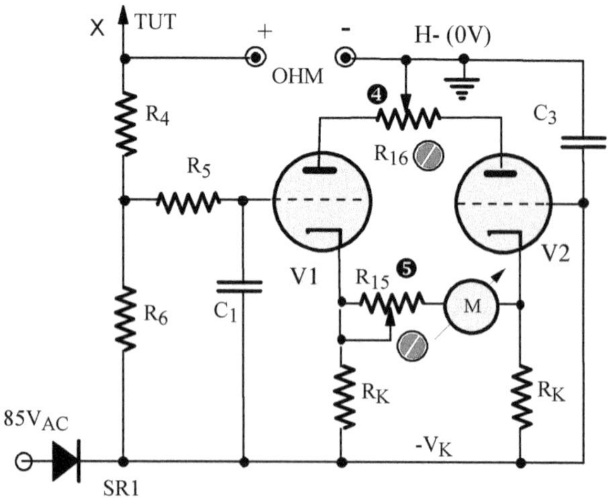

ABOVE: The VTVM at the heart of PACO 650 tube tester is a differential tube amplifier

RIGHT: 76V_{DC} is the test voltage used by the MΩ meter, as demonstrated here

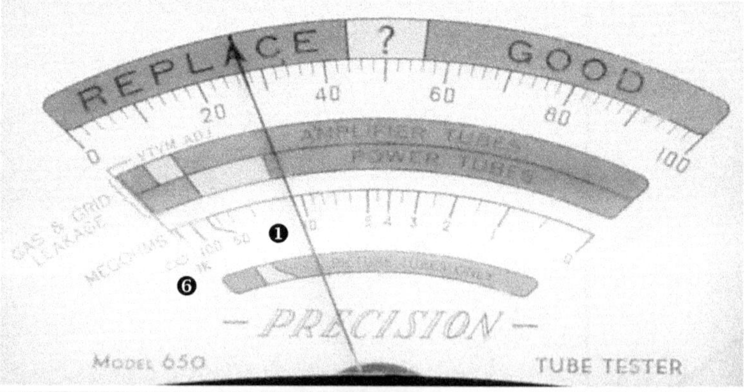

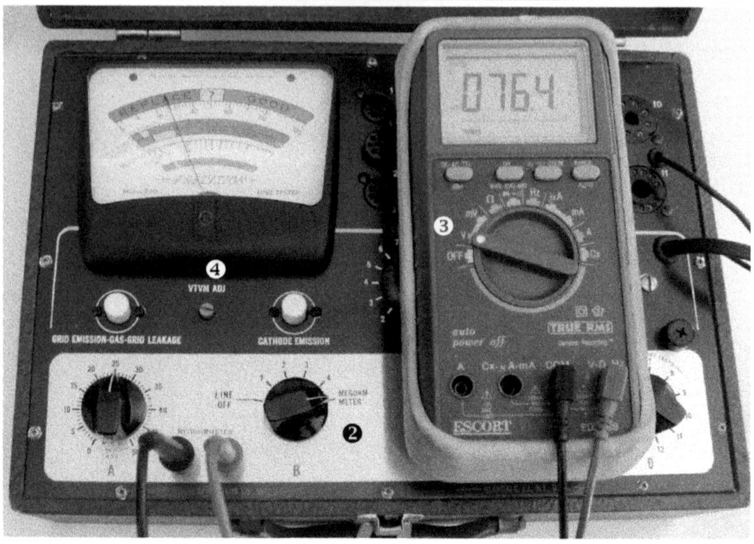

A 12AU7 duo-triode is configured as a differential amplifier, and this potentiometer R16 is in their anode circuits, roughly half acting as anode resistor for each triode. By moving its slider one way or the other, one anode resistance increases while the other decreases, thus balancing the voltage gains of the two halves of the differential amplifier and zeroing the meter. The meter is connected as a load between two cathodes through the "Cal. control" trimmer resistor R15 (5).

Triode V1 is the "active" half; the signal is brought to its grid through the voltage divider R4-R6 and a low pass filter formed by R5 and C1 (shunting high frequencies away from the grid). The grid of V2 is grounded for AC signals through capacitor C3.

Normally a positive anode voltage $+V_{BB}$ is on the anodes, and cathode resistors R_K are at the GND or 0V potential. Here, the anode resistor R16 is at GND or 0V, while the cathodes are at a negative DC voltage provided by the rectifier SR1, which half-wave rectifies one of the power transformer secondary voltages, namely $85V_{AC}$, which is then filtered and smoothed out by a 10μF capacitor C3. This reversal of polarity makes no difference; as long as anodes are more positive than cathodes, a tube will work as it should. Voltages (potential differences) are a relative term!

A differential amplifier VTVM is superior to single tube or FET circuits used in other grid testers. The meter reading is not affected since supply voltage fluctuations, drift, and aging affect both triodes equally.

Variable resistors: potentiometers and rheostats

Potentiometers are variable resistors, the ones used in the audio are of a single-turn construction (270°), but precision 10-turn pots are also widely used in applications where more precise control of resistance is required.

Potentiometers that need to be adjusted frequently, such as those that control the volume or tone in amplifiers, have a plastic or metal shaft onto which a knob is attached, either a push-on or screw-type.

Potentiometers that are adjusted rarely, only during servicing or calibration, instead of a shaft and a knob have a slot for a screwdriver. These are usually mounted inside an amplifier or an instrument, so the user has no access to them and are called trimmer-potentiometers or "trimpots."

Film potentiometers are made by depositing a conductive film onto a plastic substrate. The conductive film is usually carbon or "cermet," a unique conductive ceramic material.

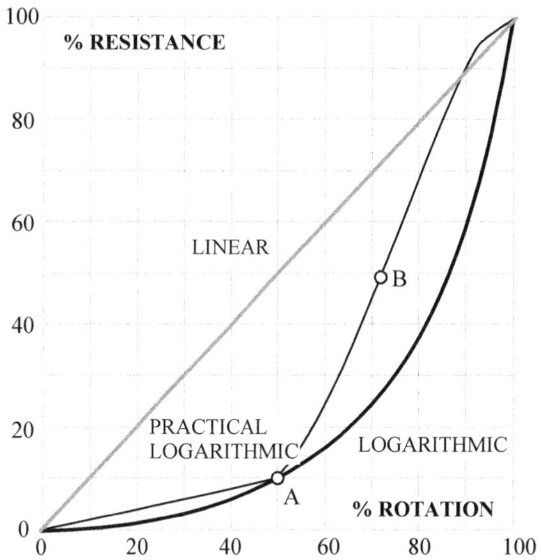

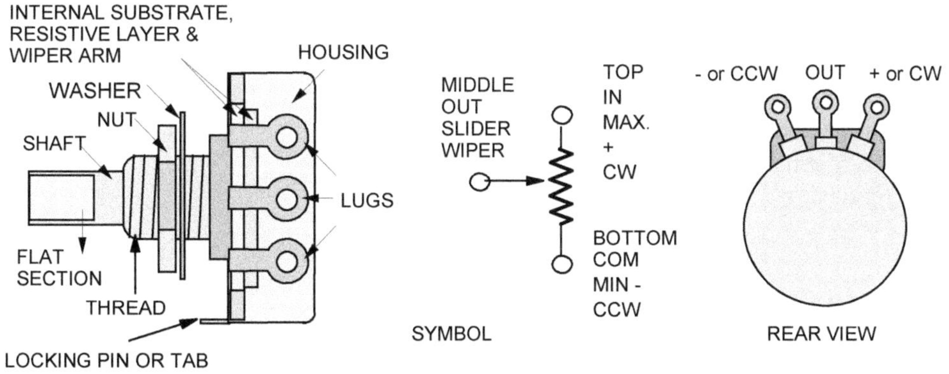

ABOVE: Percentage of resistance versus percentage of rotation curves for a linear potentiometer, nominal (ideal) logarithmic or audio potentiometer, and real (practical) logarithmic potentiometer

LEFT: Outside construction, symbol, and rear view of a typical potentiometer

TThe wiper or sliding arm makes mechanical contact with the conductive film and is the weakest link since such contacts are abrasive and gauge the conductive layer, eventually making pots "scratchy" or "crackly," even losing contact altogether.

The two main types of potentiometers are linear taper, usually marked "B," and logarithmic or audio taper, usually marked "A." The logarithmic potentiometers are used for volume controls since human ears respond to sound pressure in a logarithmic manner.

Manufacturing true log tapers by carbon or plastic deposits is very difficult, so manufacturers usually resort to approximating the log curve with two linear segments. As illustrated, up to 50% rotation, where amplifier's volume controls are set most of the time, the deviation from the actual log curve) is quite small, and at higher volumes such error is not noticeable at all.

Just like their fixed-resistance relatives, rheostats or high-power resistors are wire-wound components. The thinner the wire, the finer is the step-adjustment (the jump in resistance between adjacent turns), but the lower the current rating.

Testing potentiometers for continuity

This is one application where digital ohmmeters or rather digital multimeters measuring resistance simply cannot do. You'll need an analog ohmmeter or analog multimeter. Connect it between the wiper and one end of a potentiometer using crocodile clips and rotate it slowly from one end to the other. Observe the needle; it should move continuously and smoothly, without jumping forward towards lower resistances or falling back towards zero. Any breaks or irregularities in the resistance track will be indicated by the instrument meter's needle (pointer).

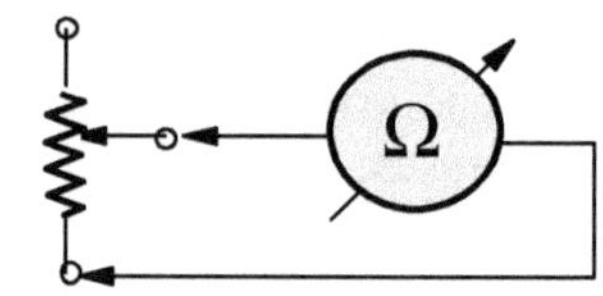

ABOVE: Connect an analog ohmmeter between the wiper and one end of a potentiometer using crocodile clips and rotate it slowly from one end to the other. The needle should move continuously and smoothly.

Testing potentiometers for tracking error

The most critical parameter of dual audio potentiometers used for volume control is the tracking between the two gangs. A significant difference in the resistance of the two gangs may manifest itself as one channel (with higher resistance between the pot's wiper and ground) of a stereo amplifier sounding louder than the other.

In this simple test, any battery can be used, standard voltages of 1.5V or 9V are acceptable, or you can use a benchtop DC power supply. Since it works as a null-indicator in this bridge circuit, the moving-coil instrument should have a center-zero so it's able to deflect to either side.

The four halves of the dual-ganged potentiometer form four legs of the bridge; the meter is connected in the bridge diagonal, between the two wipers. Ideally, the meter should stay dead center. The larger the deflection to either side, the more significant the discrepancy between the two gangs of the potentiometer! This test is only qualitative; an ohmmeter is needed to express the error quantitatively.

A two ohmmeter method is also illustrated, but a single ohmmeter and a changeover switch can also be used. In that case, the test is not continuous but taken in a dozen or so points across the potentiometer's range, from 0 to 270 degrees.

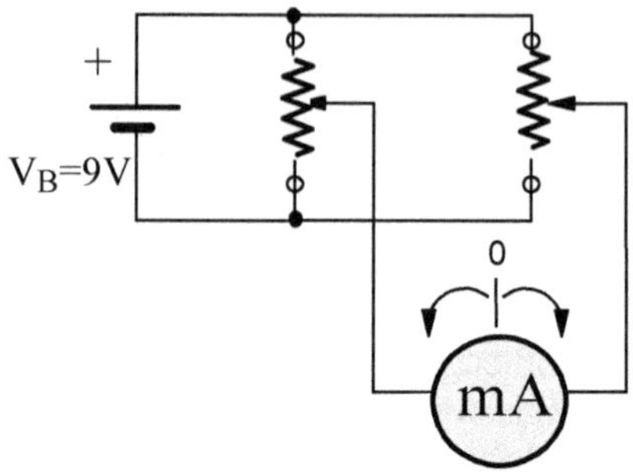

ABOVE: This simple setup to test dual-gang potentiometers for tracking error requires only a DC battery and a zero-centered moving coil meter.

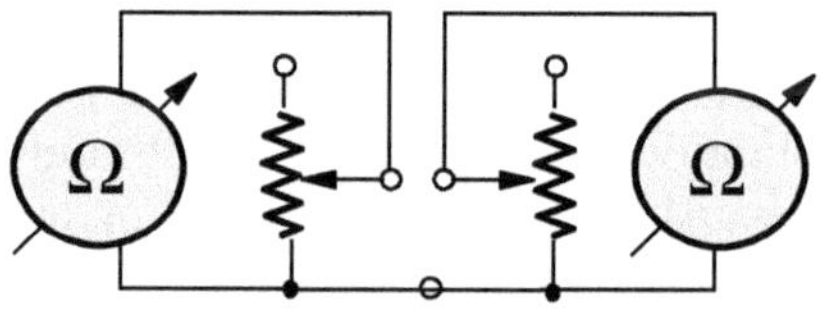

ABOVE: Two identical analog ohmmeters to test dual-gang potentiometers for tracking error

TESTING CAPACITORS

An ideal capacitor

An ideal capacitor would have an infinite resistance for DC current, thus allowing no DC current to pass. The AC current through an ideal capacitor precedes the voltage by 90 degrees. The impedance Z_C of an ideal capacitor is identical to its reactance X_C, inversely proportional to frequency f and its capacitance C:

$$Z_C = X_C = 1/(2\pi fC)$$

The higher the frequency of the signal, the lower the capacitor's impedance. This fact will have important repercussions in amplifier circuits, where unwanted or parasitic capacitances will form low pass filters with adjacent resistances and divert high frequency signals to the ground, thereby limiting amplifiers' frequency range. The same applies to multimeters and oscilloscopes.

In an ideal capacitor, the current leads the voltage by 90°, so the power factor PF is always zero. The PF is defined as a cosine of the phase angle Φ between voltage and current **PF=cosΦ**.

Except for the fact that the current lags the voltage by 90°, the same applies to ideal inductors (chokes), their PF is also zero, meaning there are no losses of any kind and no *active* power is spent or dissipated. Thus, the *ideal* L (inductor) and C (capacitor) are purely *reactive* components.

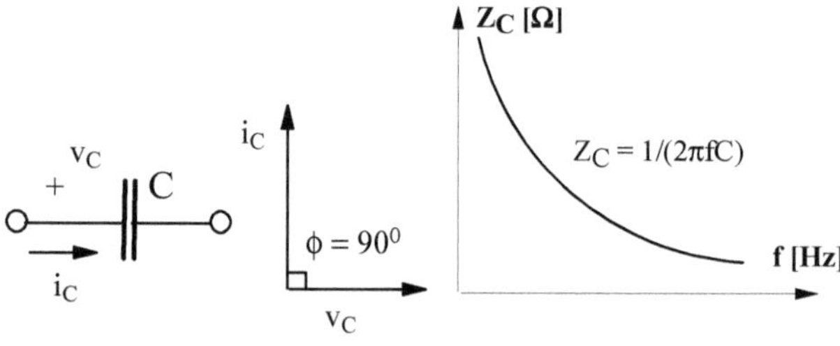

ABOVE: Ideal capacitor and its vector diagram: current leads voltage by π/2 radians or 90° The impedance of an ideal capacitor is inversely proportional to frequency.

BELOW: The phase relationship between the AC voltage across and current through an ideal capacitor.

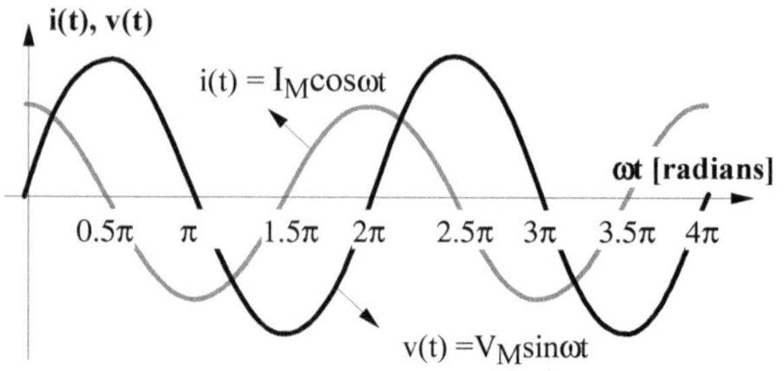

IMPEDANCE OF AN IDEAL CAPACITOR

$$Z_C = 1/\omega C = 1/(2\pi fC)$$

The AC losses (D-factor) and a vector diagram of a real capacitor

In electrolytic capacitors, the two major contributing factors to ohmic losses are the conductivity of the aluminium oxide layer and the resistance of the connections and leads. This behavior is modeled by adding a resistor of a small resistance in series with the ideal capacitor. This resistance R_S is often called ESR or the Equivalent Series Resistance. It is an essential parameter of electrolytic capacitors (ideally zero), but in reality, it can be pretty high, from 1 to 10Ω!

The vector diagram for the real capacitor features angles δ (delta) and ϕ (phi), the sum of which is always 90^0 (degrees). Delta is zero for an ideal capacitor, meaning $\phi = 90^o$.

From the right angle voltage triangle we know that $\tan\delta = v_R/v_C$. Since = $1/\omega C_S$, $v_C = i_R * X_C = i_R*(1/\omega C_S)$ and $v_R = R_S * i_R$ ($i_R = i_C$) it follows $\tan\delta = v_R/v_C = \omega R_S C_S = D$. That parameter was named D for "dissipation factor" (a ratio of resistance R_S and reactance X_C) and is zero for an ideal capacitor. We can relate Q-factor and D - they are a reciprocal of each other: Q=1/D!

The AC behavior of a real capacitor can also be represented by the parallel model. The vector diagram now features the right angle current triangle, so that $\tan\delta = i_R/i_C$. Since $v_C = v_R = R_P * i_R$ and the same voltage $v_C = v_R = X_C * i_C = 1/(\omega * C) * i_C$, it follows that $R_P * i_R = i_C/(\omega * C)$. If we express $i_R/i_C = 1/(\omega * R_P * C)$ we get $\tan\delta = i_R/i_C = 1/(\omega * R_P * C) = D$.

So, the lower the D-factor, the higher the Q-factor, and the better the capacitor, the lower its dielectric AC losses. These losses are, of course, thermal, so they raise the temperature of the capacitor. This is most noticeable on vintage electrolytic capacitors, which have high losses and are hot to touch. The top of the metal case often even bubbles up, and such a bulge is a sure sign that the capacitor has reached the very end of its life and should be replaced immediately.

Eventually, the build-up of gasses will cause the metal can to rupture, and the capacitor will explode, spewing the gunk all over the insides of the amplifier! Film capacitors do not use an electrochemical process as electrolytic capacitors do, so there is no gas buildup or the possibility of explosion.

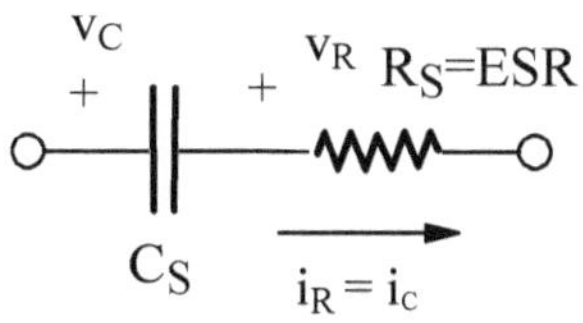

ABOVE: Series equivalent circuit of a capacitor

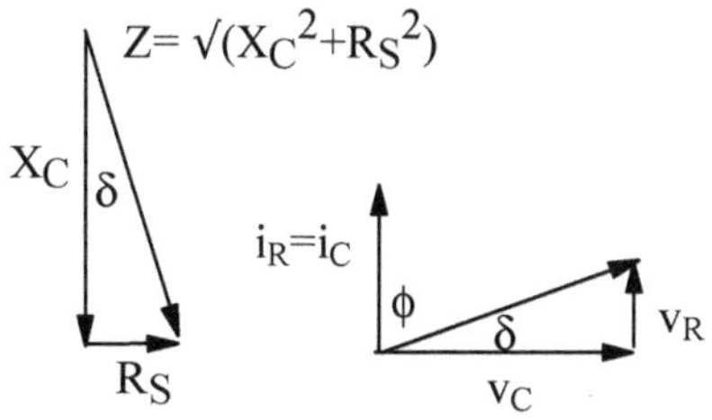

ABOVE: The impedance diagram of a real capacitor and the vector diagram of its currents and voltages (series equivalent circuit)

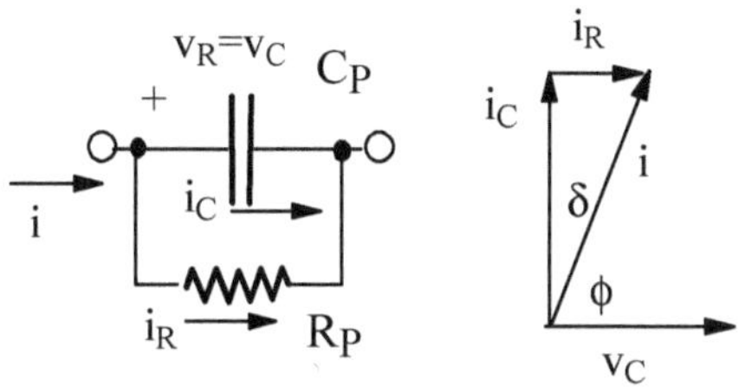

ABOVE: Parallel model of a capacitor and its vector diagram

BELOW: A more accurate Ac model of a real capacitor, with ESL added

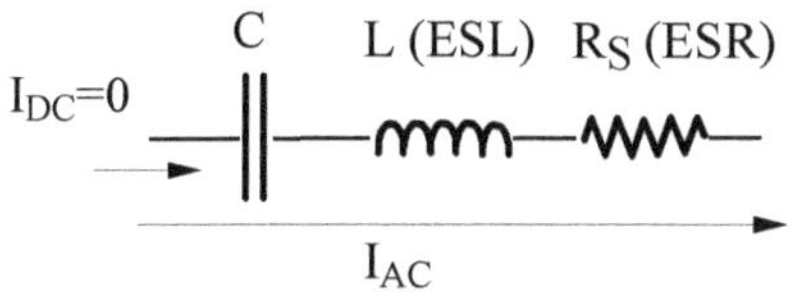

100Hz	1kHz	10kHz	100kHz
0.003	0.004	0.007	0.009
669.2 nF	666.8 nF	659.8 nF	646.6 nF

ABOVE: An example of how the measured values of "D" and "C" change with test frequency for a 680nF film capacitor

RIGHT: A simplified 3-component model of an electrolytic capacitor and the impedance versus frequency curve for electrolytic capacitors.

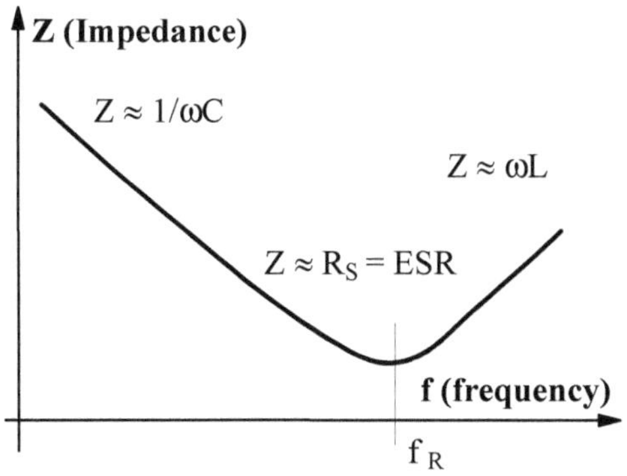

If your LCR meter has more than two test frequencies, measure the capacitance and the D-factor of the same film or electrolytic capacitor at each frequency. You will notice that D-factor will increase with frequency and that the measured capacitance will decrease (see test results in the box above). This leads us to the next, even more complex capacitor model, with added ESL or Equivalent Series Inductance.

When capacitors are wound using film & foil or metalized film, there are many "windings" in such a cap, and it can be expected that such a coil-like structure will have some inductance as well. With film caps, such inductance only becomes relevant at very high or radio frequencies.

Still, electrolytic caps start losing their capacitance very quickly with rising frequency. Above a specific signal frequency f_R (where they behave like a resistor), the inductance becomes the dominant parameter, and the capacitor loses its filtering capability! A qualitative frequency curve (above) shows these three regions. At the self-resonant frequency f_R the impedance is at its minimum and equal to the ESR.

Resonant frequency: $f_R = 1/[2\pi\sqrt{(LC)}]$ Impedance: $Z = \sqrt{[R_S^2 + (\omega L - 1/\omega C)^2]}$

The DC losses (leakage) added to the AC model

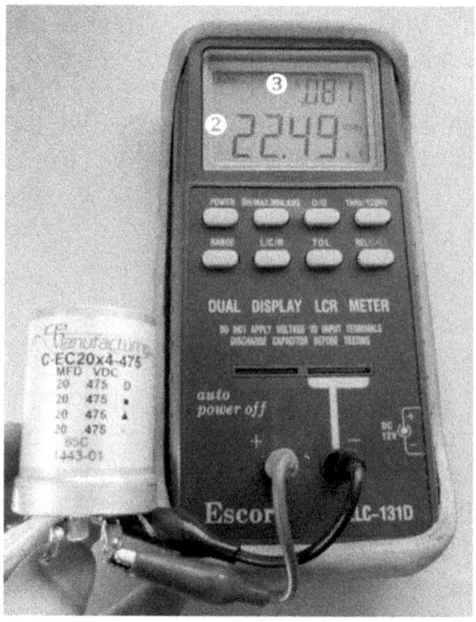

Real capacitors allow a small leakage DC current to pass through their dielectric, so the final model for both AC and DC conditions has an additional component, the parallel leakage resistance R_L. Do not confuse it with R_P in the parallel model, which is an AC resistance only, meaning only for AC signals.

LCR meters must measure ESR or D factor without charging up the capacitor. That is why they use only a small amplitude AC test signal with no DC component. Most meters do not display the amplitude of the test voltage like this Made-in-China instrument (1) but specify it in their manuals. The test voltage is under 1V, usually 500 or 600 mV_{RMS}.

Some LCR meters test components using only a series or parallel model; others have a default test mode (for instance, a series mode), but the user can manually override it and choose the parallel mode. Japanese DER EE LCR meter, model DE-5000, uses the series mode if the tested component's impedance is under 100kΩ (L_S, C_S, or R_S are displayed) and parallel mode above 100kΩ.

The default of our Escort LCR meter (10,000 count, 0.9V_{RMS} test signal) is the parallel mode for capacitance and resistance tests and series mode for inductance measurements. The photo (upper right) shows it testing one section of a multi-section elco. Using 120Hz test frequency, it measured 22.5 µF (2) and D=0.081 (3). What was the ESR value?

Since $D=R_S/X_C$, $R_S=DX_C= D/(2\pi fC)= 0.081/(2*3.14*120*22.5*10^{-6}) = 4.77$ ohms.

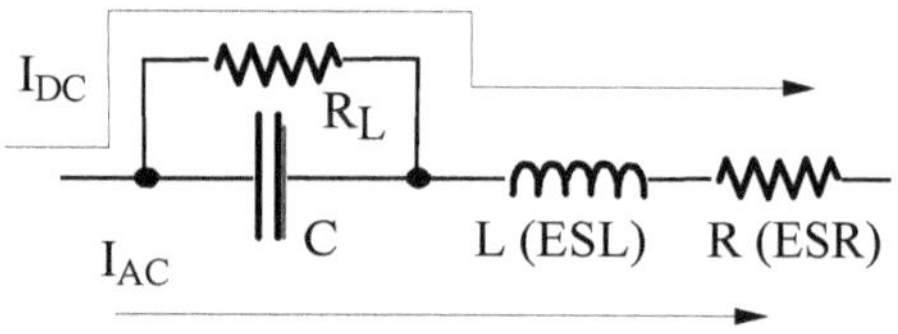

LEFT: The 4-component model of an electrolytic capacitor with a leakage resistance R_L added. This establishes a DC current leakage path through the capacitor.

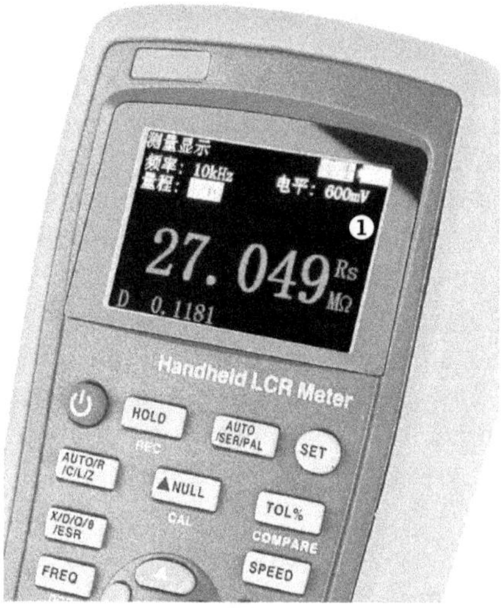

The leaky capacitor puzzle

When you measure capacitance on digital LCR meters, the measured value could be much higher than the marked value! The leaky vintage USA-made film capacitor pictured (4) should measure around 50 nF, but the LCR meter says 63nF (5). The problem is caused by the method most LCR meters use to measure capacitance, counting pulses during the time it takes to charge a capacitor to a certain value. The longer the charging time, the larger the capacitance. This si a typical example of a *systematic error,* inherent to the test method.

A leaky capacitor such as this one (akin to a paralleled resistor R_L which keeps discharging the capacitor) will increase the charging time and the number of counted pulses, which the LCR meter will interpret as higher capacitance.

Since LCR meters display the D-factor as well (.016 in the photo), a very high D-factor reading will help you identify these false measurements. Good quality, non-leaking film caps will have a D of _ _ _ (unmeasurable), .001 or .002, not .016!

Testing electrolytic capacitors with an analog ohmmeter

To test electrolytic capacitors use an *analog* ohmmeter on x1kΩ (x 1,000) range. If the capacitor is open circuit, there will be no meter indicator movement. For a good capacitor, the meter needle should move quickly towards the low resistance side of the scale (usually on the right) as the battery inside the ohmmeter charges the capacitor. Then it should slowly drop back towards infinity. The final resting point of the needle will indicate a very high resistance, the insulation resistance of the capacitor.

If the capacitor is short-circuited, the needle will move towards the low Ω region of the scale and stay there; it will not return quickly to the high end of the scale. Illustration on the next page.

RIGHT: Testing electrolytitc capacitors with an analog ohmmeter. The meter needle should move quickly towards the low resistance side of the scale (a) and then drop back towards the very high resistance end (b).

BELOW RIGHT: Cross-section of a typical shielded or coaxial cable reveals that the whole cable is in effect one huge cylindrical capacitor, with the inner insulation as the dielectric.

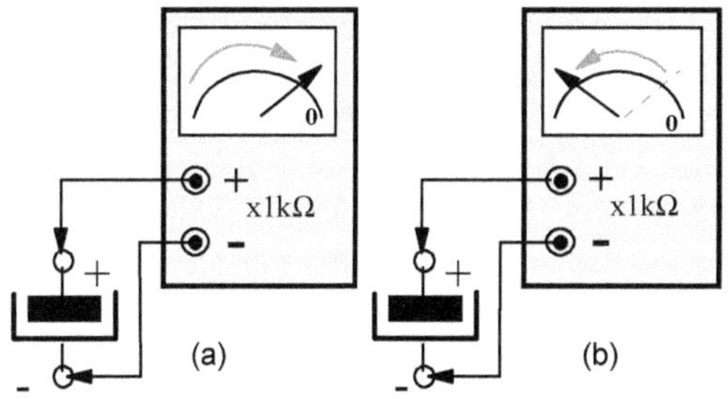

Measuring the capacitance of shielded audio cables

Every shielded cable is, in its essence, a low-pass filter with distributed parameters (distributed across the length of the cable). From a cross-section of a shielded or a coaxial cable, it is evident that the whole thing is one giant capacitor: two metal structures (the inner conductor and the shield or braid), separated by a dielectric (inner insulation). Indeed, connect an LCR meter between the active conductor and the shield and measure its capacitance.

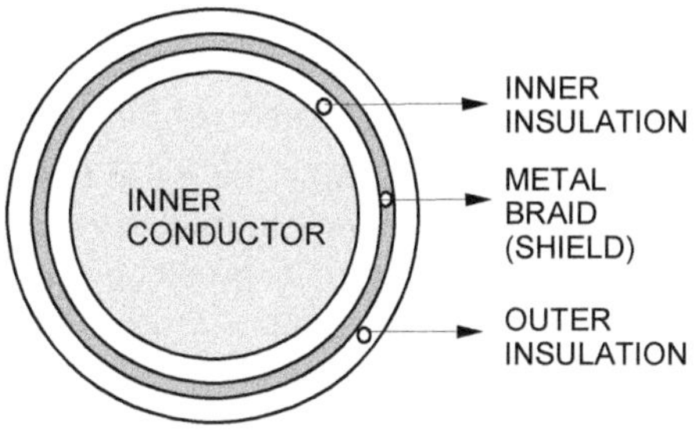

In a signal chain (right), the cable will shunt higher frequencies away from the load. For instance, the source can be a CD player, while the load can be a preamp or an amp. Or, it may be the shielded cable used inside the amplifier, from the RCA inputs at the back across the whole depth of the chassis to the potentiometer at the front.

If you measure both R_C and C_C, you can even calculate the frequency at which the cable's capacitance will attenuate the signal by -3 dB. At the angular frequency $\omega_0 = 1/R_C C_C$, the reactance $1/\omega C_C$ equals the resistance R_C, so the output voltage is 1/2 of the input voltage, and the filter transmits 50% of maximum power. Halved power is the equivalent of -3 dB signal attenuation.

Next time you do a listening test comparing cables and conclude that a certain cable lacks the "top end", measure its capacitance and see if it's really high or is it just your imagination.

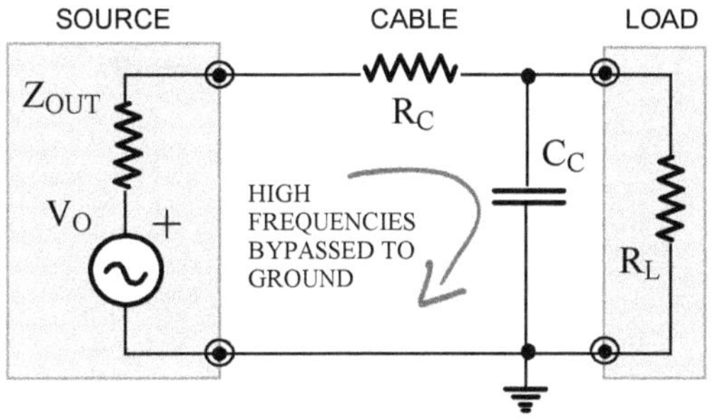

ABOVE: A high capacitance cable is a low-pass filter which will shunt higher signal frequencies to ground.

Testing and reforming electrolytic capacitors

Vintage tube-based capacitor checkers come in many types and under many names: capacitor checker, condenser tester, capacitor analyzer. Most test capacitors are out-of-circuit only, while some do in-circuit tests, a few units can do both. As capacitance meters, these old-timers are inferior to modern digital LCR meters. However, they have two significant advantages. They test capacitor leakage (scale calibrated in MΩ) at high voltages, typically up to 600 Volts. Simpson Capacohmeter goes even higher, up to $900V_{DC}$!

Sprague TO-4, TO-5, and TO-6 have an analog meter that displays the leakage current, as do the Japanese-made Lafayette TE-46 and Olson TE-189. Most others use a "magic eye" to qualitatively indicate excessive leakage.

Due to their chemical nature, elcos deteriorate more sitting on a shelf than when working in equipment. Unless permanently damaged, they can be reformed on these high voltage checkers. Leave them energized for 1 hour for each year they have been inoperative (10 hours for an elco that hasn't been used for 10 years).

The power supplies of these testers have been designed to sag significantly under a load such as a current-hungry electrolytic capacitor.

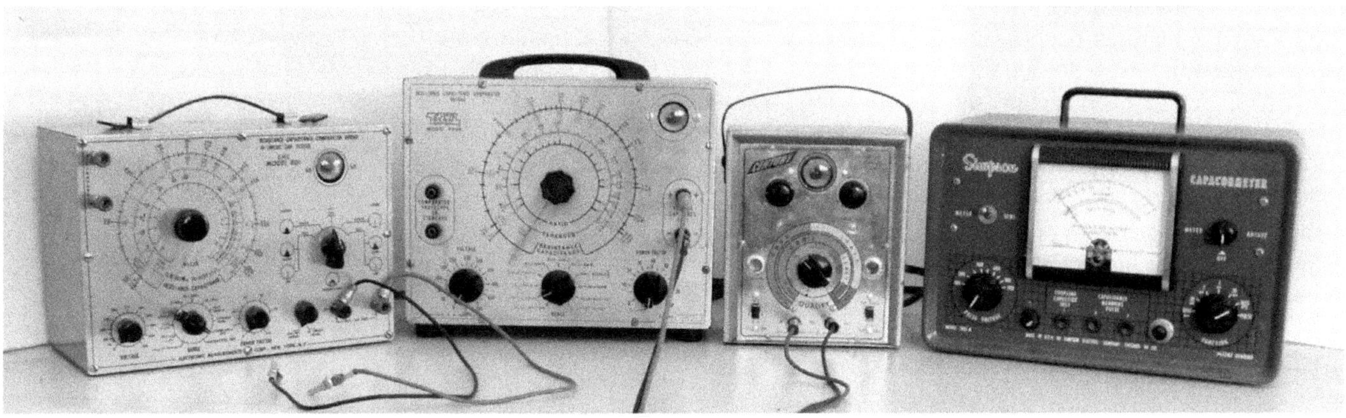

The four models pictured are (L-R): EMC model 801, Eico 950, Century CT-1 and Simpson Capacohmeter. The first three all use the same type of an AC bridge but do not display leakage current quantitatively, only as a qualitative indication on a magic eye tube.

So, although say 400V may be dialed on the tester, the higher the charging current, the more the output voltage will drop. The reforming process may start at 30 or 40 volts; in five minutes, the multimeter connected in parallel with the capacitor may show 120V, after 20 min it may be up to 210V, and so on.

The Simpson Capacohmeter does not measure resistance or inductance. It is not an RC bridge like the other three instruments, but it can test capacitors using a steady DC voltage or a pulse technique. Internal Thyratron tube produces high voltage pulses for that purpose.

Simpson "Capacohmeter"

Simpson 383A Capacohmeter is an interesting 3-in-1 instrument. As a capacitance meter, its specs are pretty limited; its five ranges cover a narrow range of 10pF to only 10μF. While AC sinewave voltage is used to measure capacitance, capacitor leakage is checked using a high DC voltage of 400V. This is a highly sensitive test; the instrument working as a megohmmeter can detect leakage as high as 1,000 MΩ (1GΩ)!

The most interesting feature is the pulse test, using voltage pulses produced by the internal relaxation oscillator with a Thyratron tube. By adjusting that tube's grid bias voltage by external control, peak pulse voltages of between 100V and 900V are available. In-circuit tests can be performed (on high impedance tube circuits only, of course). The three leads are connected across a coupling capacitor (as per the diagram on the right), with the third lead at ground potential.

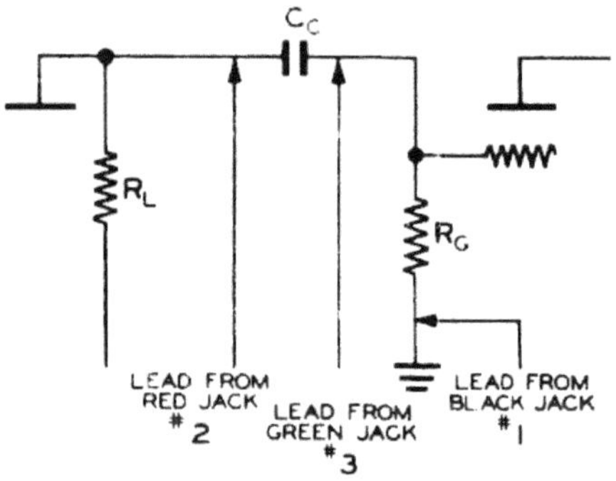

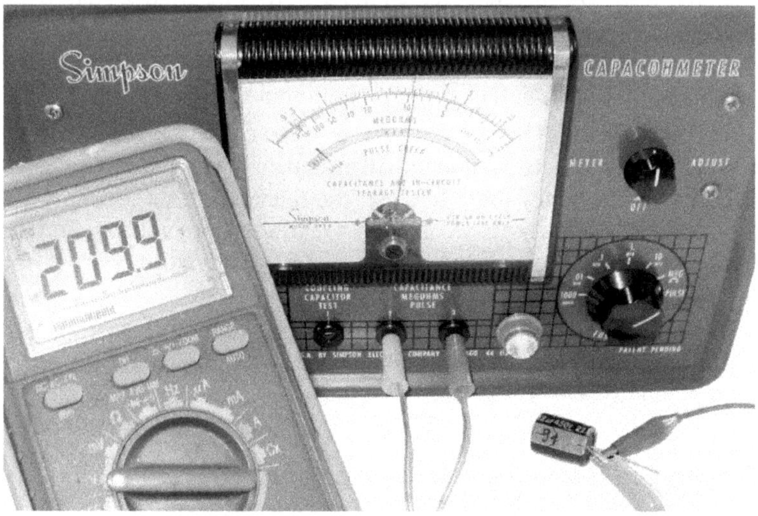

ABOVE: 3-terminal in-circuit test of coupling capacitors in tube circuits

RIGHT: Pulse testing of a 22μF elco on our test bench

Sencore LC analyzers LC53, LC75, LC76, LC77, LC102, LC103

In contrast to most T&M equipment makers who released new models once in a blue moon, Sencore constantly upgraded their gear, especially their tube testers and this range of LC analyzers. They remain highly sought after (read: very expensive) even today, more than 40 years after the first model, LC53, was released.

LC53 Z-meter features twelve switch-selectable test voltages from 1.5V to 600V. Later models, such as LC76, go even higher, up to 1kV (16 voltage choices), and have a dedicated push-button to display the ESR of the capacitor under test.

Such a high voltage testing regime is most likely the reason for their high prices on eBay and other online sales websites, from US$400 to over US$1,000.

The most modern models, LC102 and LC102, fetch a few thousand dollars and compete with laboratory-type LCR meters such as those made by HP (Agilent) and others.

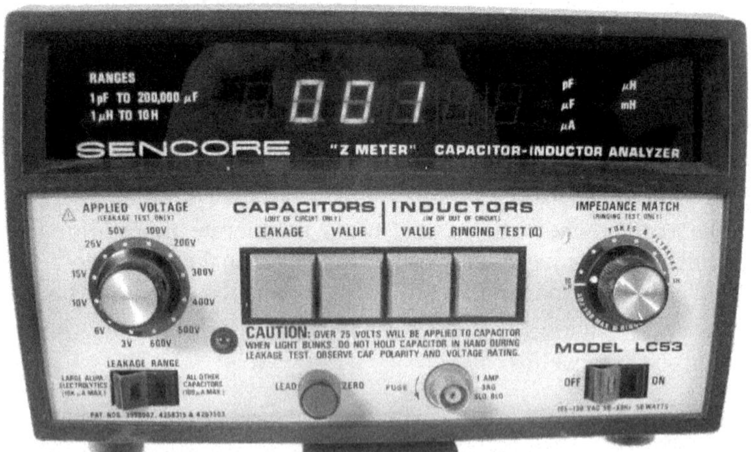

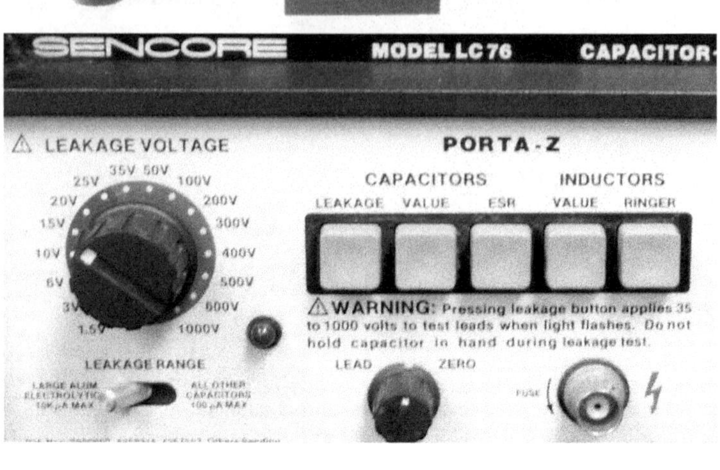

THE BRIDGE METHODS OF TESTING CAPACITORS

Although modern digital LCR meters are much more precise and faster to use than these vintage "utility bridges," many important lessons can be learned from these instruments. So let's look at how they measure resistance and capacitance, the power factor of a capacitor, its losses and leakage, and how they do one thing that digital LCR meters cannot and never will be able to do, "reform" electrolytic capacitors. However, we must first digress and look at various DC and AC bridges and their operating principles.

The evolution of various capacitance bridges

Dozens of variations of AC bridges exist, but they all originate with the DC resistance bridge, known as the Wheatstone bridge.

There are four arms of the bridge and four nodes. DC voltage is connected to nodes U & V, and a null-detector is in the bridge diagonal, between nodes X and Y. The null detector indicates when the bridge is "balanced" when there is no voltage and thus no current flowing between points X and Y, i.e. when they are at the same potential.

DC bridges usually use a moving coil meter with a center zero to deflect in both directions, depending on the polarity of V_{XY} voltage. Circuit analysis will show that for the bridge to be "balanced," $R_1/R_2=R_X/R_3$ or the ratio of resistances (or impedances) to the left of the null detector must equal the ratio of resistances (or impedances) to its right.

If three of the four resistor values are known (precision resistors used), the bridge can be used to measure the resistance of the unknown resistor R_X. Since there is one equation with one unknown, it is enough to make only one resistor (R_3 in this case) variable (adjustable), and its dial can be calibrated in the values of R_X!

By simply replacing the DC power supply or battery with an AC voltage source such as an audio oscillator or even a secondary of a mains transformer, we get an AC bridge, which can measure not only resistance but also capacitance and inductance. The basic capacitance bridge has two resistors in the upper arms and two in the lower arms, one unknown capacitor C_X and one standard or precision capacitor, C_3.

The bridge is balanced when the ratio of the two capacitive reactances equals the ratio of the two resistances: $X_{C4}/X_{C3}=R_1/R_2$. Since $X_{C4}=1/\omega C_4$ and $X_{C3}=1/\omega C_3$ the angular frequencies ω cancel out and we get $C_3/C_4= R_1/R_2$.

Headphones are used as an audio null-indicator. There will be no AC signal or tone in the headphones when the bridge is balanced. The detector or indicator can also be visual, an oscilloscope, an AC voltmeter (or DMM), or an eye tube.

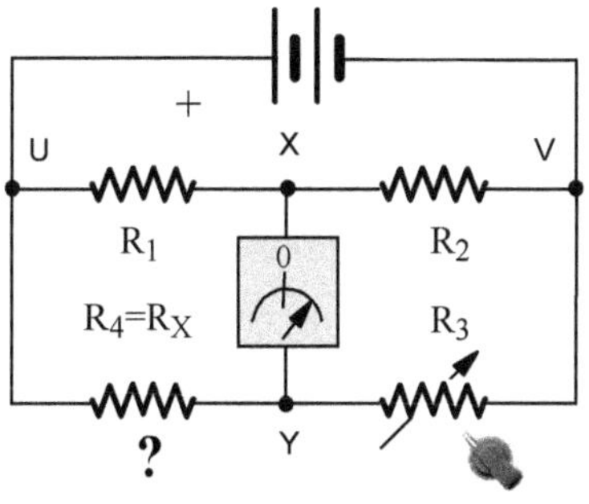

ABOVE: The Wheatstone bridge is balanced when $R_1/R_2=R_X/R_3$ or $R_1*R_3=R_2*R_X$ i.e. the product of resistances in opposite arms must be equal!

BELOW: The basic capacitance bridge

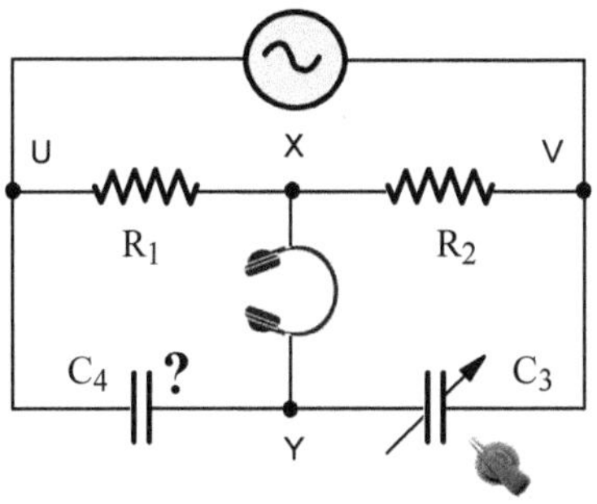

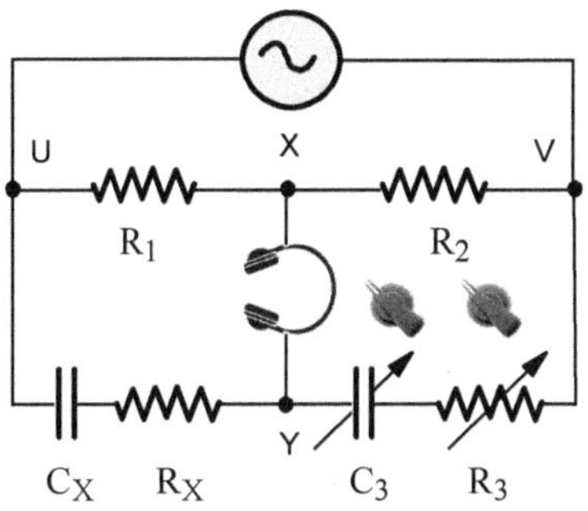

ABOVE: The basic capacitance bridge modified to test lossy capacitors

BELOW: Eico 950B bridge

In gray frame is the magic eye detector (partial drawing not showing the anode and its supply circuitry)

Adding a provision for lossy (leaky) capacitors

This simple bridge will work only for capacitors with zero or low losses. Any significant series or shunt resistance (remember the models of a leaky capacitor) cannot be balanced out. Eico 950B and other simple service-type resistance/capacitance bridges only use this arrangement when testing low-value film and mica resistors, in Eico's case, the lowest 10pF-5nF range, since they usually have the lowest losses (zero or near-zero power factor).

Thus, the next, more complex version has an addition of a variable resistor R_3 in the C_3 arm. That resistor must be adjustable to balance out the R_X of the capacitor under test. On Eico 950B and other commercial vintage bridges, that potentiometer's dial is marked "Power Factor" and calibrated 0-80%. Power factor is the ratio of resistance to impedance. Remember that D (dissipation factor) is a ratio of resistance R and reactance X_C, so for small values of PF, it is approx. equal to D, or $PF\approx\omega R_S C_S = D$. $PF=\cos\phi$ and since $\phi=90^o$ when there are no losses, PF=0 for an ideal capacitor!

Instead of using expensive and bulky tuning capacitors, service type bridges use a potentiometer "P," whose dial is then calibrated in the resistance and capacitance values (since it is used for both R and C measurements). R_C is a current limiting resistor, and the additional resistor R_A is only switched into the R_2 arm at the highest capacitance range (50μF-5μF for Eico 950B).

In the gray frame is the detector, a 1629 "magic eye" tube. This partial drawing does not show its anode and related circuitry; see the circuit diagram of Lafayette TE-46 and Olson TE-189 (page 83), which used the same detector arrangement as Eico.

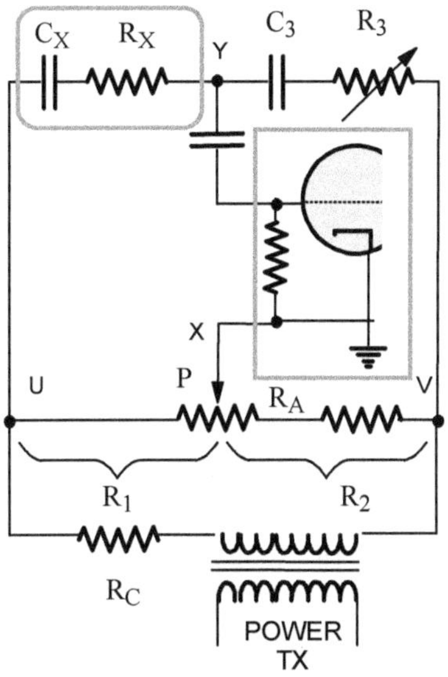

Electron-ray or "magic eye" tubes as tuning and bridge balance indicators

Since the most common use of electron-ray tubes was as tuning indicators in radio receivers, they became known as "magic eyes." They were also often used in service type (meaning cheaper) test instruments as cheap and sturdy replacements for analog moving coil meters. Meters could easily be mechanically damaged or get overloaded and burn out. Magic eye tubes could sustain overloads and were smaller and less troublesome.

There are two triodes in one glass envelope. The first is a voltage amplifier; the output from its anode drives the deflector electrode of the second assembly, whose circular anode dish is coated with fluorescent material and is known as the "target ."The ray-control electrode is shaped as a vertical wire parallel to the cathode. It is typically 150-200V negative with respect to the target, so it repels some of the electrons and thus alters the path of the electron beam. This results in an unlit conical or trapezoidal segment on the target called a shadow. The more negative the ray-control electrode is with respect to the target, the wider the shadow becomes.

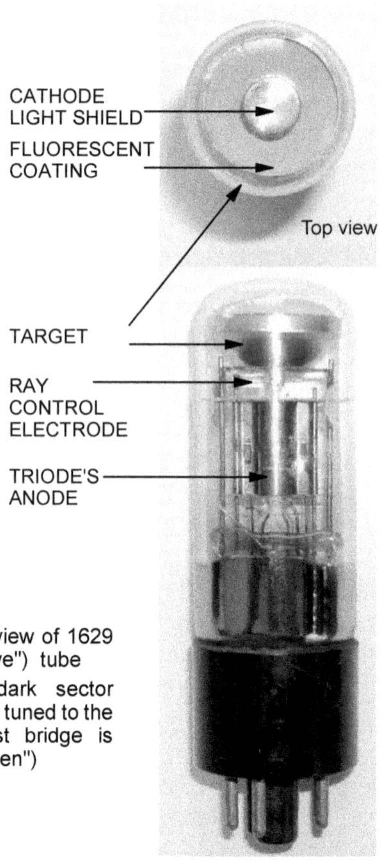

Bridge closer to the balance point (dark sector appears)

Bridge past the balance point (dark sector narrowing - eye closing again

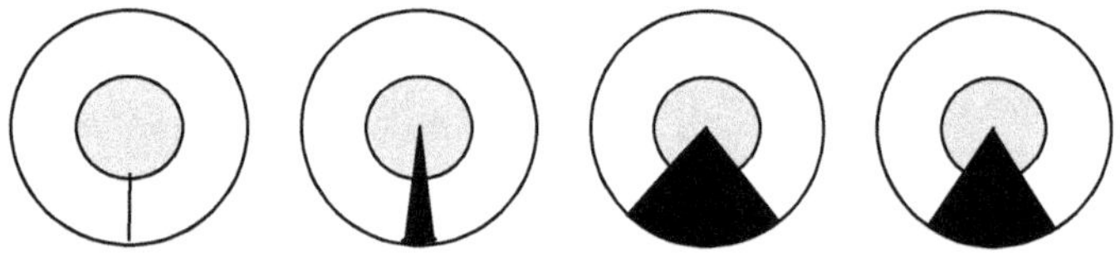

Bridge unbalanced (eye closed or overlapping)

Bridge balanced (dark sector the widest - eye fully open)

RIGHT: Side and top view of 1629 electron ray ("magic eye") tube

LEFT: The widest dark sector indicates that a radio is tuned to the station or that a test bridge is balanced (eye fully "open")

The Schering bridge

The comparison capacitance bridge discussed so far cannot accurately measure capacitors with very low dissipation and power factors. For that to happen, the standard internal capacitor C3 must have a much lower dissipation factor (lower losses) than the measured capacitors, which is next to impossible to achieve, especially in cheap service-type bridges. By the way, if you get one of those, they will be 50+ years old, so all capacitors and most resistors inside will have to be replaced with precise values. Otherwise, the instrument, whose accuracy is +/- 3% at best (even with precisely selected C and R values), will be useless.

Laboratory-type bridges use the Schering bridge to measure capacitors with very low losses. The balance conditions require that the sum of phase angles of opposite arms 2 and 4 equals the sum of the phase angles of arms 1 and 3. Since the standard capacitor C_1 is in arm 1, the sum of the phase angle of arms 1&3 will be $90+0=90^o$. To obtain 90° needed for balance, a small variable capacitance C_2 is added in parallel to R_2 to give that arm a slight capacitive angle and thus to cancel the tested capacitor's resistance R_X!

Adding DC polarization (bias) for high voltage capacitor testing

Electrolytic capacitors have significant losses (large ESR and D-factor), which vary with the DC voltage across the capacitor, as does its actual capacitance. For more accurate measurements of both, the AC bridges discussed so far can be modified, and an HV DC power supply (0-450V) can be hooked up externally to provide DC bias. The blocking capacitor CB is needed to prevent a large DC current through the null-detector. The DC source can be added in parallel with CX (through a large inductor L which will prevent AC signal from being diverted through the DC source) or in series with the AC source, which then must be coupled through a transformer TX, so there is no DC current through it (the AC source).

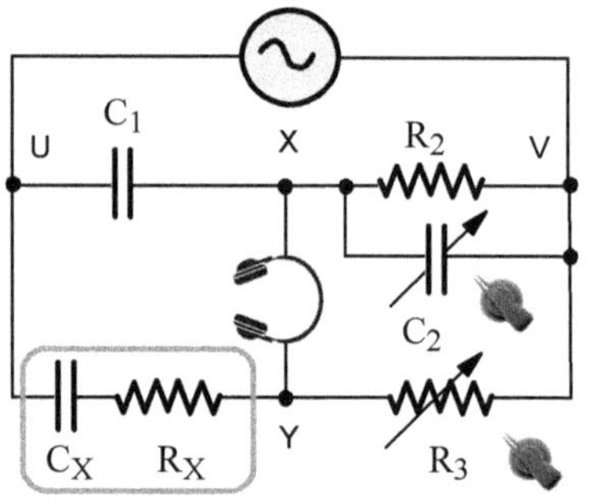

ABOVE: The Schering bridge

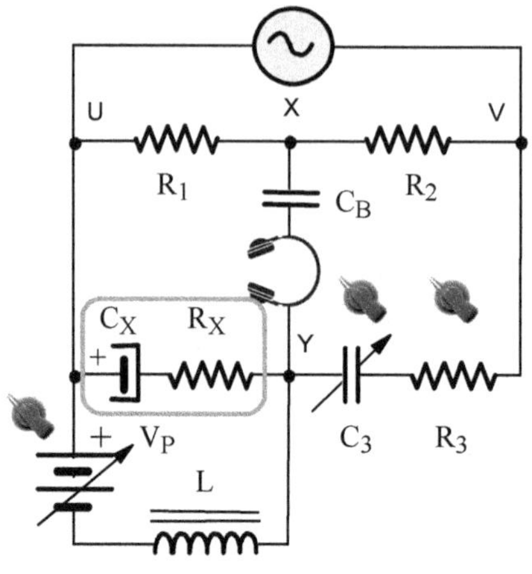

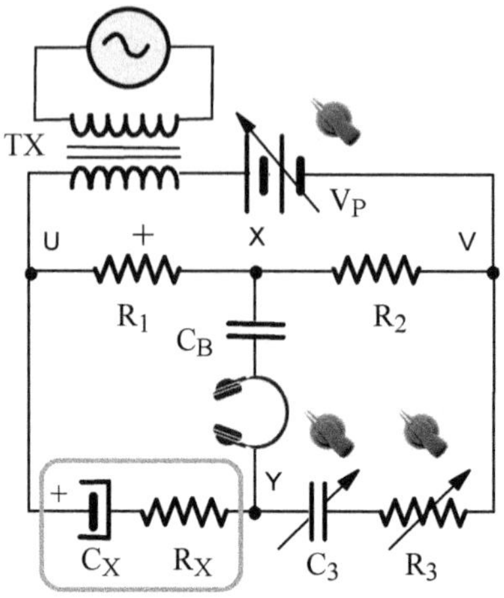

RIGHT: Two ways of adding DC bias (polarization) VP for testing HV electrolytic capacitors in a service-type bridge

Vintage utility RC bridges: Heathkit C-3, Eico 950B and PACO C-20

The range selector switch on the Heathkit selects one of two resistance ranges (four on Eico 950B), four capacitance ranges (ditto for Eico 950B), and five test voltages for the "Leakage Test," 25, 150, 250, 350 and 500 V_{DC}. This function on Eico is performed by a power rheostat calibrated from 10 to 500V, meaning that the test voltage can be adjusted precisely and continuously. However, the power rating of that rheostat is only 4 Watts, and it often burns out, as it did on our unit. Luckily, we had a similar 100k/5Watt rheostat from another salvaged piece of test gear.

Eico 950B also has another pair of binding posts for connecting a referent capacitor or resistor so that the instrument can serve as a comparator of the unknown R or C with the referent value.

R11 is the calibrated bridge balance potentiometer (#5 on the photos above), R14 is the power factor rheostat, and R15 adjusts the negative DC voltage during leakage tests. The bridge operates on $55V_{AC}$ from its own secondary winding, with R10 as the current limiting resistor.

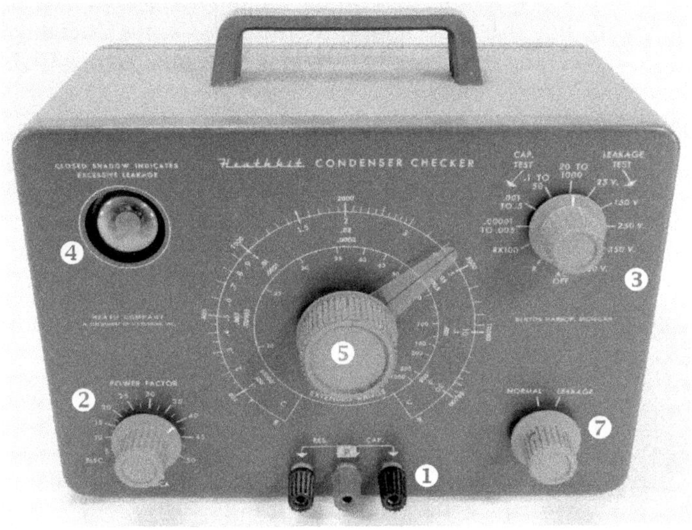

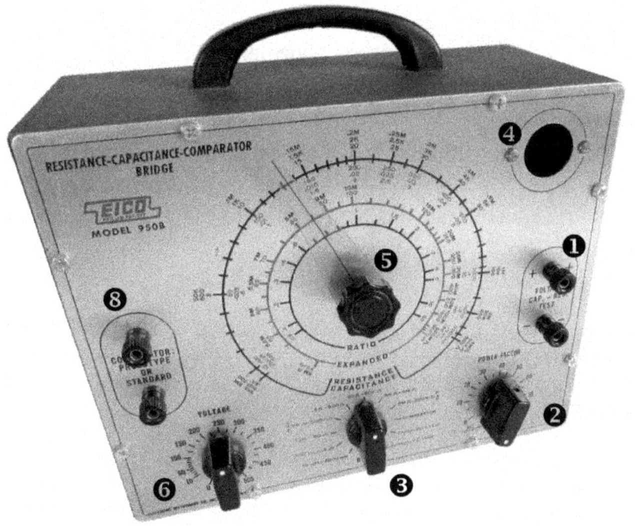

ABOVE: 1) R_X or C_X terminals (binding post type) J1-J2 2) Power factor rheostat 3) Function switch 4) Magic eye indicator 5) Calibrated bridge balance potentiometer 6) Leakage voltage selector switch 7) "Normal-Leakage" selector switch (Heathkit only) 8) Referent R or C terminals J3-J4 (Eico only) BELOW: PACO C-20 is identical to Eico 950B; its diagram is neater and better marked.

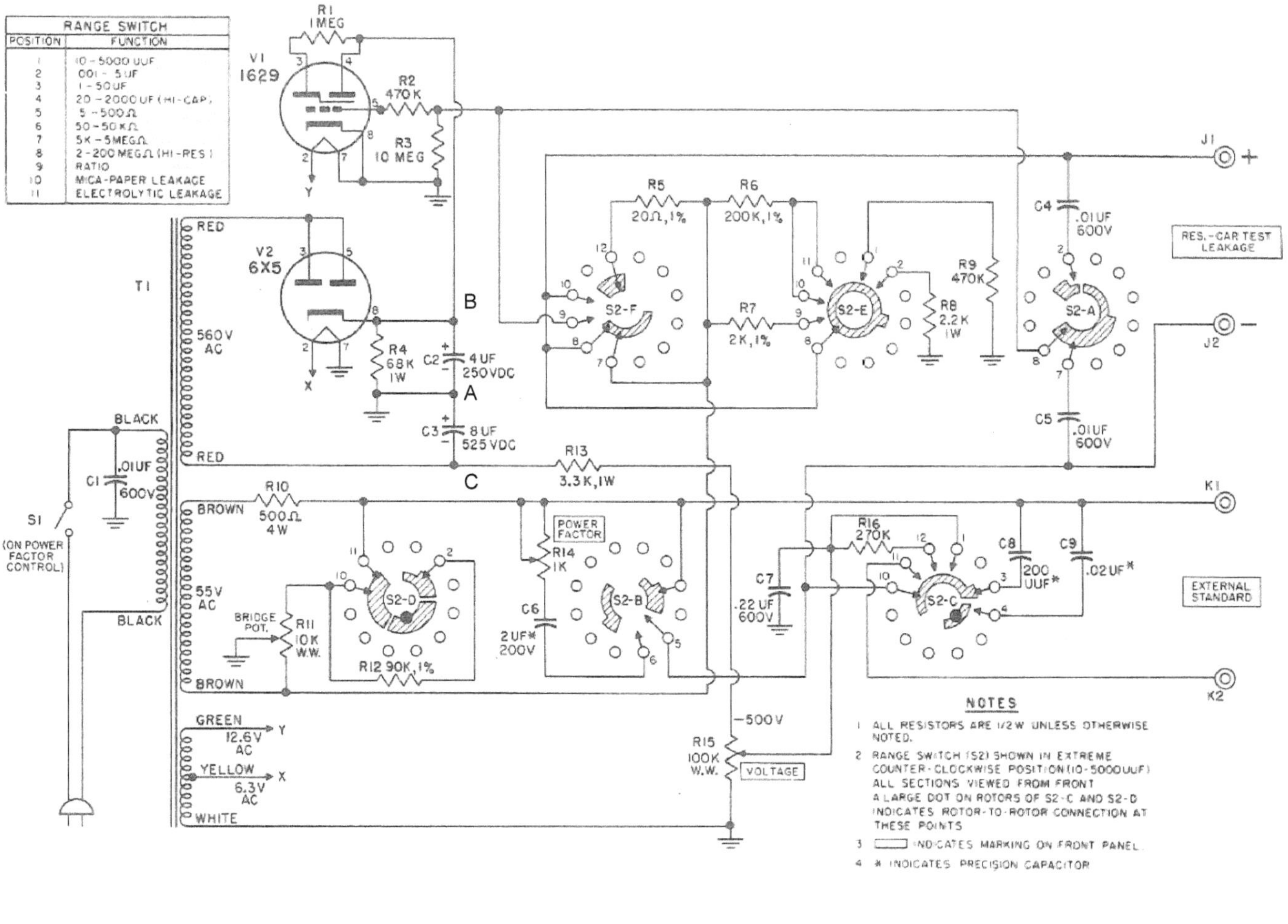

Notice how the DC voltages are obtained. The top secondary's $560V_{AC}$ is half-wave rectified (both diodes inside the 6X5 tube in parallel). There is $-520V_{DC}$ in point C and $+200V_{DC}$ in point B (anode supply for the 1629 magic eye tube). The total rectified voltage is $V_{BC}=720V_{DC}$, but since point A is grounded (earthed), the two voltages mentioned are of the opposite polarity with respect to point A, the referent point (0 Volts).

PRACTICAL APPLICATION: Measuring transformer's turns ratio

When the "RANGE" switch (3) is the "COMPARATOR" position (1), the bridge compares the impedances of two components, the tested device connected to terminals J1 & J2, and the referent component on terminals J3 & J4. One application of that mode of operation is determining the turns ratio of unknown power or audio transformer. When inductances or resistances are being compared, the known/standard/referent value must be divided by the indication obtained on the dial (2). Notice the symmetry of the dial, with the range from 0.05 (5) at one and 20 (4) at the other end of the scale, with "1" or 1:1 in the scale's middle (6).

The unknown transformer is connected with its primary at ref. terminals J3 & J4 and its secondary at unknown terminals J1 & J2, as per diagram. The tuning eye shows balance at the dial reading of close to 20, about 18. The transformer is actually a 300B amplifier output transformer with N1:N2 turns ratio of 17.7, meaning the impedance ratio is 17.7 squared or 312.5. With an 8 ohms secondary load, the primary impedance will be 8*312.5 or 2,500 ohms.

However, the scale isn't precise between 10 and 20. Many audio output transformers have turns ratios above 20:1, so more precise measurements based on other methods must be used, as outlined in the transformer measurements chapter. While this test method is valid and can be very precise if properly implemented, the vintage instruments of the Eico 950B kind are "cheap & cheerful," far from the laboratory quality and precision level.

RIGHT: Comparator bridge arrangement for determining turns ratio of a transformer. J3-J4 are comparator or standard terminals (8) on previous page photo, J1-J2 are R_X or C_X binding posts

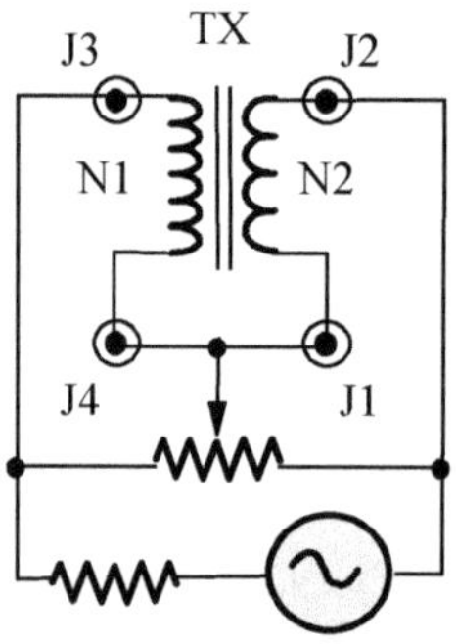

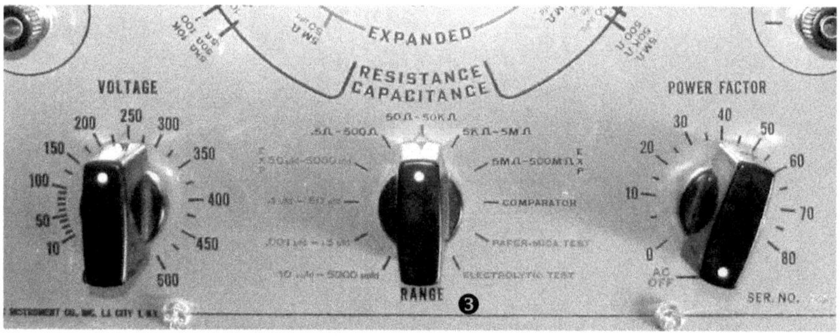

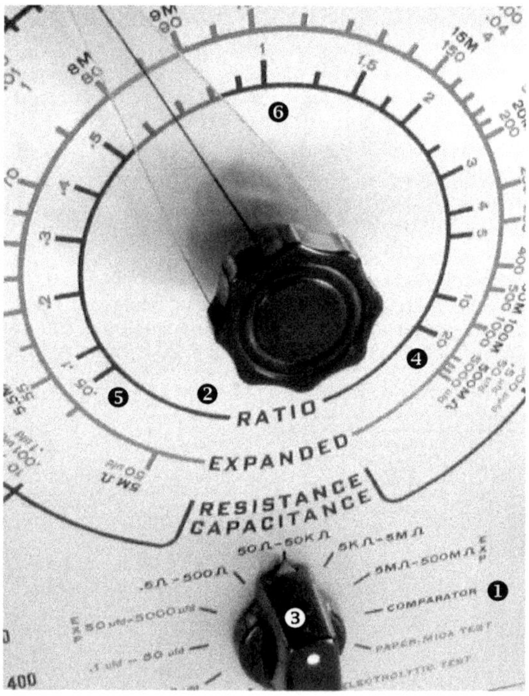

Vintage utility RC bridges: Lafayette TE-46 and Olson TE-189

Apart from capacitance, resistance, leakage currents and power factors of electrolytic capacitors, these instruments can also measure transformer turns ratios and general impedance ratios of two components.

The 900V center-tapped secondary voltage (1) is rectified by the 6X4 rectifier (2) and filtered to obtain around $630V_{DC}$. In position #5 of the "METER" switch (4) this voltage is used to directly test insulation resistance of electronic components. In positions #3 ("10mA leakage") and #4 ("1mA leakage"), the moving coil meter (10) is shunted to display the selected leakage current range. For this test one chooses a test DC voltage by "VOLTAGE ADJ." switch (3), the choices are 3V-6V-12V- 25V- 50V-150V-250V- 300V- 350V- 400V- 450V- 600V.

To reduce target wear, on leakage and insulation resistance tests the indicator tube (8) is switched off, it is only active during R, C and ratio measurements, when the METER switch is position #2, as indicated on the diagram ("Meter off").

The two leakage positions can also be used for reforming electrolytic capacitors. Capacitors are connected to terminals B & C and the voltage is gradually raised to the rated DC voltage of the capacitor. As the capacitor's electrolyte and plates are "reformed" the leakage current indication on meter "M" will slowly decrease.

The meter has no protection against overload so excessively large leakage currents can burn it out. Thus, two silicon diodes connected in anti-parallel arrangement should be shunted across the meter terminals.

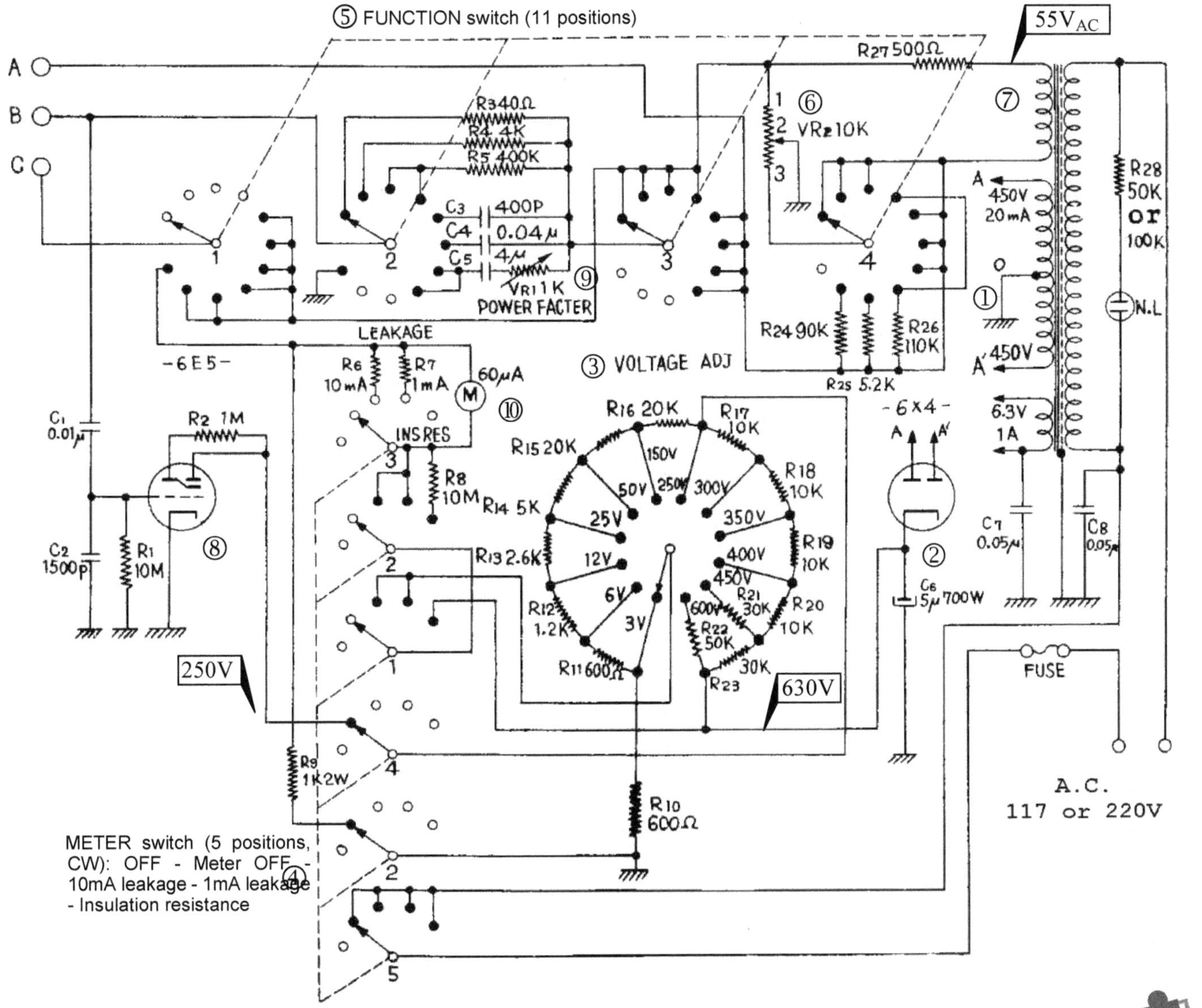

The measuring bridge is balanced by potentiometer VR1 (6), whose large dial is calibrated in various scales in C, R, and turns ratio values. The "bridge balance" potentiometer (6) is connected across the $55V_{AC}$ secondary transformer winding (7), and the "magic eye" tube is used as a null indicator, as on all similar bridges. With the bridge out of balance, its target will be evenly lit. Closer to the balance point, a dark triangular segment will appear at the bottom of the target and will widen as the balance is approached. Pass the balance point, the target will start narrowing again (see page 80).

R3-R5 and C3-C5 are reference components in the bridge and must be of exact values, as specified. The three capacitors should have no leakage at all.

Lafayette TE-46 and Olson TE-189

- Capacitance measurement: 2 pF - 2,000mF
- Resistance measurement: 2Ω - 200 MΩ
- Transformer turns ratio 1:1 - 200:1
- Impedance ratio - 1:1 - 40k:1
- Electrolytic capacitor leakage at 3-$600V_{DC}$, ranges: 0-1mA, 0-10mA
- $600V_{DC}$ insulation resistance test 0-300 MΩ
- 6X4 (rectifier) and 6E6 ("magic eye" indicator) tubes

INDUCTANCE MEASUREMENT METHODS AND INDUCTANCE BRIDGES

An ideal inductor

An ideal inductor has no resistance for DC current and no power losses, so there is no energy loss. The voltage on the choke precedes current, or you can say that the current lags voltage by 90 degrees, the opposite of a capacitor. The reactance of a choke is linearly proportional to its inductance L: $X_L = \omega L = 2\pi fL$. Thus, the higher the frequency of an AC signal, the higher the reactance of a choke and the higher the choke's attenuation of the signal.

An inductor is named after a physical phenomenon called electromagnetic induction, while a "choke" is its popular name due to its effect on alternating currents, which it suppresses or "chokes." A choke resists any change in current that flows through it by inducing an opposite voltage which then tries to push the current in the opposite direction to neutralize the current that caused the induction in the first place. That mechanism is called *self-induction.*

A choke has a low impedance for DC currents and high reactance to AC currents. This desirable combination of properties makes chokes useful in power supplies, as ripple filters, and in audio circuits as inductive loads.

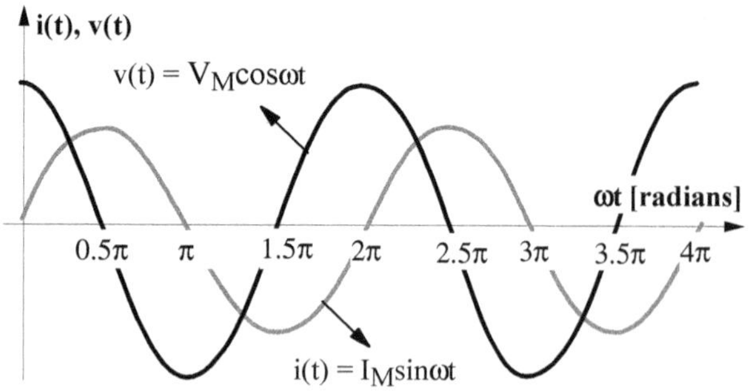

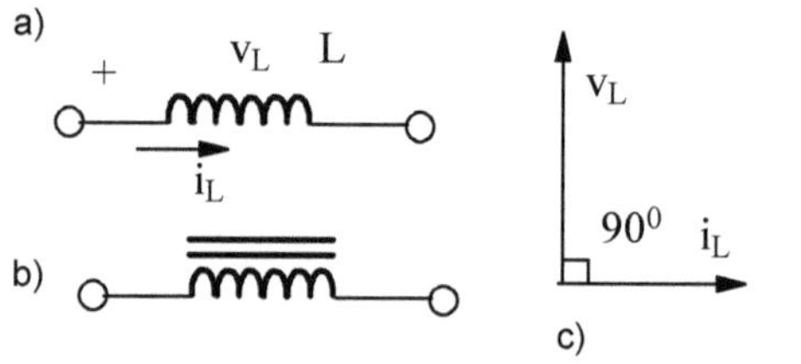

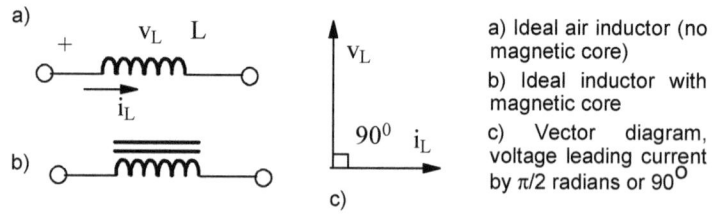

a) Ideal air inductor (no magnetic core)

b) Ideal inductor with magnetic core

c) Vector diagram, voltage leading current by $\pi/2$ radians or 90^{0}

LEFT: The phase relationship between the AC voltage across and current through an ideal inductor.

Real inductors, their losses (Q-factor) and vector diagram

Since inductors are made of many turns of wire on a bobbin (plastic frame), such a considerable length of relatively thin wire (typically 0.2-0.6 mm diameter) has a measurable resistance. We model real chokes as a series RL circuit by adding an ideal resistor in series with the choke's inductance.

The vector diagram for the real choke also features angles δ and ϕ. The same current flows through the resistance and the inductance, but there is a phase shift of 90 degrees or $\pi/2$ radians between the voltage drop on the resistance (which is in phase with the current) and the voltage drop on the inductance (which advances 90^{0} ahead of the current).

From the right-angle voltage triangle, we know that $\tan\delta = v_R/v_L$. Since $i = v_R/R = v_L/X_L$ and since $X_L = \omega * L$, we get $v_R/R = v_L/\omega * L$, $tg\delta = v_R/v_L = R/\omega L$. The quality factor or Q-factor for short, is the inverse of $\tan\delta$: $Q = 1/\tan\delta = \omega L/R$. The lower the tangent delta or the higher the Q-factor, the more the choke approaches the ideal inductor.

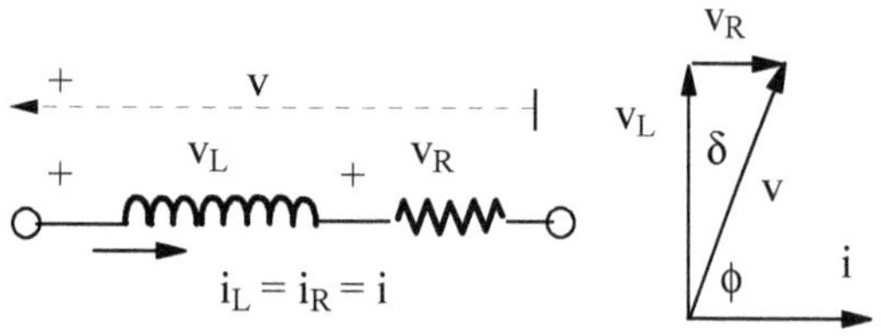

LEFT: Real inductor's voltage and current vector diagram shows voltage leading current by $\phi = \pi/2 - \delta$ radians

RIGHT: The impedance vector diagram for a real inductor

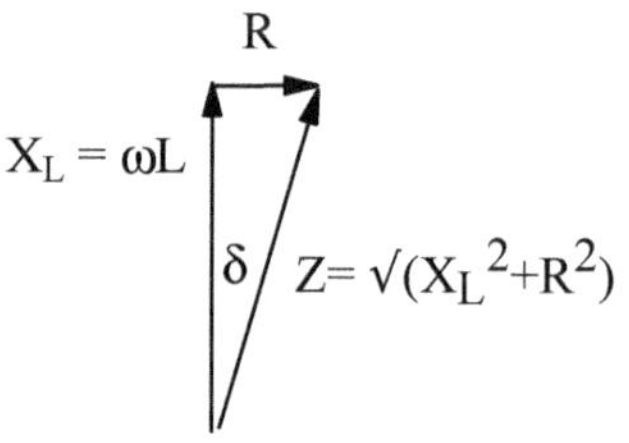

Inductance bridges

Unknown inductance can be determined by a comparison-type bridge, where the unknown inductor (choke or coil) is "balanced" by a variable inductor in the adjacent branch of the bridge. While simple to understand, this method requires precisely calibrated standard inductors, which are large, expensive, and sensitive to outside electromagnetic fields.

Luckily, the reactance of a coil can also be balanced by a capacitive reactance in the diagonally opposite branch of the bridge. Capacitors are small and cheap and not susceptible to external fields and interference.

The two most commonly used bridge types are the Maxwell (also called Maxwell-Wien) bridge and Hay bridge. As with other bridges, these do not measure the actual impedance but the components of that impedance, in this case, inductance L and quality (or storage) factor Q. The resistive component of the impedance is not measured and indicated directly but through Q and D values.

Variable capacitors aren't needed at all; the inductive reactance can be balanced out by a variable resistor R_2, while the impedance's losses or resistive component can be determined by adjusting R_1. As can be seen from the schematics, the only difference between the two bridges is that a parallel R_1C_1 connection is used in the Maxwell bridge, while the Hay bridges feature a series R_1C_1 arrangement.

The "CRL" and "DQ" controls refer to commercial impedance bridges' markings since they can also measure R, C, and D-factor. The Maxwell bridge's C, R, and L values are frequency-independent.

Still, the readings of D and Q will depend on the oscillator test frequency, so the instrument must be precisely calibrated by a trimmer capacitor in its phase-shift network for the fine-tuning of the oscillator frequency. The L-reading for the Hay bridge also depends on the test frequency.

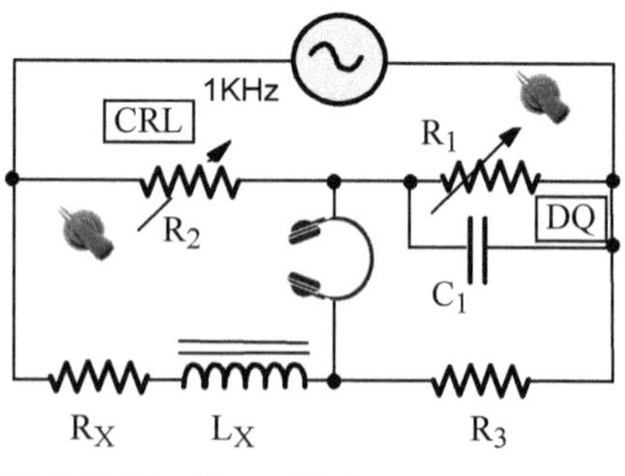

ABOVE: The Maxwell inductance bridge for low Q coils (<10)

BELOW: Hay inductance bridge for high Q coils (10-1,000)

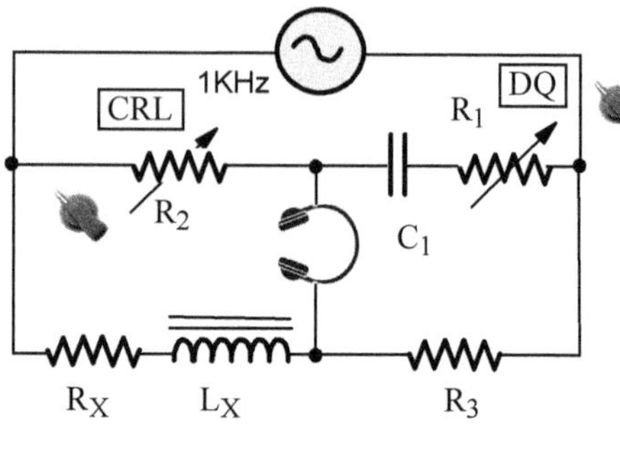

Vintage impedance bridges

Heathkit IB-28 was designed to compete with much more expensive laboratory-grade impedance bridges of its day. It's a 4-in-1 bridge with an electronic detector (a 2-stage tube amplifier and a moving coil meter), a Wheatstone DC bridge that measures resistance, a capacitance comparison bridge for capacitance, while Maxwell and Hay's bridges give the user a choice when measuring low Q (below Q=10) and high Q (10-1,000) coils respectively.

A 100-0-100 μA meter is the null indicator, but there is a connection for an external indicator such as headphones or an oscilloscope. The bridge has its internal 1kHz sinewave oscillator, with a provision for connecting an external generator for measurements at other audio frequencies. The oscillator's output is fed to a level control and then to a buffer amplifier (cathode follower), which prevents the changes in load impedance from affecting the oscillator by causing shifts in its frequency and thus affecting the accuracy of the instrument. An impedance matching transformer (see the photo below) is employed to enable optimum power transfer between the buffer amplifier and the bridge itself.

The measurement ranges are 0.1Ω to 10MΩ, 0.1nF to 100 μF, 0.1 mH to 100H, dissipation factor (D) from 0.002 to 1 and storage factor (Q) from 0.1 to 1,000. A bridge of this kind isn't the best choice for testing capacitors because there is no possibility of connecting an external DC polarizing voltage for electrolytic caps, and the capacitance test range is quite limited. Unfortunately, unless you modify the instrument yourself, there is no provision for DC current biasing of chokes and single-ended audio transformers, whose inductance varies with the amount of DC current.

LEFT: The above-chassis view of IB-28 . Below the chassis are the smaller RCL components. The bridge's restoration required the replacement of *all* capacitors and the selenium rectifier bridge.

1) D-Q dial
2) CRL dial (dual concentric scales)
3) Function switch
4) Range switch
5) Generator level control and generator switch
6) External generator connection (binding posts)
7) Detector switch
8) External detector connection (binding posts)
9) R_X, C_X, L_X terminals

Leader LCR-740 bridge controls:

1) DQ dial + "x1-x10" multiplier switch
2) C-R-L dial
3) Normal/"+1.00" switch
4) RCL (parameter) selector and range multiplier switches
5) Parameter selector switch (AC bridge for RCL and DC bridge for R)
6) On-Off switch and Sensitivity control
7) External oscillator input
8) Headphones or oscilloscope (external detector) output
9) R_X, C_X, L_X and "External +DC" terminals (binding post type)

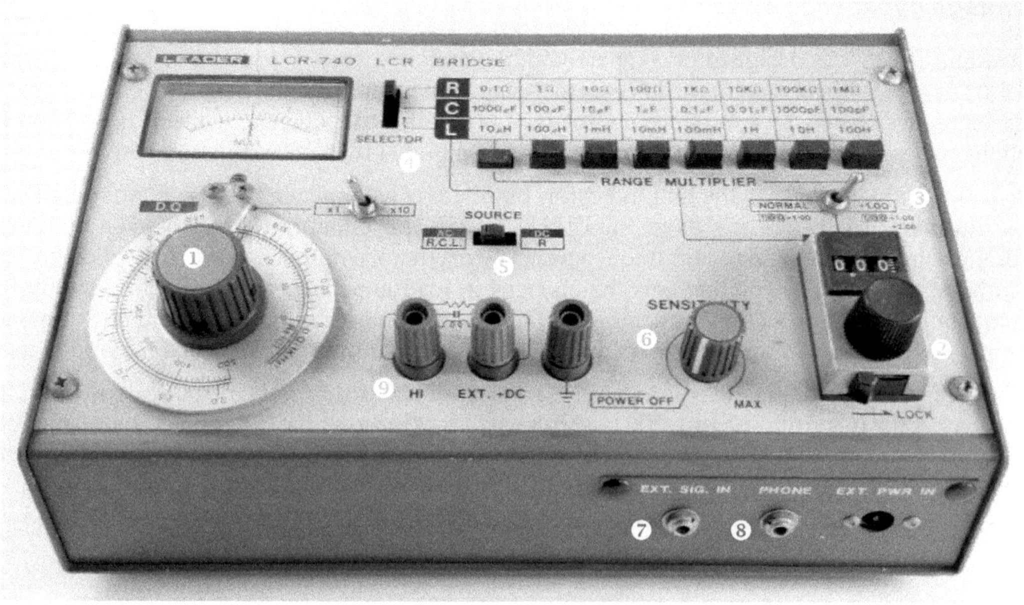

Made in Japan, Leader LCR-740 uses only a Maxwell bridge, while Heathkit IB-28 features both Maxwell and Hay bridges. Thus, while LCR-740 can only measure Q up to the value of 30, the Hay bridge in Heathkit IB-28 enables it to measure Q up to 1,000! Naturally, being a more modern solid-state instrument, LCR-740 is a fraction of the Heathkit IB-28 size, but the basic blocks are functionally equivalent, such as a 1 kHz oscillator (RC phase shift type with a single pentode in IB-28 versus 2-stage bipolar transistor in LCR-740) and the analog meter driven by a two-stage pentode amplifier in IB-28 versus three bipolar transistor stages in LCR-740.

Upon its release in 1968, when General Radio's 1650-B replaced its predecessor, model 1650-A, this professional bridge (photo below right) cost US$450. Fully solid-state, its specifications were much better than Heathkit IB-28 and similar bridges. For instance, it can measure the capacitance from 1 pF to 1,100 μF, inductance from 1mH to 1,100H, and resistance from 1mΩ to 1.1 MΩ. Powered by four D-cell batteries, this portable bridge is very compact and can be tilted at any angle.

Except for the binding posts for the unknown component (9), all connections are on the side (external oscillator, external DC bias up to 600V, and external null-detector, among others). DC current may be supplied to inductors by various methods described in an informative user manual to measure incremental inductance. Entry-level professional bridges of this class had an accuracy of +/- 1%, medium-priced bridges +/- 0.1%, with the highest priced ones only managing to halve that to +/- 0.05% accuracy, the limit of this method and the available technology.

1) DQ dial
2) C-G-R-L dial
3) Parameter selector switch
4) Multiplier switch
5) Function switch (Bat. check-AC external-AC internal-power off-DC internal-DC external)
6) Oscillator level
7) Detector sensitivity
8) Ortho-null adjustment
9) R_X, C_X, L_X terminals (binding post type)

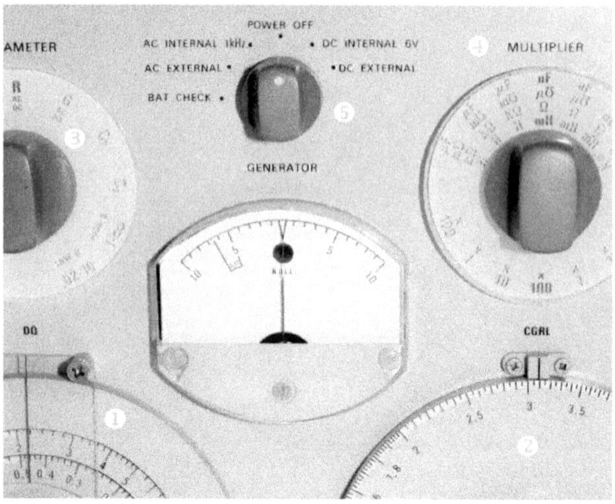

BELOW: 1650-B Impedance Bridge by General Radio

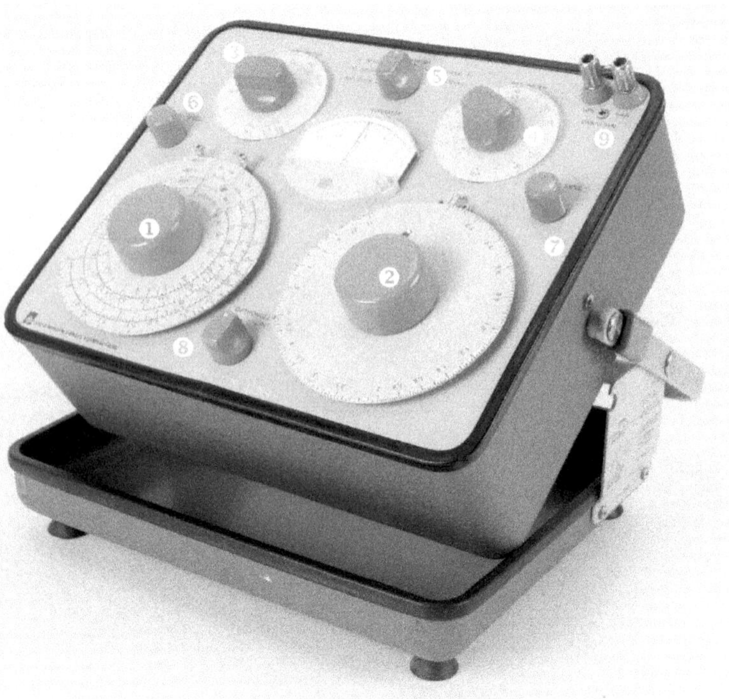

DIGITAL LCR METERS

A digital LCR (inductance-capacitance-resistance) meter is an essential instrument for audio equipment builders and fixers. Older models from the mid-late 1990s measure parameters at two frequencies, 120Hz, and 1kHz. The 120Hz frequency was chosen since it's double the 60Hz mains frequency in the USA, and that is the frequency of the fully rectified AC ripple in a power supply. More modern versions, such as B&K Precision LCR meter model 879, can also measure parameters at 10kHz, which is a handy feature when measuring transformer leakage inductance. Recently released handheld LCR meters even feature 100kHz test frequency as well.

Older models have slots into which component leads can be inserted, a feature that seems to have fallen out of favor lately. Alternatively, you can use test leads terminated with crocodile clips, included with both instruments.

All handheld LCR meters have the main parameter selector switch (1), marked "L/C/R" and "Freq." switch (2), often marked 1kHz/10/kHz on models with only two test frequencies.

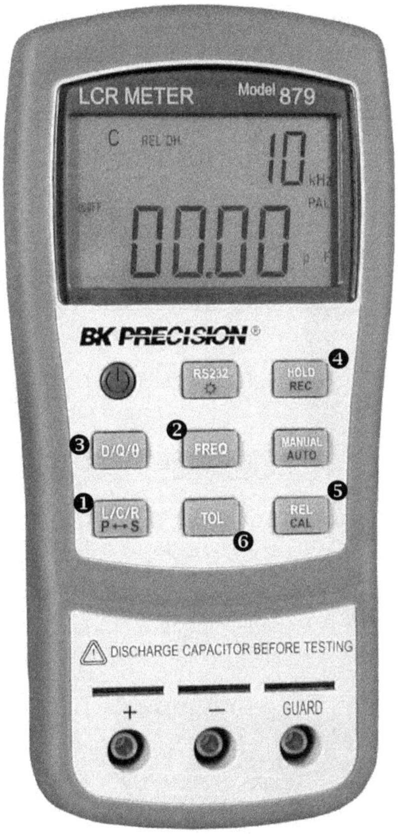

This toggles the test frequencies through a few choices (120Hz, 1kHz, 10Khz, and 100kHz). Some models with "auto-detect" mode automatically choose the most appropriate frequency for the component connected.

For instance, when a high inductance inductor (choke) is detected, it's automatically tested at 120Hz, or when a low-value film or ceramic capacitor is being tested, a 1kHz or 10kHz frequency is used, in contrast to high-value electrolytic caps, which are most likely to be used in power supplies. Just like filtering chokes, those are also tested at 120Hz.

Likewise, most LCR meters have a default mode of measurement, usually the series mode for inductance (where resistive losses are significant) and parallel mode for capacitance measurements. This can be overridden, and a different mode can be manually selected by the user for each test. On model 879, the "L/C/R" button is used for that purpose as well (lower functions in black lettering are selected by pressing the button for two seconds or longer).

The secondary parameter to be displayed is also selectable (3), the toggle button marked "D/Q" or "D/Q/q".The "Rel" button sets the display to zero and stores the displayed reading as a referent value. Afterward, all subsequent readings will be shown in terms of their relative value above or below the reference figure.

The static recording mode is entered by pressing the "HOLD/REC" button for 2 seconds. As the recording is performed, maximum, minimum, and average values are stored, and a beep tone sounds as acknowledgment.

The tolerance mode (6) is handy for quick component sorting. 1%, 5%, 10% (and 20% on some models) tolerance limits are selectable. A "beep" or tone indicates that the component is within tolerance, and three "beeps" a component out of tolerance.

Auto-balancing bridge method LCR meters

The auto-balancing bridge is the most accurate method, with an accuracy of 0.05% easily achievable, and it offers the widest measurement range.

A high gain amplifier automatically adjusts the gain level so that the current I_2 through the resistor R is equal to the current I_1 flowing through the DUT (Device Under Test). Since it is connected to the inverting input to the operational amplifier (I-V converter), the low potential side (L) of the DUT is always at the virtual ground level. The impedance of the DUT is calculated using the voltage V_1 measured at the high terminal (H) and the voltage across R (V_2).

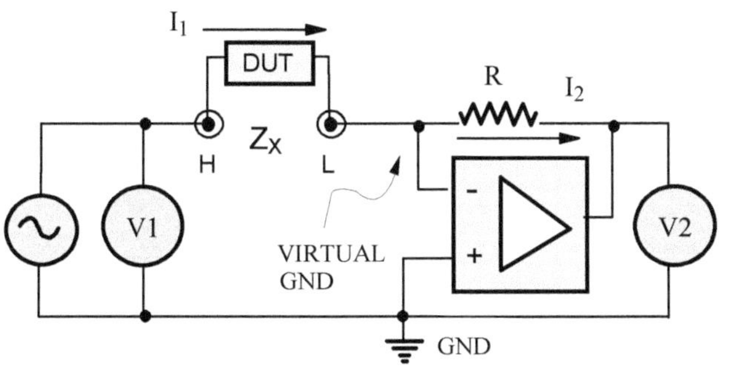

ABOVE: How auto-balancing bridge LCR meters work

$V_2=I_2*R$ or $I_2=V_2/R$

$Z=V_1/I_1 = V_1/I_2= R*V_1/V_2$

In practice, this simple configuration with an op-amp is only used at low frequencies, typically below 100 kHz, while at higher frequencies, a more complex arrangement is needed (due to the limits of a simple op-amp). This usually requires a null detector, phase detector, integrator (loop filter), and vector modulator. Resistor R, also called the range resistor, is switchable, so various values can be selected, thus maintaining the measurement accuracy across a wide range of frequencies.

From Q-meters to resonant-type LCR meters

Q-meters were among the most important test instruments of the 20th century, especially in high frequency (radio and TV) component testing. The conventional impedance bridges we have studied recently were limited to low or audio frequencies. Although RF (radio frequency) bridges were available, they were costly and out-of-reach of most DIY enthusiasts, radio amateurs and equipment builders, so many used Q-meters instead.

The instrument operated on a resonant principle. A series resonant circuit is formed when an unknown coil (inductor, choke) is connected in series with capacitor C.

A sinewave signal of a stable and known amplitude V_S and variable frequency f is fed into the circuit through a precision low-ohm shunt resistor R_{SH}. That source voltage is measured by voltmeter V1.

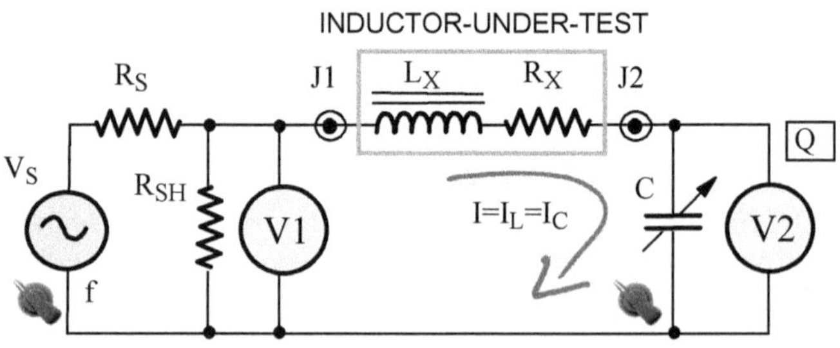

ABOVE: How Q-meters work

BELOW LEFT: M4070 handheld LCR meter

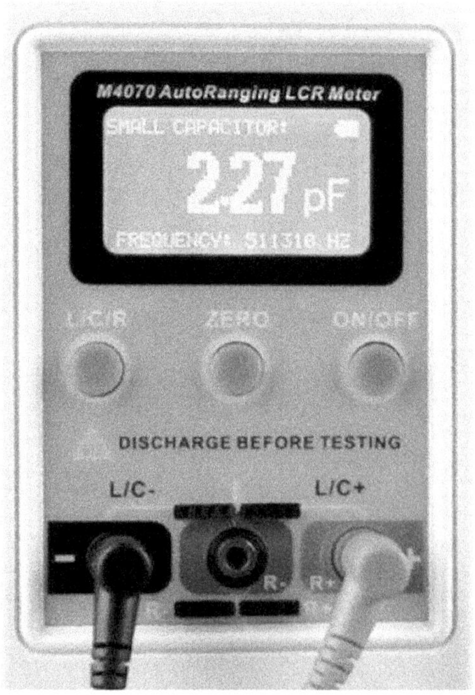

The frequency of the wideband oscillator is typically in the 50kHz to 50 MHz range but can extend to 300MHz or even higher. The capacitance C and/or the frequency f are tuned until a sharp peak (increase) in the voltage V2. This indicates the resonance condition where $X_L=X_C$, so effectively, the inductive and capacitive reactances cancel each other at that specific frequency f_R and only R_X figures in the circuit.

Since the impedances of electronic voltmeters V1 and V2 are so high that they can be neglected, the same current flows through the tested coil and the resonating capacitor ($I_L=I_C=I$), so we have $V1=I*R_X$ and $V2=I*X_C=I*X_L$.

Finally, $Q=X_L/R_X = X_C/R_X= V2/V1$. Since V1 is known, the scale of voltmeter V2 can directly be calibrated in the values of Q!

L_X can be calculated from $X_L=X_C$ so $L_X = 1/(2\pi fR)^2C$ [H].

Some modern LCR meters, both bench/laboratory types and those of handheld variety, utilize the same operational principle.

M4070 LCR meter tunes the test frequency until resonance is achieved. For a low value of capacitance, as displayed (2p27), such frequency is relatively high, 511.310 kHz. Since $\omega L=1/\omega C$, we can calculate the internal inductance the instrument had to use to achieve resonance at this frequency, $L= 1/\omega^2C$ or, in this case, 42.7mH.

The obvious disadvantage of such an instrument is that the user has no control over the test frequency used.

6 TESTING AUDIO AMPLIFIERS AND PREAMPLIFIERS

- VOLTMETER (MULTIMETER) AND LCR METER AMPLIFIER TESTS
- OSCILLOSCOPE TESTS
- MEASUREMENTS ON TUBE CIRCUITS (INSIDE TUBE AMPLIFIERS)

> "One may even suspect that there is more to reality than measurements will ever reveal."
>
> Michael Crichton

VOLTMETER (MULTIMETER) AND LCR METER AMPLIFIER TESTS

A voltmeter, multimeter, FET, or vacuum tube voltmeter can test quite a few parameters of interest to amplifier builders: maximum output power, the frequency range (upper and lower -3dB frequencies f_U and f_L), crosstalk, and the output impedance (from which amplifier's damping factor can be calculated).

Checking the maximum output power with an AC voltmeter

Assuming an 8Ω amplifier's output, terminate the amp with an 8Ω dummy load (R_L), a power (usually wire wound type) resistor of suitable power rating. With 1 kHz frequency and 1 V_{RMS} sine wave signal at the amp's input, measure the maximum output voltage at the amplifier's speaker terminals. The maximum power is then $P=V^2/R$, where R=8Ω. Say you measure 12.3 V_{RMS}. The power is then $P = 12.3^2/8 = 18.9W$.

connected in parallel with the multimeter or AC voltmeter, will be very distorted, so make sure the instrument used is a True-RMS type; otherwise, a significant error will be present.

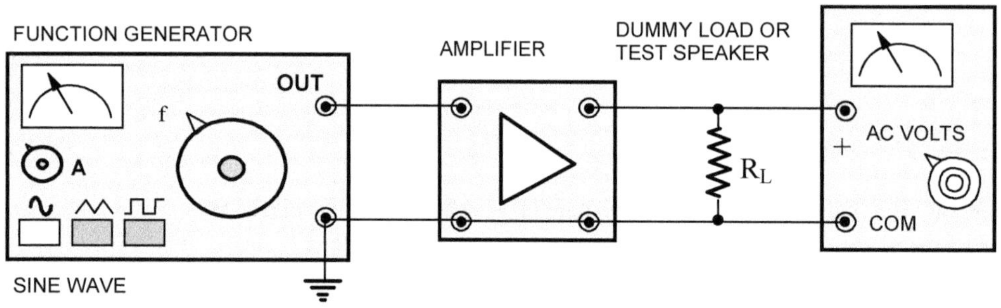

LEFT: The basic test setup for voltmeter measurements on amplifiers, preamplifiers, filters, crossovers, etc.

Determining the frequency range of (pre)amplifier using an AC voltmeter or an oscilloscope

On page 17, we have explained how to measure the upper and lower -3dB frequencies of a test instrument such as a digital multimeter. The same procedure is used to find these frequencies for an audio amplifier or preamplifier.

Start with the frequency of 1 kHz on the function generator (sine wave) and adjust the reading on the voltmeter or scope to be 2.83 V_{AC} on the voltmeter or 4 V_{PP} on the scope. Reduce the frequency until the RMS output voltage of the amp drops to 2.00 V_{AC} or until the peak-to-peak voltage on the oscilloscope drops to 2.8V. This is the lower -3 dB frequency or f_{L0}. The voltage dropped to approximately 71% of its value at 1 kHz.

Now increase frequency back to 1 kHz and recheck the indication on the voltmeter or the scope; it should return to the original 2.83 V_{AC} on the voltmeter or to 4.0 V_{PP} on the scope. Keep raising the frequency until the voltmeter's reading again drops to 2.0 V_{AC} (or 2.8 V_{PP} on the scope). That is the f_{U0} frequency.

Make sure the output amplitude of the function generator or low distortion oscillator stays constant. If it varies with frequency, you must adjust it to the same level as at 1 kHz during this test. Also, the AC voltmeter used should have its own f_U higher than the f_U of the tested amplifier.

Measuring the power bandwidth of an amplifier

Power amplifiers' half-power or -3dB frequencies depend significantly on the output power level. As we have just seen, it is customary to perform these measurements at two power levels, usually at 1 Watt (2.83V_{RMS} into 8Ω load) and at the amplifier's rated power output.

The frequency range at the rated power output P_R will always be narrower than the -3dB range at 1 Watt P_0. In tube amplifiers, this is primarily due to the imperfections of the output transformer.

In tube amps, the output transformers' performance is affected at low frequencies due to the limited primary impedance and magnetic core saturation and at high frequencies due to parasitic capacitive effects.

In the example illustrated here, the single-ended triode amplifier rated at P_R=15W was tested and the half-power frequencies (-3dB frequencies) were found to be f_{LR}= 10 Hz and f_{UR} = 49 kHz.

However, when the same amplifier was tested at a lower power output level of 1Watt, the -3dB frequencies were f_{L0}=5Hz and f_{U0}=57 kHz, meaning the frequency range was much wider. That is why quoting a frequency range without specifying the power level and the load impedance is meaningless.

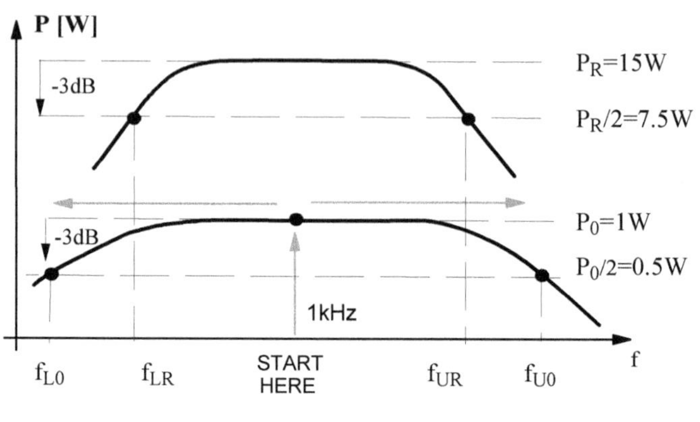

Testing high power amplifiers

While very few tube amplifiers exceed the 100 Watt per channel power rating, many solid-state amps are rated at 200, 400, or more Watts per channel. Sourcing 400 or 500 Watt power resistors for their dummy loads is hard, but again you can use a combination of identical resistors of a lower power rating, say four 100 Watt resistors as pictured.

The output power is four times the measured output power or $P=4V^2/R$, where R is the resistance of one of the four identical resistors, 8Ω in this example.

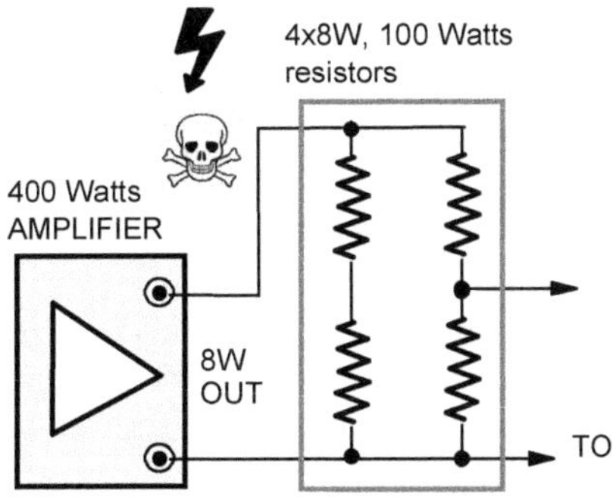

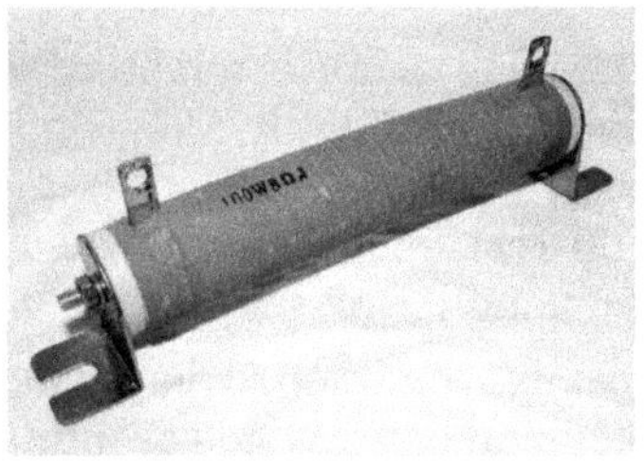

ABOVE: Using four 100 Watts 8Ω resistors to construct an 8Ω 400 Watts dummy load. WARNING: High AC voltages may be present!

Amplifier input sensitivity test

To produce its nominal (declared by the manufacturer) output power, an amplifier requires a signal of a certain amplitude at its input (V_{IN}). To determine the value of such voltage or amplifier's sensitivity, calculate the RMS voltage that will produce the rated power across the nominal load impedance. Say we are testing a 100 Watt (per channel) 8Ω amplifier. Since $P=V_{OUT}^2/R$, $V_{OUT}=\sqrt{(PR)}$, or, in this case the voltage on the load must be $V_{OUT}=\sqrt{(100*8)} = \sqrt{800} = 28.28V_{RMS}$!

Set the function generator to a sine wave output of a certain frequency (1), typ. 1 kHz. Increase (2) the amplitude of the function generator's voltage output (feeding the input of the amplifier being tested) until the multimeter or AC voltmeter across the load indicates 28.3V (3).

Read the function generator's output voltage on its meter, or, for signal sources without the metered output (4), measure the RMS voltage at the amplifier's input with the same or another multimeter or AC voltmeter, or even with an oscilloscope. That is V_{IN}. The amplifiers with volume control must have their volume control at the maximum.

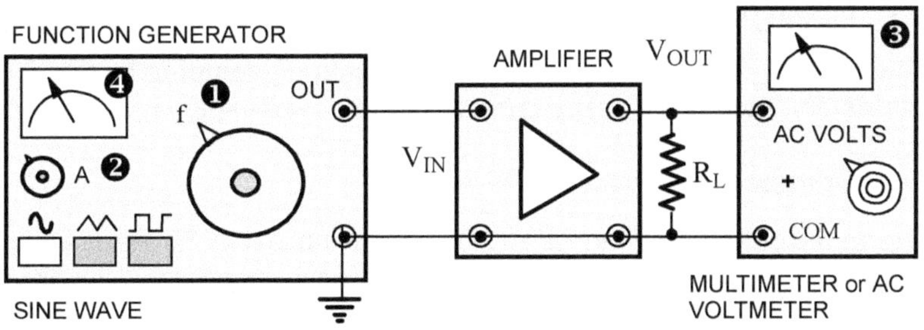

Crosstalk or channel separation

Crosstalk is the unwanted signal introduced from one channel into the other. Its inverse is channel separation. The lower the crosstalk figure, the better the channel separation. Measure crosstalk directly in dB by connecting a sine wave signal to one channel only and setting the attenuation on the voltmeter or the volume control on the amp to get 0 dB reading, and then switch the voltmeter (without touching any controls) to the other channel (without the signal at its input). The dB scale will indicate the crosstalk signal level. The outputs should be terminated with a rated load, and the input without the signal (excitation) should be short-circuited to the ground, so no spurious signals are picked up at its input.

If your voltmeter doesn't have a dB scale, measure the signal at the output of the right channel and the crosstalk on the left channel's output in Volts, divide the crosstalk with the signal, take a logarithm and multiply by 20. It does not matter if you measure the RMS, average, or peak values since their ratio matters here.

Crosstalk can be different between channels. Assuming a two-channel or stereo amplifier, LR (Left-to-Right) crosstalk (signal fed into the left channel, crosstalk measured at the output of the right channel) may be -40 dB, while the RL (Right-to-Left) crosstalk (signal into the right channel, crosstalk measured in the left channel) may be -37 dB.

Crosstalk is usually frequency-dependent. Since you can measure it at any frequency you like, perform the measurements at a few typical test frequencies, such as 100Hz, 1 kHz, 5 kHz, and 20 kHz. Test results of one SET power amplifier using EL34 tubes are tabulated. The crosstalk figures were very poor, ranging from -39 dB at 100 Hz to only -23 dB at 20 kHz. The crosstalk increased as we raised the test frequency; higher frequencies tend to "jump around" more!

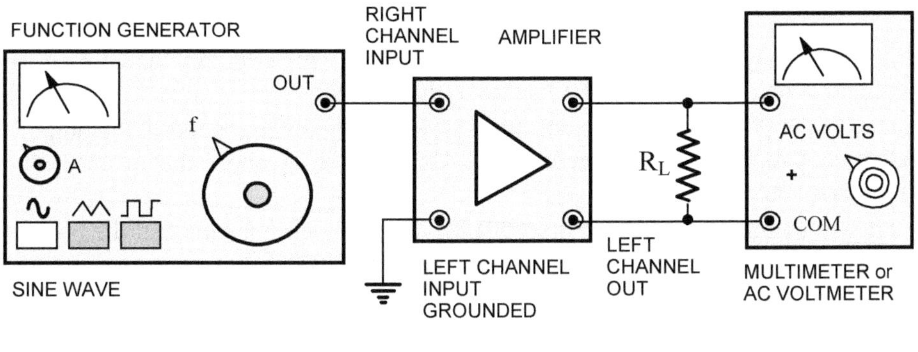

Frequency	Crosstalk [dB]
0.1kHz	-39
1kHz	-27
10kHz	-25
20kHz	-23

LEFT: Measuring channel separation, or its opposite, crosstalk

ABOVE: Crosstalk test results of a low fidelity Made-in-China tube amplifier

Output impedance and damping factor of a tube amplifier

Damping factor is a ratio of load impedance and amplifier's internal or output impedance: DF= R_L/Z_{OUT} . It helps us predict how well a particular amp would control the speaker cone. It is crucial at low frequencies (bass frequencies below 100 Hz) for the control of the woofer's cone and tight bass. The higher the damping factor, the "tighter" the bass.

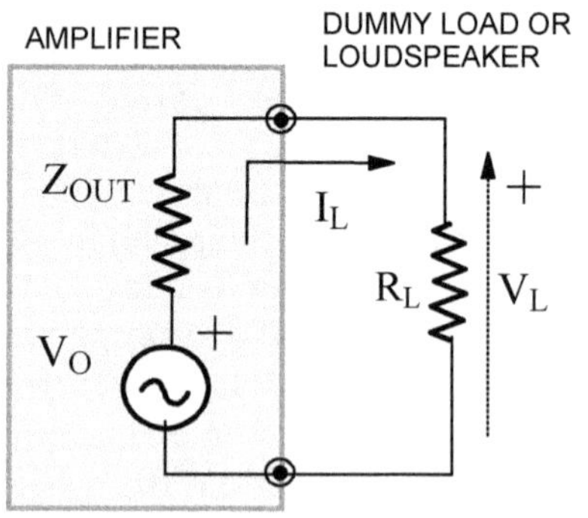

The Thevenin's equivalent circuit of the amplifier's output stage and load

Ideally, the internal impedance of the amplifier Z_{OUT} would be zero, and its damping factor would be infinite. In practice, the load current I_L creates a voltage drop on the internal or output impedance of the amplifier Z_{OUT}, so some output power is dissipated on the Z_{OUT} and wasted into heat.

Connect a 50-100 Hz signal from a function generator to the amplifier's input and adjust the volume control and/or function generators' output level until you get V_L=2.83V on the 8Ω dummy load. Leave the voltmeter connected to the speaker terminals and briefly disconnect the dummy load.

Note the voltage jump on open speaker terminals (V_O) and quickly reconnect the load resistor. Best is to use banana plugs for quick & easy unplugging and plugging.

If you write the voltage equation for this series circuit you will get: $V_O=I_L*Z_{OUT} + I_L*R_L$ Since $V_L=I_L*R_L$, once you eliminate I_L and express the R_L/Z_{OUT} ratio you will get **DF = $R_L/Z_{OUT} = V_L/(V_O-V_L)$.**

We tested two single-ended tube amps with EL34 output tubes strapped as triodes. The V_O of our own amplifier with no feedback was 3.6V, so DF=2.83/(3.6-2.83)=3.675, meaning the output impedance of our amp was 3.7 times lower than the load of 8 ohms, or 2.18 Ω. In comparison, one Made-in-China 300B amplifier (already mentioned on the previous page) had an incredibly poor damping factor of 1.11, despite the use of negative feedback (NFB raises the DF), so its Z_{OUT} was 8/1.11= 7.2Ω!

WARNING: Don't use this method of determining damping factors on solid-state amplifiers - you may destroy the output transistors. Incredibly, many transistor amps, even expensive "hi-end" brands, don't have any output protection!

Output impedance versus frequency

Due to the output transformer's leakage inductance that increases with frequency, the output impedance of a tube amplifier rises with frequency, especially in the absence of negative feedback. Even with relatively mild feedback, the same amplifier exhibits a much lower rise in the output impedance than the rise when NFB is removed.

This could be one of the possible factors that have made us conclude (after extensive listening evaluations) that mild feedback is usually beneficial and preferable to no NFB at all.

This significant rise of Z_{OUT} at high frequencies may seem intolerable, but luckily, there are mitigating factors. The most significant rise is in the 10-20 kHz band, and very little power is needed in that frequency band to drive a typical tweeter in a dynamic loudspeaker system.

Very few older audiophiles can hear frequencies above 15 kHz so even a significant drop in the upper frequency power levels may go unnoticed and actually "mellow" the amplifier's sound!

Last but not least, the impedance of a typical dynamic loudspeaker also rises at these frequencies, so while the output impedance of a typical amplifier Z_{OUT} rises in the absolute sense, the rise in the load impedance tends to offset it to some degree. Remember, DF is measured with a fixed resistive load, not with real loudspeakers!

Repeat the DF test using higher test frequencies of say 5kHz, 10kHz, and 20kHz (the four test points marked on the Z_{OUT} curve on the right).

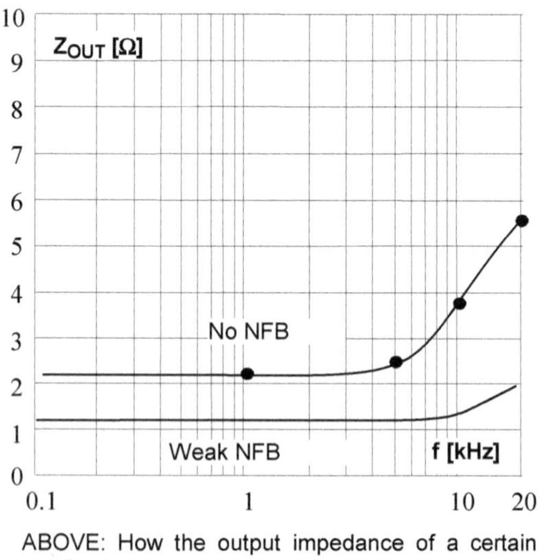

ABOVE: How the output impedance of a certain 300B single-ended triode amplifier (8Ω output) varied with frequency when no negative feedback was used and with a mild NFB.

Quick noise/hum tests

To measure the residual hum or noise of a preamp or power amp, terminate the power amp with a dummy load. Preamps have an internal resistor across the output; there is no need to connect a load to a preamplifier.

Turn the equipment on and short the input(s). The AC millivoltmeter will show the residual hum and noise at the output. Single-ended tube amps with directly-heated tubes such as 300B will measure 0.5-2mV; anything above that means back to the drawing board! Preamps should be much quieter.

After the voltage, measure the frequency. It will be 50Hz (60Hz) or 100 (120) Hz. 50/60Hz hum means that the power transformer or wiring radiates the mains frequency AC signal, which is getting picked up by the tubes or the wiring. Better shielding or transformer positioning is needed.

A 100/120Hz hum means that DC power supply filtering & smoothing is inadequate, that the AC ripple on DC lines is too high and is getting amplified by the circuit. A filtering choke could also be radiating magnetic flux and needs to be shielded and/or repositioned.

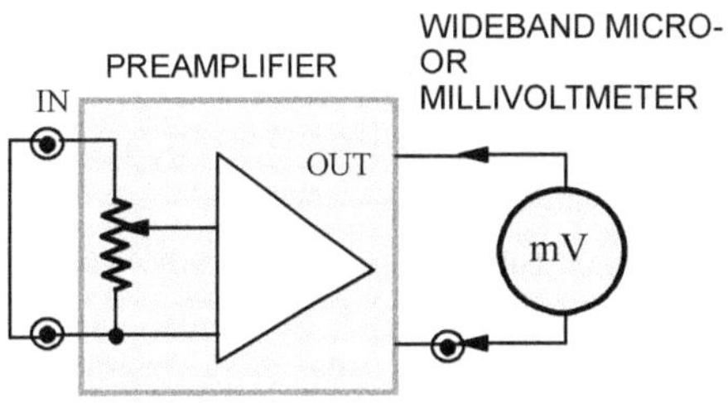

Measuring the hum or noise at an amplifier's (BELOW) or preamplifier's (ABOVE) output.

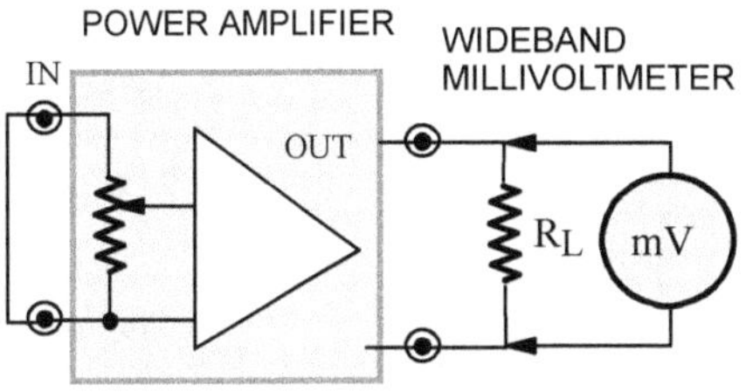

Bob's Black Box (BBB) versus an electronic wattmeter: How to determine the power factor (PF) of the load

This plug-in digital Watt and kWh meter (1) sells online for less than AU$15 and can also be programmed to display the cost of the power consumed. The voltage and current draw can also be read by pressing a few buttons. How does its reading compare with that of BBB (2)?

In this example, the power consumption of our EL12N push-pull amplifier at idle was 141.5 Watts. The true RMS meter measured $6.39V_{AC}$ across the 10Ω resistor in series with the load (inside BBB), meaning the AC current was 0.639A. The mains voltage was measured as 248V by a True-RMS multimeter (not shown), so the total or *apparent* power drawn was S=V*I = 248*0.639 = 158.47 VA.

Remember from the right-angle power triangle and Pythagoras Theorem that apparent power is $S^2=P^2+Q^2$, where Q is reactive and P is active power. Since the *active* power is $P=S*\cos\Phi$, we can calculate that the power factor of that amplifier was $\cos\Phi=P/S= 141.5/158.5 = 0.8927$. The digital wattmeter measured 0.7A & 239.4 V, so S=167.6 VA and PF=0.844.

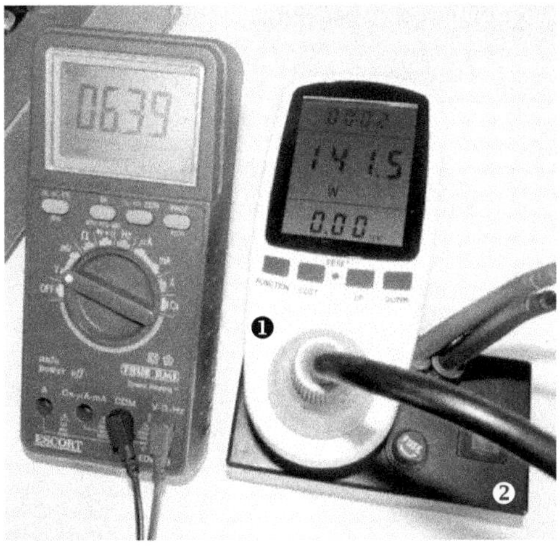

Input impedance of an amp or preamp

LCR meters are great for measuring the input impedance of amplifiers and various other audio devices. You can do the test either in a de-energized state or while the device-under-test is powered up, since all the LCR meter does is inject a small audio-frequency signal into the input and measures the response, so you cannot damage anything.

Measure input capacitance, reactance, and resistance, calculate capacitive reactance as $X_C=1/\omega C_{IN}$ and inductive reactance as $X_L=\omega L_{IN}$, and, finally, calculate the modulus of AC impedance at the test frequency as

$$Z = \sqrt{(X_L-X_C)^2+R_{IN}^2}$$

For one of our 300B SET amplifier we got the following results (1kHz LCR test frequency): C_{IN}=92.7pF, L_{IN}=0.952H, R_{IN}=100.1 kΩ.

Thus, $X_C=1/\omega C$ = 1,717 Ω, $X_L=\omega L$=5,982 Ω and

$$Z = \sqrt{(X_L-X_C)^2+R_{IN}^2} = \sqrt{4{,}265^2+100{,}100^2} = 100{,}191\ \Omega.$$

The nominal value of the input resistor was 100kΩ (measured as 100,100 Ω), so the contribution of the reactive components of the input impedance was only 91Ω or 91/100,191 *100% = 0.091%! In other words, the reactive aspects of Z_{IN} can be neglected, and $Z_{IN} \approx R_{IN}$!

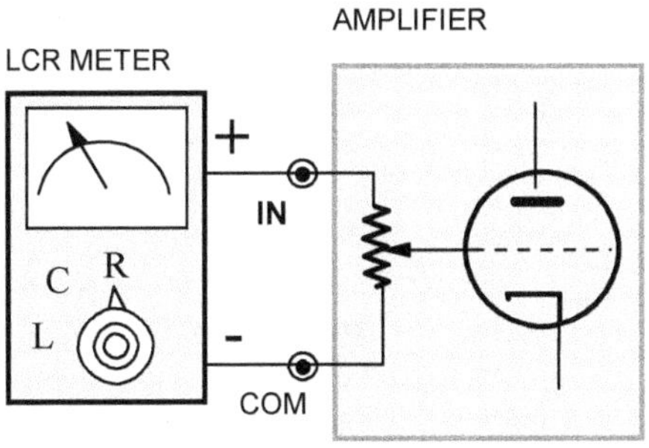

ABOVE: Measuring the input resistance of an amplifier or preamplifier with an LCR meter

Output impedance of a tube preamplifier

Power the preamplifier up and set the output voltage to a desired value with the open output. That is voltage V_O. Then, connect a 1 kΩ or 5 kΩ potentiometer as per the drawing and adjust its slider until the voltage V_{OUT} (as continuously measured by an RMS voltmeter) drops to half of the open circuit value V_O.

Turn the preamp off, disconnect the potentiometer and measure its lower section R_X (between the wiper and the fixed bottom end) with an ohmmeter. That is the value of the preamplifier's internal or output resistance because at that point, half of the voltage V_O is across Z_{OUT}, and half is across the load R_X, which are thus equal in magnitude.

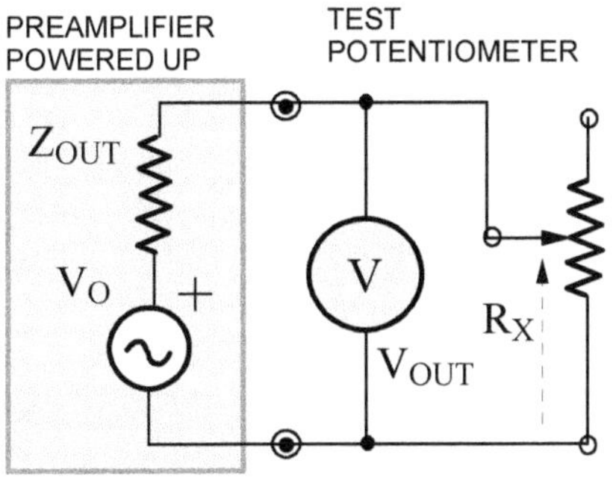

ABOVE: The potentiometric measurement of the output impedance of a preamplifier

Testing RIAA curve accuracy of phono stages

There are two ways to check the RIAA filtering accuracy of a phono stage. The direct method uses a sine signal of a known frequency and amplitude at the input of the phono stage, and the output voltage is measured by a precise AC voltmeter. The test is repeated for various frequencies (20-30 points are sufficient), the gain (or attenuation) is calculated and compared to the standard RIAA curve.

The error method uses an inverse-RIAA filter at the input of the phono stage, which, in theory, should cancel the phono stage's filtering, and a flat frequency characteristic should result at the output. The signal source (function generator set on sine wave output) must have a constant amplitude output. Adjust the amplitude of the function generator's output or the preamp's volume control (if any) until the AC voltmeter reads 0 dBm or 0.775 V_{AC}.

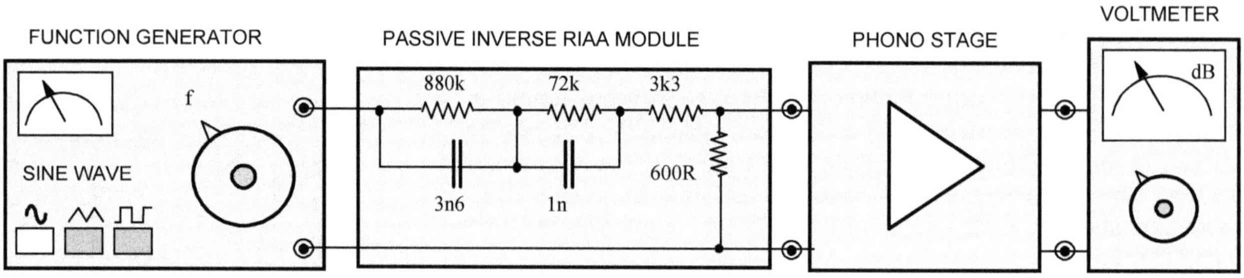

ABOVE: Test setup for measuring the accuracy of RIAA filtering curve of phono stages (turntable preamplifiers)

ERROR MEASUREMENT		DIRECT VOLTAGE MEASUREMENT	
FREQ f	ERROR dBm	VOLTAGE Vac	dBm
30	1.43	2.895	11.56
40	1.81	3.171	12.34
50	1.85	3.294	12.68
60	1.84	3.324	12.76
70	1.76	3.298	12.68
80	1.63	3.226	12.48
90	1.57	3.128	12.2
100	1.47	3.021	11.85
200	1.18	2.008	8.38
300	0.86	1.477	5.68
400	0.61	1.218	4.04
500	0.43	1.066	2.86
600	0.26	0.972	2.01
700	0.14	0.902	1.36
800	0.06	0.846	0.83
900	0.01	0.811	0.43
1000	0	0.774	0
2000	-0.14	0.575	-2.54
3000	-0.16	0.447	-4.73
4000	-0.17	0.366	-6.48
5000	-0.17	0.306	-8.03
6000	-0.18	0.261	-9.41
7000	-0.18	0.226	-10.66
8000	-0.19	0.198	-11.8
9000	-0.21	0.178	-12.72
10000	-0.22	0.161	-13.86
12000	-0.53	0.129	-15.5
14000	-0.62	0.108	-17.03
16000	-0.73	0.092	-18.5
18000	-0.85	0.077	-19.93
20000	-0.96	0.065	-21.39
22000	-1.11	0.054	-22.97

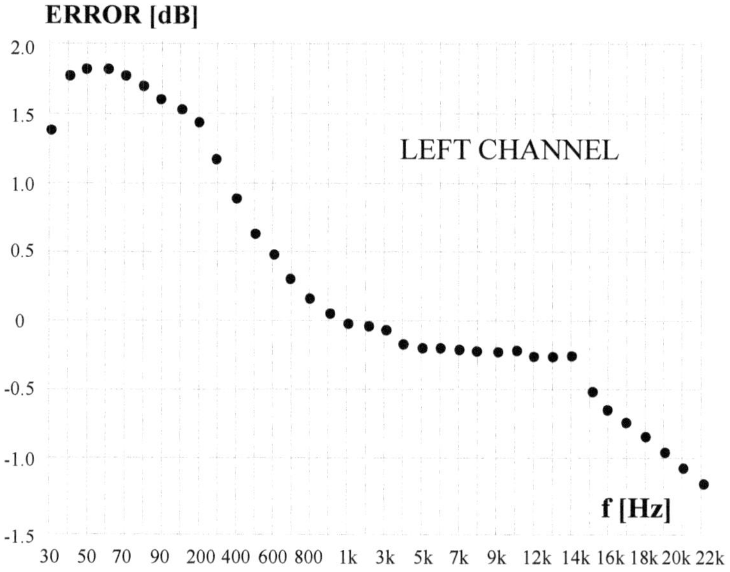

BELOW: Alternative version of the passive inverse RIAA filter

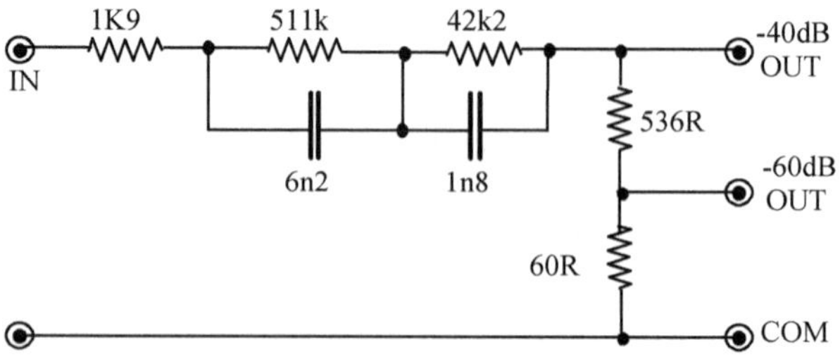

The table shows the measured values of our Accurus MM phono stage (one channel only). The second column is the error in dBm, the 3rd column is output voltage measured by a direct method (without the inverse RIAA module), and the 4th column is the corresponding measurement in dBm.

While the numbers in a table don't mean much, the graph of the error versus frequency tells the whole story at one glance. A picture is indeed worth a thousand numbers! Notice the positive error at low frequencies, peaking around 50-90 Hz, that was done deliberately to boost the low bass. The error is around -0.15dB in the midrange, then increasing to around -1% at 20kHz. Slightly attenuating the treble, as in this case, produces a smoother top end and removes traces of harshness.

OSCILLOSCOPE TESTS

Estimating the maximum "undistorted" voltage and amplification factor

A multimeter on "AC volts" range can only measure the effective value of input and output voltages of an amplifier. To see the waveform, how "clean" or distorted it is, you need an oscilloscope.

In this test, the input signal is brought to CH2 (Channel 2) on the scope, while the signal from the output of the amplifier under test is brought to CH 1. You can connect them the other way around, of course, it does not matter which signal goes into which channel on the CRO, since both oscilloscope channels are identical.

With 1 kHz frequency and 1V_{RMS} at the input of the amp, start with a small signal at the input and increase its amplitude until you observe visible rounding of the top or the bottom of the sine wave. Stop increasing the signal's amplitude and measure the maximum power by reading the PP (peak-to-peak) amplitude on the scope, then divide the reading by 2.8 to get the effective (RMS) volts.

Although we call that maximum "undistorted" power, once you notice slight rounding of the sine wave, the signal is already significantly distorted. An oscilloscope is not a precise instrument for measuring distortion. Or, to be more succinct, oscilloscopes are precise, but our eyes aren't. You can perform this test at any frequency. At low test frequencies of say 50 or 100Hz, you will notice that the maximum undistorted output will be lower than the maximum undistorted output at 1kHz, due to the limitations of the output transformers.

RIGHT: Two channel oscilloscopes make quick measurement of the amplification factor and output power easy.
The input signal to the amplifier under test is displayed on one channel and the output signal on the other.

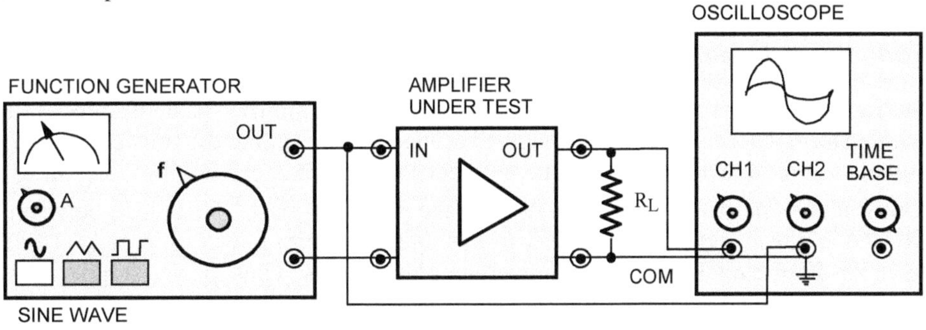

Square wave response of an amp or preamp

Since a square wave is comprised of an infinite number of sine waves, this test is the most revealing of all. Only a few frequencies are sufficient to indicate the quality of an amp or preamp. Typically, square waves of 100 Hz, 1 kHz and 10 kHz frequencies are used.

Again, just as with frequency range measurements, we usually start with a 1kHz square wave test signal and observe the fidelity of the waveform. At 1kHz, the output of a good quality amp should be an almost perfect square wave (waveforms drawn in gray). Yet, many expensive amplifiers by reputable manufacturers show serious weaknesses in these tests. Three *typical* responses are illustrated here.

The top is the underdamped response, the output voltage rises too slowly, so the amplifier will most likely sound sluggish and lacking in dynamics. This refers to a 1kHz test, of course. If a 10kHz response looked like this, we would proclaim this to be a very good amplifier indeed!

The middle response is generally better, faster rise time but a slight overshoot. There is no oscillation; the amplifier is stable. This amplifier would have a more detailed and crispier top-end than the previous one.

Finally, the bottom response is of a marginally stable amplifier, with pronounced "ringing". The signal takes too long to settle. Amplifiers with this type of response usually have a harsh sonic signature, just like many cheap and nasty sounding CD players, whose square wave response looks similar.

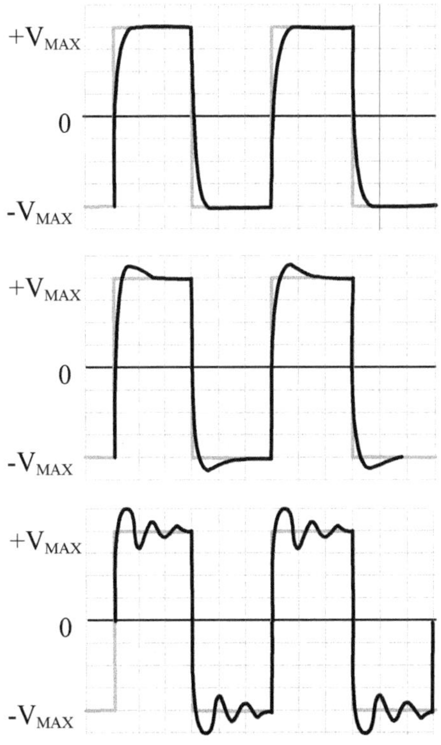

The snapshot of the oscilloscope's screen on the right shows the response of a certain tube hi-fi preamplifier to a 210 kHz square wave signal. Notice the quick rise time, a very small rounding off the top corner, an excellent result at such a high frequency. The upper -3dB frequency was 350 kHz!

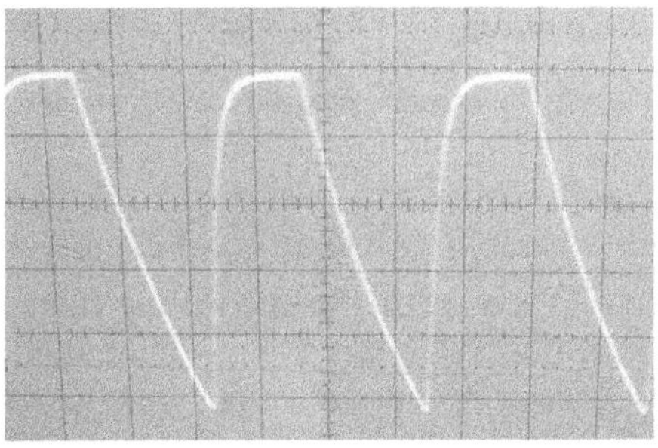

However, the falling edge is "trailing", the fall time is much longer. The designer obviously did not bother to optimize the circuit in that regard. Nevertheless, the preamp sounded very good indeed.

Let's look at oscillograms of the SET EL34 "Look" amplifier (below). The 100 Hz response is slanted, which is expected. The slope indicates how low in frequency the amp goes; the flatter the 100 Hz response (the less slanted the response), the lower the -3 dB frequency of the amp!

The 1 kHz response is poor, there's significant "ringing" due to the resonant frequency of the output transformer, usually around 60-100 kHz. Also, the rise time of the leading edges is very long; the amp is slew-rate limited, or simply "slow", it cannot cope with sudden changes in signal levels. Likewise, its 10 kHz response shows a very long rise time, an overshot plus ringing, and a very long fall time.

At 20 kHz, the output is triangular instead of square - the whole amp behaves like a giant capacitor, "integrating' the signal. This indicates that the upper -3dB frequency of this amp is very low. Indeed, we measured f_U=21 kHz!

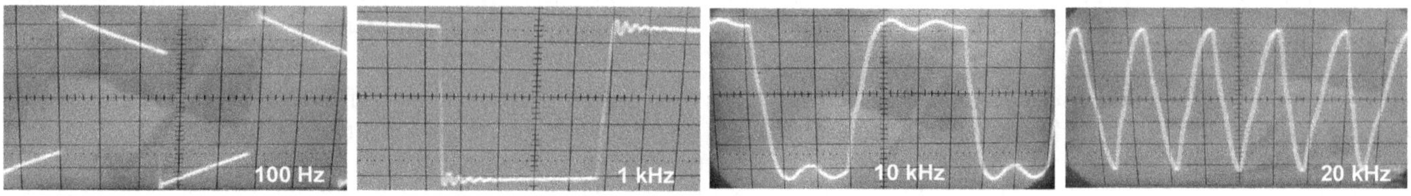

ABOVE: Square wave response of a poor-quality single-ended tube amplifier at few typical test frequencies

Tone-burst dynamic headroom (dynamic power) test

Static tests using sine and square waves are useful in amplifier basic design and troubleshooting, but they don't tell the whole story. These signals are used not because they resemble the music signals (in fact, they are as much unlike music signals as possible!) but because these signal sources are cheap and because these tests are easy to perform and interpret.

Sinusoidal signal has a small peak-to-average power ratio; the difference between the peak and average power is only 3 dB, while music signals have a peak-to-average short-term power ratio of 14-18 dB. This means that music signals often have 25-50 or even higher peak-to-average power ratios. Thus, during a quiet musical passage, the power demand on an amplifier may be 1W, but during loud crescendos, the peak power demand shoots up to 50-100 Watts!

I still remember my encounter in 1996 with a high-end dealer in Perth, Australia. I asked him to evaluate the prototype of our 10W single-ended triode amplifier, with a view of them selling it once in full production. The guy was no fan of low-powered triode amps; he hooked it up to the first pair of (inefficient) speakers he could find and played a passage from 1812 Overture by Tchaikovsky. Needles to say, the famous cannon fire sounded like somebody popping open a can of soda, the amp flopped, and the sound fizzled; it simply could not provide so much peak power in such a short time. His evaluation lasted a full minute!

We have noticed that two amplifiers of similar power output but using different power tubes and different circuit designs may both sound very good at low to moderate volumes, yet when pushed to their limits, one sounds a little harsher but still OK while the other distorts terribly and sounds awful. The explanation may be in their different overload abilities.

A tone-burst signal is controllable and repeatable, but, most importantly, it resembles music signals much better than simple waveforms. Vintage tone burst generators such as General Radio Company 1396-B are rarely available for sale online, but keep your eyes open.

The name is misleading because it is not a generator by itself; an external signal source is needed. The unit then triggers and "modulates" the external signal by adjusting the amplitude and duty cycle and feeds the tone-burst signal into the amplifier under test, as per the setup below.

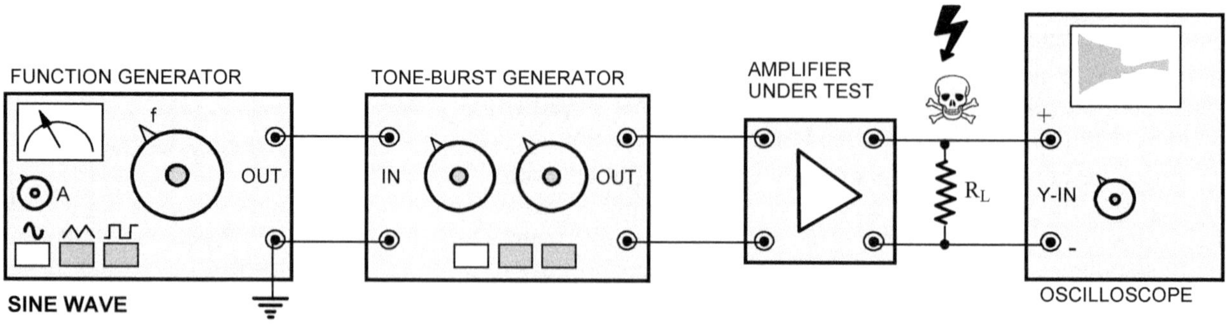

ABOVE: Tone-burst test setup requires a tone-burst "generator", or more precisely "modulator"

The (a) waveform (next page) shows the test signal, a sinewave tone burst. The larger amplitude is adjusted to take the tested amplifier into an overload of between 2 and 5dB. Responses of three different amplifiers are illustrated.

In b), the amplifier uses coupling capacitors driving the output grids, as does the amplifier in c), so there is a noticeable rise time, but both amps cope with the high amplitude signal. The amp in b) recovers faster from overload. The amplifier in d) is inferior because it cannot cope with the overload signal (sagging top), and the recovery is long and laborious.

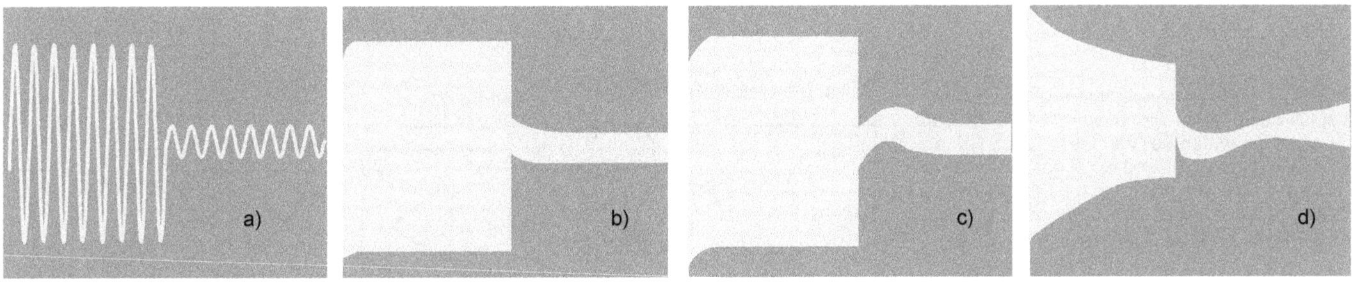

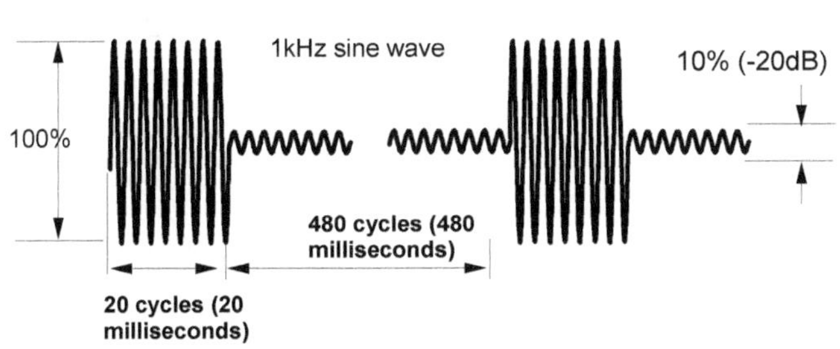

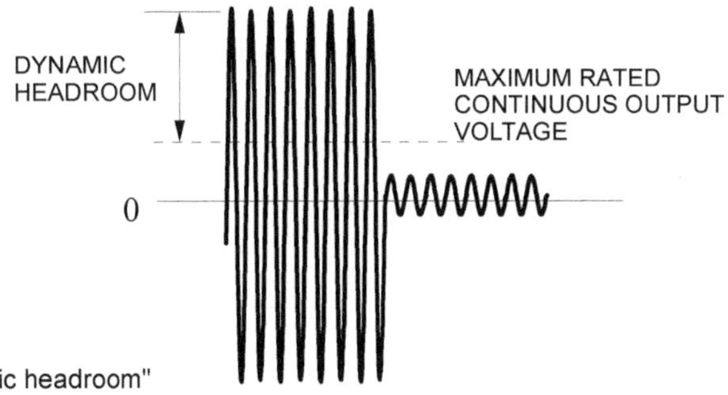

ABOVE LEFT: The burst test signal ABOVE RIGHT: The definition of "dynamic headroom"

In 1949 the Radio-Electronics-Television Manufacturers Association (RETMA), together with the Electronic Industries Association (EIA), published "Engineering Specifications for Amplifiers for Sound Equipment" (EIA/RETMA SE-104), the first attempt to standardize amplifier specifications. In 1958, the Institute of High Fidelity (IHF) expanded on the RETMA document and released IHF-A-200, a standard further amended in 1966 (IHF-A-201).

In the 1970s, ANSI (American National Standards Institute) developed a document titled "Measuring Audio Amplifier Power Output Rating for Institutional Audio Visual Equipment," but it never became a standard. Meanwhile, in West Germany, the Deutsches Institut für Normung (DIN) developed standards for hi-fi equipment in general, and for power amplifier ratings in particular, the most recent being 61305-3:1995. CEA2006 is the standard for amplifiers designed for mobile applications (cars). It is mentioned here for its dynamic power measurement, using a 1kHz burst tone method similar to the old IHF method.

The test tone is 20 cycles of 1 kHz sinewave signal (20 ms duration), followed by 480 cycles (480 ms) of smaller amplitude (10% of the previous burst or -20dB down). The distortion threshold for this test is 1% THD. The test signal amplitude is increased until this level of distortion is reached, the burst amplitude is measured, and the dynamic power maximum calculated (in Watts). The equivalent figure, expressed in dB, was called "headroom" in the old IHF standard.

ABOVE AND LEFT: General Radio 1396-B tone-burst generator, front and rear views

V_{OUT} - V_{IN} (transfer characteristic) of an amplifier, preamplifier or crossover

Using an oscilloscope in the X-Y mode, connect the output of the amplifier under test (voltage across a dummy load) to vertical or Y input and the voltage signal at the input to the amplifier to the X- or horizontal input. At a chosen midrange frequency (1 kHz, for instance), adjust the sensitivity of one or both oscilloscope inputs to get a 45^{0} line, in the example (next page), 6 vertical by 6 horizontal divisions.

To observe the dynamic range of an amplifier, increase the input voltage (the output amplitude of the test generator or the volume control of the amplifier). At some point, one or both ends of the line will start becoming rounded, which means the amplifier is distorting, rounding, or even clipping the peaks of the sine wave. Instability (oscillations) can also be detected by this method, easily identified by thickened or "smudged" part(s) of the transfer curve.

RIGHT: The V_{OUT}-V_{IN} (transfer characteristic) display using an oscilloscope in X-Y mode can be used to estimate various aspects of an amplifier (or any other device, for instance, a preamp or a crossover), such as its frequency range, phase shift at various frequencies and the onset of clipping at any frequency.

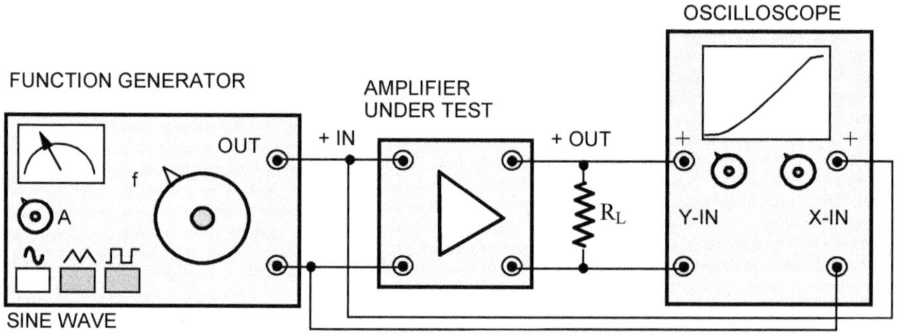

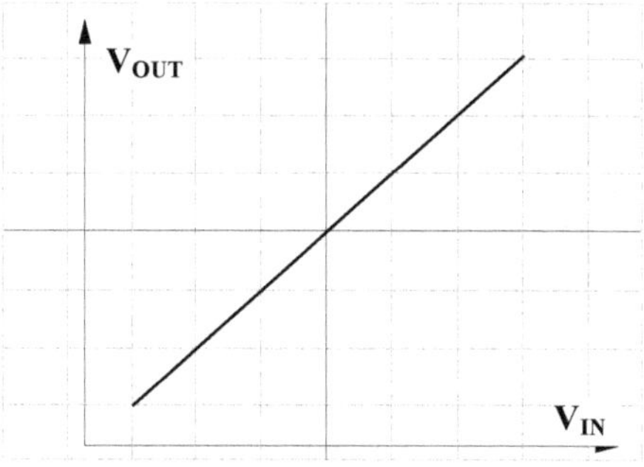

The ideal transfer curve is a straight line

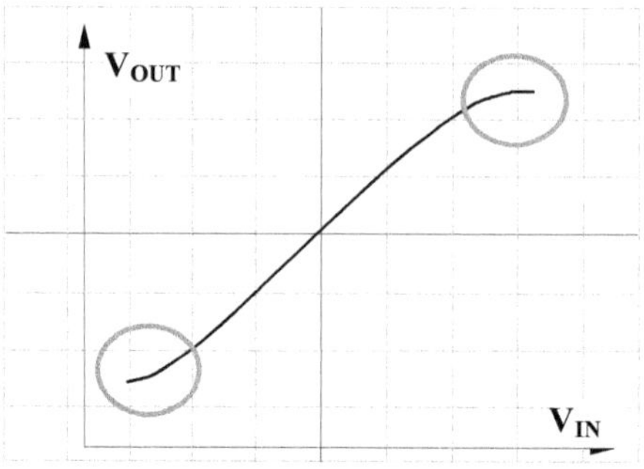

Nonlinearity (clipping) at either or both ends

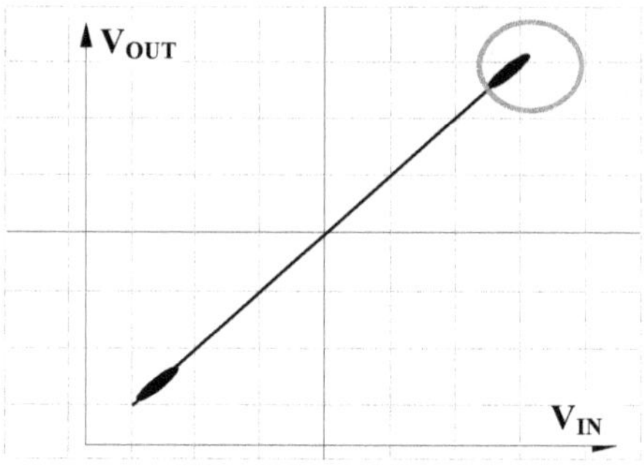

The thickening indicates HF oscillations

Gain change at frequency extremes

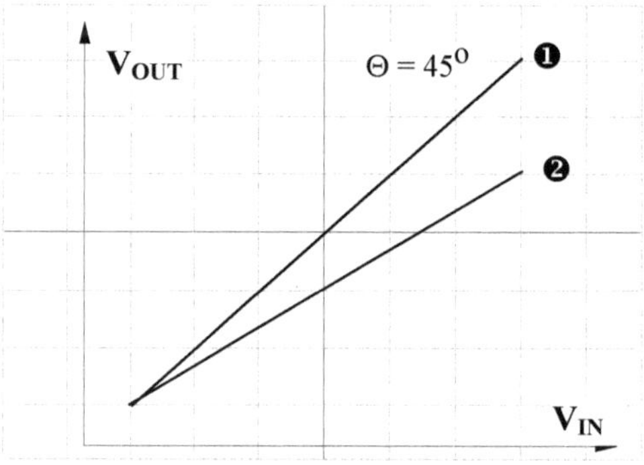

While still in the linear (non-distortion) range of the amplifier (say at 1 kHz signal frequency) with a straight 45o line display (1), we can use this method to estimate the change in amplifier's gain at various frequencies. As we increase the frequency of the test generator, the gain will start dropping, and so will the output voltage and thus the vertical deflection. The line slope will start reducing, the line becoming more horizontal. The second trace (2) dropped here from 7 down to 5 vertical divisions, a reduction of 5/7=0.714 or -2.923 dB. Read the test frequency on the test generator, and you have just found the amp's upper -3dB or half-power frequency f_U!

Measuring phase shift with an oscilloscope - method 1

This method involves bringing two signals (in this case, the input and output signal of an amplifier) to two vertical inputs and using the oscilloscope's time base for the horizontal sweep. So a 2-channel oscilloscope is needed.

Connect the input signal to the measured device (amplifier, filter, etc.) to vertical input 1 (Channel 1) of the CRO, connect the output to vertical input 2. Select Channel 1 vertical attenuation of the CRO and the time base to get one full period across the whole screen. Make sure the waveform is centered on the center line, so the positive and negative peaks are equal distance from the center line.

Center vertically and adjust Channel 2 vertical attenuation to get the same amplitude on the screen. Read the phase shift (the horizontal distance between points A and B). In our case the phase shift is 1.2 major divisions. To calculate the phase shift in degrees or radians, we need to know the oscilloscope's horizontal sensitivity - its "time base" setting.

In this case, we read that the time base selector switch is in the 10 μs/div position. That means each major horizontal division is 10 microseconds (10^{-6} seconds), so the whole period is 8 divisions or 80 μs!

The signal frequency is thus $f=1/T = 1/(80*10^{-6}) = 1/80*10^6 = 0.0125*10^6 = 12{,}500$ Hz. We didn't need to calculate the frequency to determine the phase shift; that was simply to show you how to do it.

Going back to the phase shift, we know that 8 horizontal divisions represent 360 degrees or 2π radians, so 1div = 360/8 = 45 degrees. Finally our phase shift angle is $\alpha = 360/8*1.2 = 45*1.2 = 54$ degrees. Since 360 deg = 2π = 6.28 rad, 1 deg = $2\pi/360$ rad, so $\alpha = 2\pi/360*54 = 0.94$ radians.

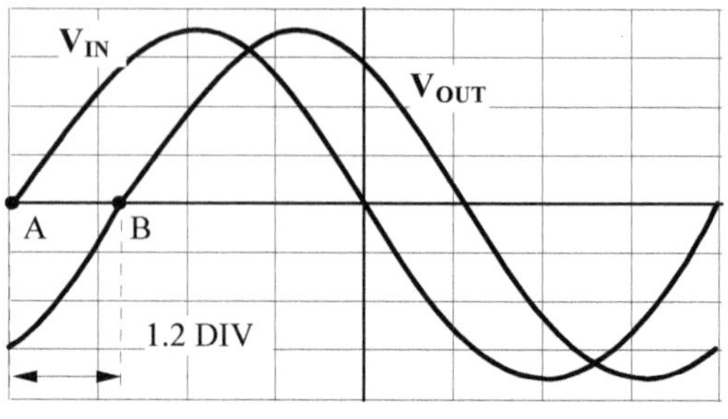

ABOVE: Direct phase shift measurement using the oscilloscope's time base

Measuring phase shift with an oscilloscope - method 2

The second method of measuring phase shift involves using the X-Y mode of the oscilloscope, where its internal timebase is disabled. One signal is brought to the input of the horizontal or X-amplifier inside the scope, the other signal to the vertical or Y-deflection amplifier.

Depending on the ratio of the frequencies of the two signals, a Lissajous curve of a certain shape is obtained on the screen, as illustrated on the right. The amp doesn't change the frequency of the test signal, so the figures are always for the 1:1 frequency ratio. The curves become more complicated for other ratios, but the frequency shift then loses its meaning.

The phase shift between the input and output signal is Φ=arcsin(A/B). Alternatively, the phase shift can be determined by measuring the ellipse's long- (L) and short-axis (S): Φ=L/S*90°. Here it is more difficult to determine L and S precisely, so the first option is recommended.

For S=0, we have a line, so Φ=0°. For S=L, the ellipse turns into a circle, and the phase shift is Φ=90°. In the example below A=2.6 and B=3, so Φ = arcsin(A/B) = arcsin0.8667 = 60.1°!

This shows you the imprecision of both methods due to reliance on visual estimation. The phase shift could be anywhere between 54° and 60°!

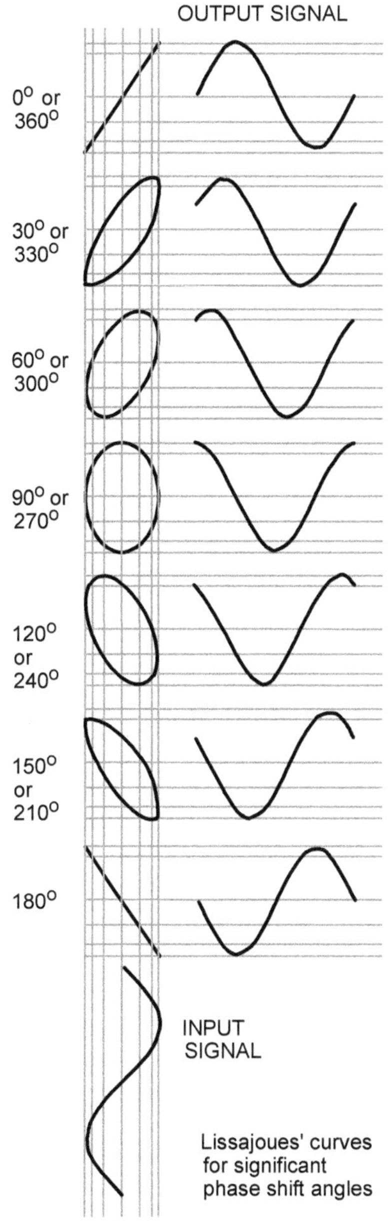

Lissajoues' curves for significant phase shift angles

RIGHT: Method #2, phase shift measurement using Lissajoues' curves (X-Y mode). The curves were named after French mathematician Jules Antoine Lissajous who investigated them in 1857.

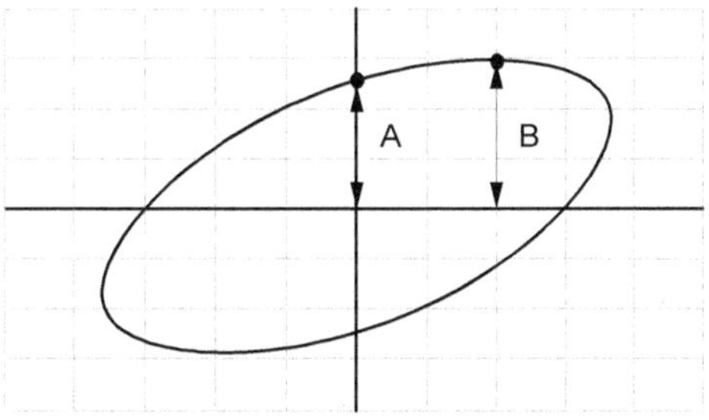

OPTION 1: Phase shift can be determined by Φ = arcsin(A/B)

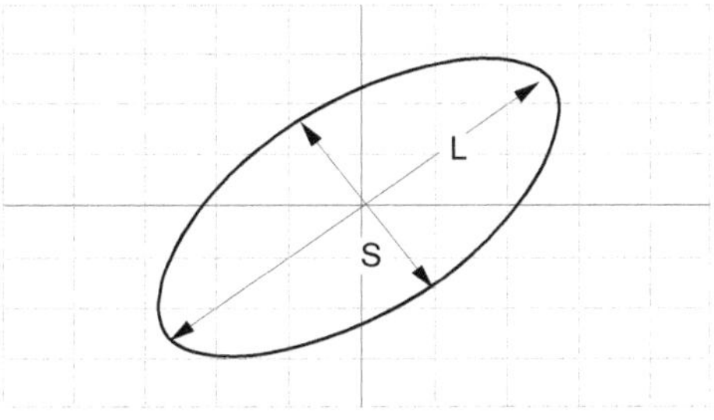

OPTION 2: Phase shift can be determined by Φ = L/S*90°

Amplitude-vs-frequency characteristic (response) of an amplifier or preamplifier

With its sweep function turned on, a sweep function generator will produce a sawtooth wave, either linear or logarithmic, on its "sweep out" output terminals. If such a low-frequency periodic signal is brought to the X-input of an oscilloscope and its timebase switch is in the X-Y mode (meaning the internal timebase sawtooth voltage is switched off), the trace on its screen will deflect in the horizontal direction in proportion to the amplitude of such sweep voltage.

At the start of the sweep (V_{MIN}) the output frequency of the sweep oscillator is at its minimum f_{MIN}, and when the sweep voltage reaches its peak (the set amplitude V_{MAX}), the output sine wave frequency is also at its maximum f_{MAX}.

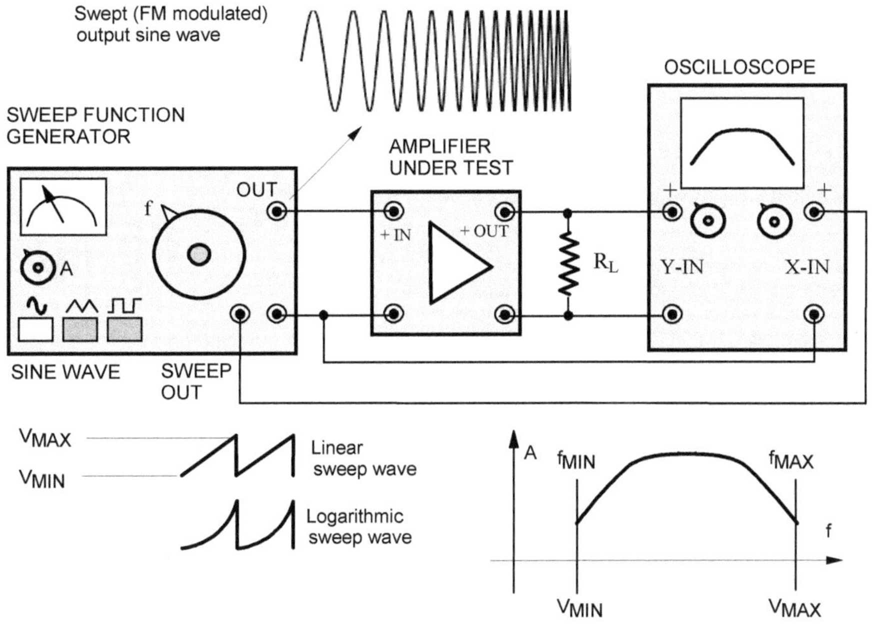

Therefore, the sine signal fed into the input of a tested amplifier is constantly changing its frequency, being "swept" across the range f_{MIN} to f_{MAX}.

Assuming the amplitude of such test signal is constant, the vertical deflection of the scope will depend exclusively on the amplification factor (gain) at various frequencies. Thus, the curve on the scope's screen will be the amplifier's A-f or amplitude-frequency characteristic!

If the sweep waveform rises linearly, the horizontal deflection will be a linear sweep of frequencies (lin scale). To get a log scale, we must use the log sweep. Strictly speaking, it is an exponential sweep since the signal rises in an exponential rather than logarithmic fashion.

The same result can be achieved by using two oscillators or function generators, providing one of them (#1 in the drawing) has a VCF input.

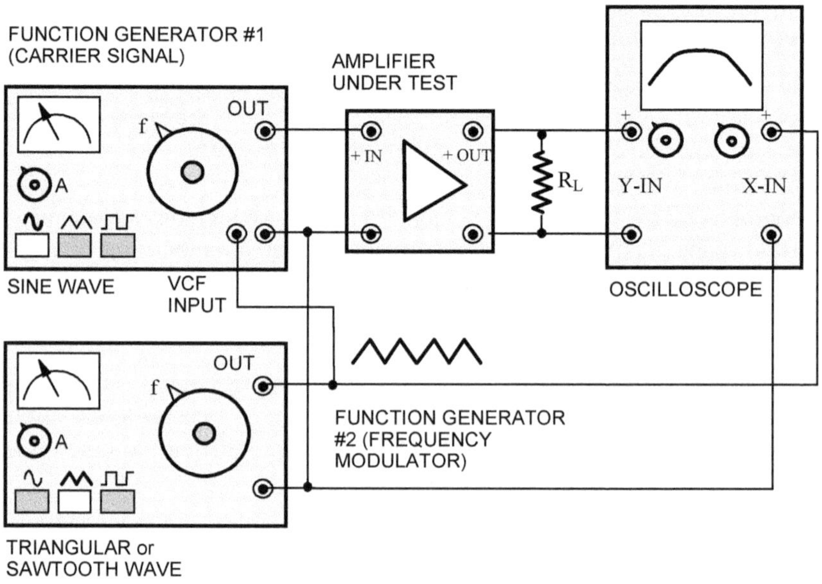

ABOVE: How to display the A-f response curve of an amplifier, method 2, using two function generators

Its output can be frequency modulated by the output of the other (#2). However, function generators don't usually have an "exp" or "log" waveform. Another limitation of this setup is that the A-f curve obtained is only qualitative in the frequency sense. Assuming all controls have been calibrated and are known, the amplitude or amplifier's gain can be determined from the scope screen, but the -3dB or other frequencies of interest cannot easily be located on the X-axis, if at all!

MEASUREMENTS ON TUBE CIRCUITS (INSIDE TUBE AMPLIFIERS)

So far, we have been discussing testing of the audio amplifiers and preamplifiers "from the outside," using what could be called a "black box" approach, only measuring and analyzing their behavior at the input and output connections. Internal checks and tests are equally important, especially for DIY constructors and builders of audio equipment.

WARNING: The AC mains voltage and high DC voltages usually aren't present when measuring components and performance parameters of audio equipment, but they are certainly a danger once live equipment is opened up, and internal tests are carried out, so keep the safety precautions in mind and follow them at all times (see the hazards on page 11)!

EXPERIMENT: Measuring AC ripple on high voltage DC power supply lines

The use of two CLC filters in the high voltage section of this tube line preamplifier reduces the AC ripple from 2.5V_{PP} to only 1mV_{PP}, as measured by a calibrated oscilloscope. Although the chokes were identical, one measured 15H and the other only 13H on a digital LCR meter. The attenuation of the 1st CLC filter was A_1=0.035/2.5 = 0.014 or A_1=20*log0.014 = -37.1 dB. The sign is negative since it is attenuation (gain lower than 1), not amplification.

The attenuation of the 2nd CLC filter was A_2=1/35= 0.02857 or A_2=20*log0.02857 = -30.9 dB, meaning the total ripple attenuation of the filtering chain was thus A=A_1*A_2= 1/2500=0.0004 or in dB A=-68dB, the same result as we would get by simply adding two dB values, A=-37.1-30.9=-68dB!

	X	Y	Z
DC VOLTAGE [V]	280	275	271
AC RIPPLE [mV_{PP}]	2,500	35	1

RIGHT: A typical DC power supply chain of a tube amplifier or preamplifier . The measured DC voltage values and ripple (AC component) in points X, Y & Z are in the table (ABOVE)

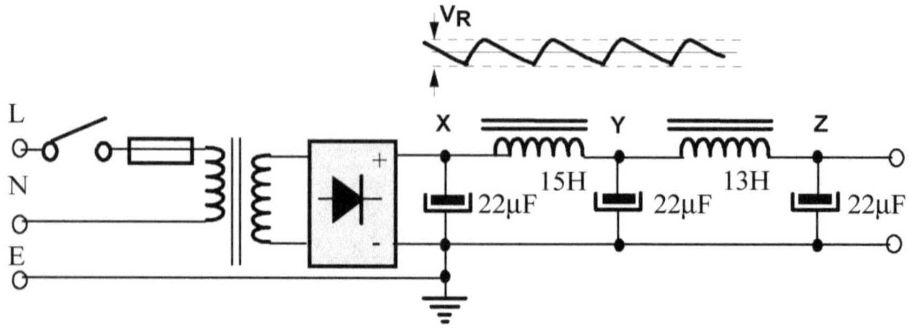

In-situ testing of coupling capacitors

Capacitor leakage is a common cause of all sorts of problems, from improper DC conditions (bias) to distortion and even tube destruction. When a cathode bypass capacitor becomes leaky (allows a small DC current to flow), the consequences are not serious, and the condition may not be discovered at all.

Looking at a typical 2-stage triode circuit (next page), say the leakage resistance of cathode capacitor C_K is 300kΩ. Together with the 2k2 cathode resistor, it will form a new parallel combination of 2k2*300k/(2k2+300k) = 2.184kΩ. As a result, the cathode DC bias of V2 will not change at all.

What about the same leakage resistance of 300k in parallel with the coupling capacitor C_C?

The DC voltage V_B of 250V will be divided according to the voltage divider formed by the anode resistor of 47k, the leakage resistance of 300k, and the grid resistor of 330k.

Even without the exact calculation, since 47k+300k is approx. equal to 330k, you can conclude that the DC voltage in point X (the grid of V2) will be slightly less than half of the 250V supply voltage, or around 120V!

The nominal cathode voltage is 4.4V so V2 will be so positively biased that a full anode current will flow through it, the anode will overheat and the tube will be destroyed!

Even a relatively small leakage (the equivalent of 3MΩ leakage resistance), would result in the positive grid bias voltage of V_G=250*330/(330+47+3,000) = 250*0.1 = 25V, which is still more than enough to fully "open" and destroy tube V2.

Measuring a D-factor on digital LCR meters is not conclusive because such meters use a very low test voltage (less than 1 Volt). The real test is under the operating voltage in an amp. One way to do that is illustrated above.

Before powering up a suspect amplifier, unsolder the coupling capacitor C_C in the point marked X (the grid of V2), then power the amp up without any signal. If you don't want to do unsoldering, the other option is to pull the V2 out, R_{G2} (330k) can stay in the circuit.

Use a good quality DMM, FET voltmeter, or VTVM with high input impedance. Measure DC voltage between point X and ground if you unplugged V2, or between the unsoldered end of C_C and ground - you should get no DC voltage at all. The higher the DC voltage you measure, the leakier the C_C capacitor is!

Checking DC imbalance (one possible cause of distortion)

Depending on what type of voltmeter you have, you'll have to perform either one or two measurements to test a push-pull stage for DC balance. Power the amp up with a speaker or dummy load connected, but bring no signal in.

If neither of the terminals of your voltmeter is grounded (both are fully "floating," as with all hand-held, battery-powered DMMs), measure the voltage difference between the two anodes. Ideally, the voltage will be zero or very low. The higher the voltage measured, the greater the DC imbalance.

If the negative terminal of your voltmeter is grounded, measure the voltage between each anode and ground and subtract the lower voltage from the higher. Again, ideally, both voltages should be the same.

WARNING: Make sure you observe the correct polarity! Connect the meter's + (red probe) to the anode and the - (black probe), which is grounded, to the ground. Reversing their polarity will ground the anodes and create a dead short circuit across the amplifier's high voltage supply!

You will detect some DC imbalance in most cases, even if you use matched tubes! The reason is that most output transformers use a faster and cheaper winding method for the primary (as mentioned earlier), where sections are wound on top of each other.

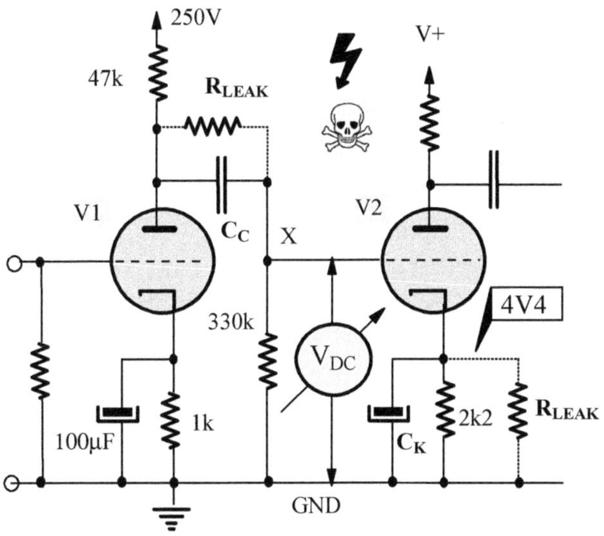

ABOVE: Since very low DC voltages are involved, leaking cathode caps aren't that serious. However, leaking coupling caps can result in a catastrophic failure of the following tube.

BELOW: If one terminal of a test instrument (voltmeter or oscilloscope) is earthed (grounded), then two measurements must be performed, between A1 and GND and between A2 and GND.

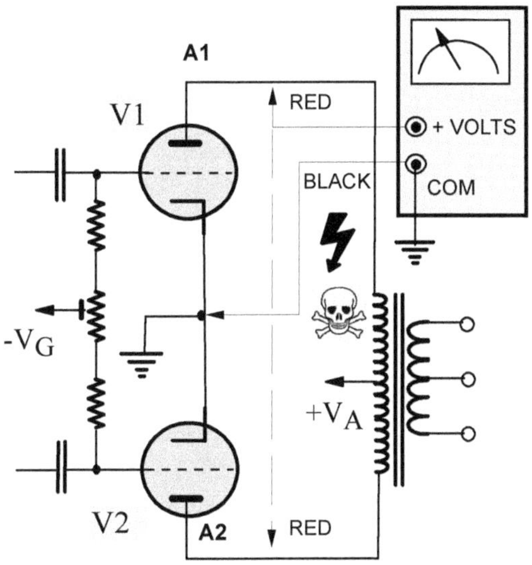

BELOW: One measurement can detect DC imbalance between A1 and A2 if a floating (differential) voltmeter is used. Battery-powered multimeters are by their design always "floating".

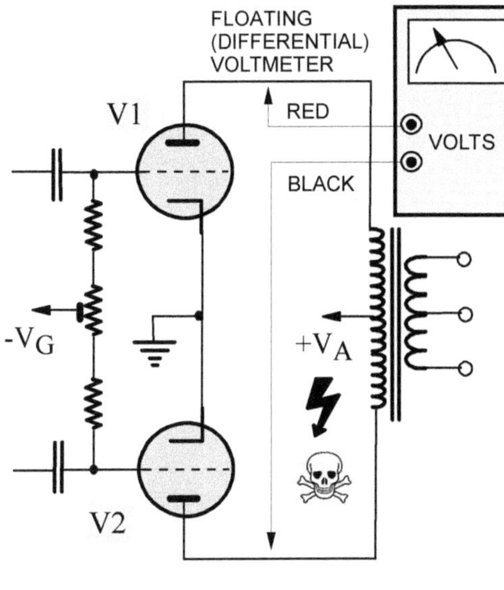

High frequency oscillations

In contrast with low-frequency oscillations (motorboating), HF oscillations are not always obvious. The human ear cannot detect ultrasonic instability (above 20,000 Hz), but instruments can.

Method #1: High-frequency oscillations are indicated on an oscilloscope by the widening or smudging of the sine wave (signal voltage) at the output of an amplifier.

Method #2: NFB and total current draw. Too much negative feedback can cause HF instability and oscillation. Sustaining such oscillations (no matter if they are of the low- or high-frequency kind) requires energy from the amplifier's power supply, so the presence of oscillations increases the total power draw from the amp's power supply.

Turn the amplifier off. Insert a mA meter (DC) in the high voltage supply CT (Center Tap). Turn the amplifier on. Without any signal at the input, record or remember the mA current draw of one channel.

Momentarily disconnect the feedback loop from the amplifier's output as indicated on the drawing. Compare the mA reading. If the mA reading goes down, the amplifier oscillated with the NFB applied.

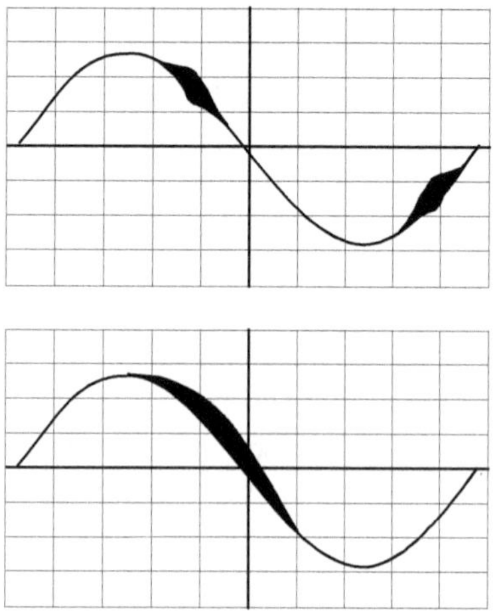

ABOVE: The "thickening", spreading or fuzziness of the waveform on oscilloscope screen indicates HF oscillation and/or noise

BELOW: NFB can cause HF oscillations, which can be detected by disconnecting the NFB loop and observing any drop in anode current draw.

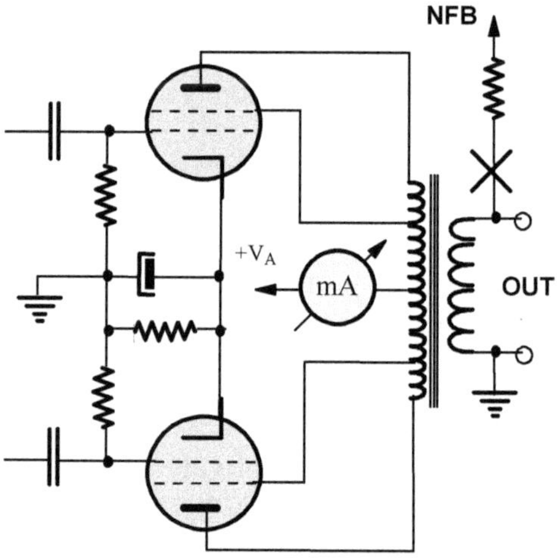

7 DISTORTION MEASUREMENTS

- DISTORTION IN AUDIO AMPLIFIERS
- TOTAL HARMONIC DISTORTION (THD) METERS
- SPECTRUM ANALYZERS AND FFT
- MEASURING INDIVIDUAL HARMONICS WITH A SELECTIVE VOLTMETER (WAVE METER)
- MEASURING INTERMODULATION DISTORTION

> If you can't measure it, you can't fix it.
>
> Dhanurjay "DJ" Patil

DISTORTION IN AUDIO AMPLIFIERS

While distortion may be a desirable feature of guitar amplifiers, it is generally avoided in high fidelity ones. I say generally because not all distortion is bad or unpleasant sounding, and also because certain measures that reduce distortion may also negatively impact the other aspects of sound quality. A typical example is the negative feedback.

Harmonic distortion (THD)

Assuming a simple sine wave signal at the input of a real amplifier (in contrast to an "ideal amplifier" that does not distort at all), the output will also be a sine wave but slightly distorted. Such a distortion cannot normally be detected visually, by observing the waveform on an oscilloscope, for instance, except in the cases of severe distortion. Nevertheless, should such a signal be brought to the input of a distortion or spectrum analyzer, the presence of new harmonics will be detected. Real amplifiers generate harmonics due to their nonlinear input-output or "transfer" characteristics. For the transistor fanatics in our midst, bipolar transistors are even less linear than vacuum tubes. The most linear of all amplifying devices is a triode!

Usually, the second harmonic (of twice the original signal's frequency) has the highest amplitude, followed by the third, the fourth, and so on. The relative amplitudes of various harmonics depend on the type of amplifying device (triodes distort differently from pentodes and beam tubes), the power level at which the measurement is taken, the design of the circuit (single-ended or push-pull), and many other factors.

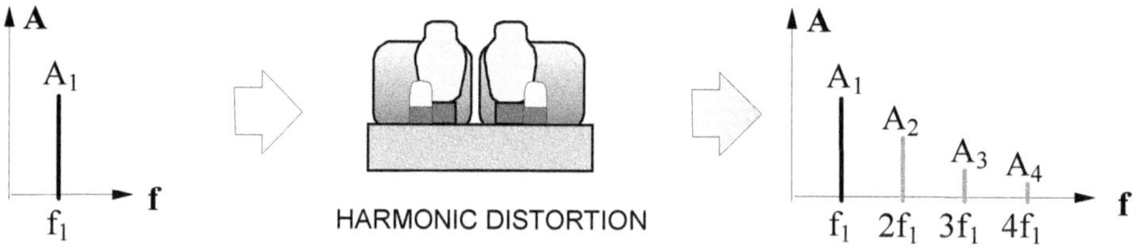

Intermodulation (IM) distortion

To explain this type of distortion, let's look at how it's measured. Two pure sine wave signals are mixed and fed into an amplifier. One is of a lower frequency f_1 (usually the mains frequency, 50 or 60Hz), the other of fifty times higher frequency f_2 (assuming f_1=50Hz, f_2=2,500Hz). Various commercial analyzers use different frequencies, but their operation's principle is the same. The amplitude of the lower frequency signal is adjusted to be four times higher than the amplitude of the higher frequency signal ($A_1=4A_2$).

An ideal amplifier would amplify both signals equally, but a real amplifier will generate two unwanted signals (or "sidebands") of the higher frequency signal. One will be of f_2-f_1 frequency (or 2,500-50 = 2,450Hz in our case), the other will have a frequency f_2+f_1 (or 2,500+50 = 2,550Hz)!

In severe cases of distortion, second sidebands will be also be generated, f_2-2f_1 and f_2+2f_1. The situation is analogous to AM radio, although frequencies in question aren't in the radio but the audio range.

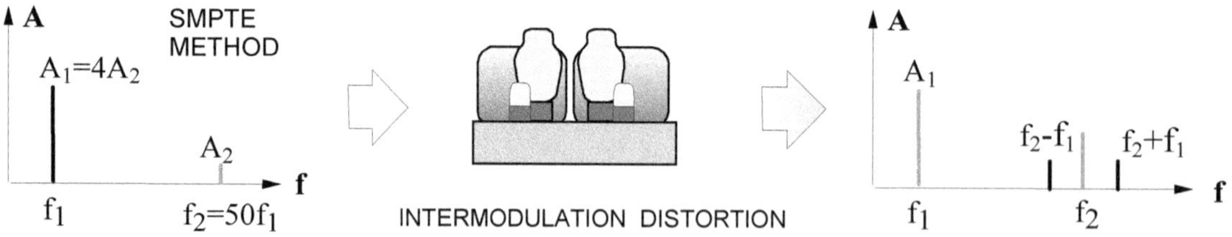

The output signal will be an amplitude-modulated carrier, the f_2 signal is the high frequency (HF) carrier, while the low frequency (LF) signal (f_1) modulates the LF signal's amplitude.

IM distortion is more unpleasant to human ear than harmonic distortion and thus its reduction should be even higher on the priority list of an amplifier designer than the reduction of harmonic distortion.

Going back to the desirability of a wide bandwidth debate, a poor quality amplifier may generate distortion tones at say 28kHz and 31kHz. Due to IM distortion these will then produce sidebands, one of which, 31-28= 3kHz, will fall in the audible range, dead smack in the midrange where human ear is the most sensitive!

Delay distortion

The phase angle between the input and output signal of a typical real amplifier stays constant through the midrange frequencies, but changes significantly at frequency extremes. The illustration (next page) shows a complex input audio signal comprising of the a fundamental and its 3rd harmonic.

If the third harmonic lies in the frequency region where the phase angle changes from its midrange value, it will be shifted or "delayed" in phase by the angle q. Their sum, the output signal, will differ from the input signal's waveform.

This kind of distortion is called a phase or delay distortion. Looking at how different the shape of the resultant output voltage is from the input waveform, you'd think that this kind of distortion would be the most serious (malign) of all. Paradoxically, our ears do not object to this kind of distortion; in fact, they don't even detect it.

The gain-frequency characteristic of a typical audio amplifier is illustrated below. The midrange gain, in this case, is 30, and the phase shift is 180 degrees, meaning the output signal lags behind the input signal by 180°; in other words, the amplifier inverts absolute polarity, the input and output signals are "out-of-phase."

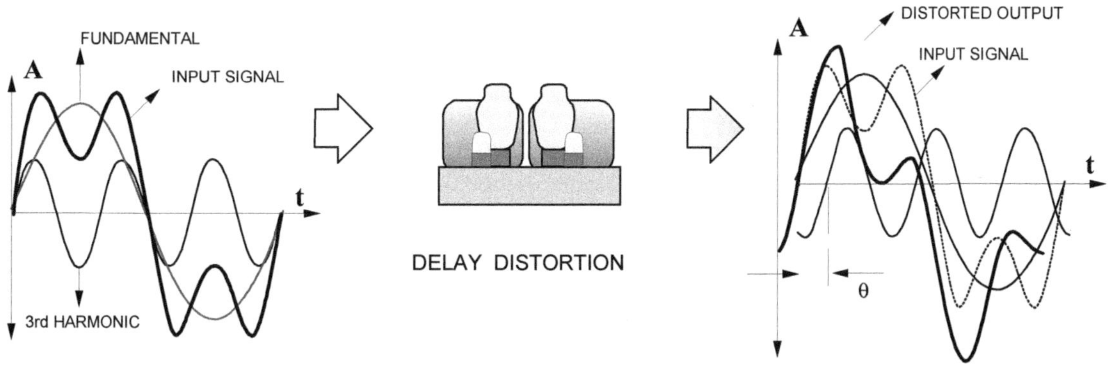

The phase characteristic of an ideal amplifier would be a straight line - there would be no phase shift between the input and output signals (or it would be 180°, as illustrated), and such a phase shift would be constant, the same for all signal frequencies. As with signal's amplitude, the phase relationships of real amplifiers also change at frequency extremes.

At low frequencies, the phase lag is increased, and at high frequencies the lag is decreased. Notice that at our half-power or -3dB points, frequencies f_L and f_U, the phase changes $+/-45^o$ from the midrange phase. At f_L the phase is $180^o+45^o = 225^o$, and at f_U the phase is $180+45 = 225^o$.

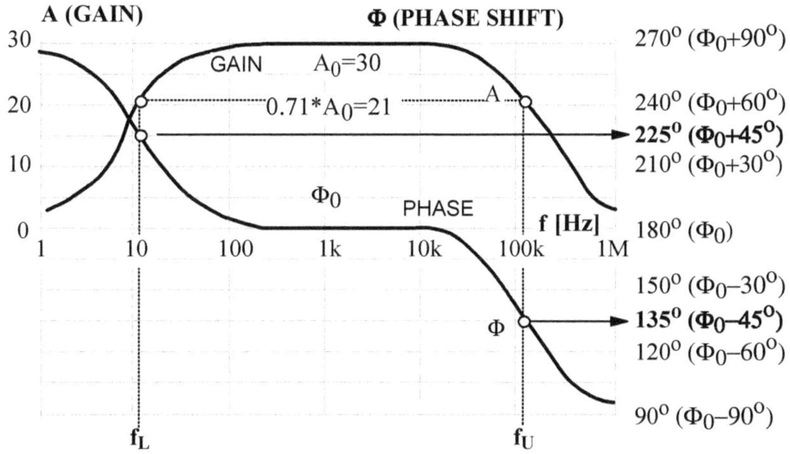

ABOVE: Typical amplitude versus frequency and phase versus frequency characteristics of a real amplifier and the meaning of -3dB points

TOTAL HARMONIC DISTORTION (THD) METERS

Operational principle

Most analog THD meters from the 1970s to 1990s are very similar, three instruments in one, an electronic DC and AC millivoltmeter and voltmeter, a THD meter, and a signal-to-noise meter. As such, they represent a great value for money since you don't have to buy another voltmeter or multimeter for audio tests for which two voltmeters are needed.

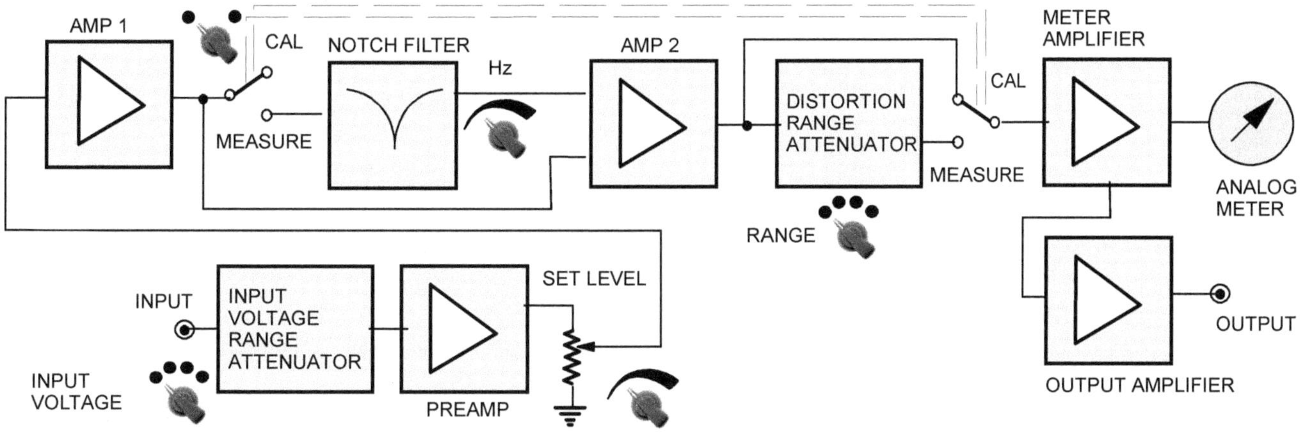

ABOVE: A simplified block-diagram of a typical manual THD meter such as Leader LDM-170

In THD mode the instrument subtracts the fundamental (1st harmonic) from its distorted input signal, the difference being the distortion added by the device under test. A highly selective balancing network (usually a tunable Wien-bridge notch-filter) eliminates the fundamental frequency in the 20Hz-20kHz range; the distortion harmonics are amplified by a wideband amplifier (AMP 2) and fed to "METER AMPLIFIER" which drives the analog meter calibrated in % THD.

In voltmeter and S/N modes, the notch filter and the distortion range attenuator are switched out, so 0.1mV-300V signals in the 20Hz-200kHz range are measured directly.

A useful feature of all three analyzers pictured above is the output that can be connected to an oscilloscope so that you can see the waveform of the distortion components by themselves. All three can also measure noise in Volts and dB and S/N (signal-to-noise) ratios in dB.

While HP8903A and HP8903B are quite common on eBay, they are still priced at $800-$1,200, way too expensive. Unless you can find one locally and test it before you buy it, the risk of a unit being or arriving faulty is very high due to its complexity (microprocessor control and a very high number of printed circuit boards and discrete components), and it's unlikely that you would be able to fix it yourself! Plus, the international postage or courier charges would run into hundreds of dollars due to their size and weight.

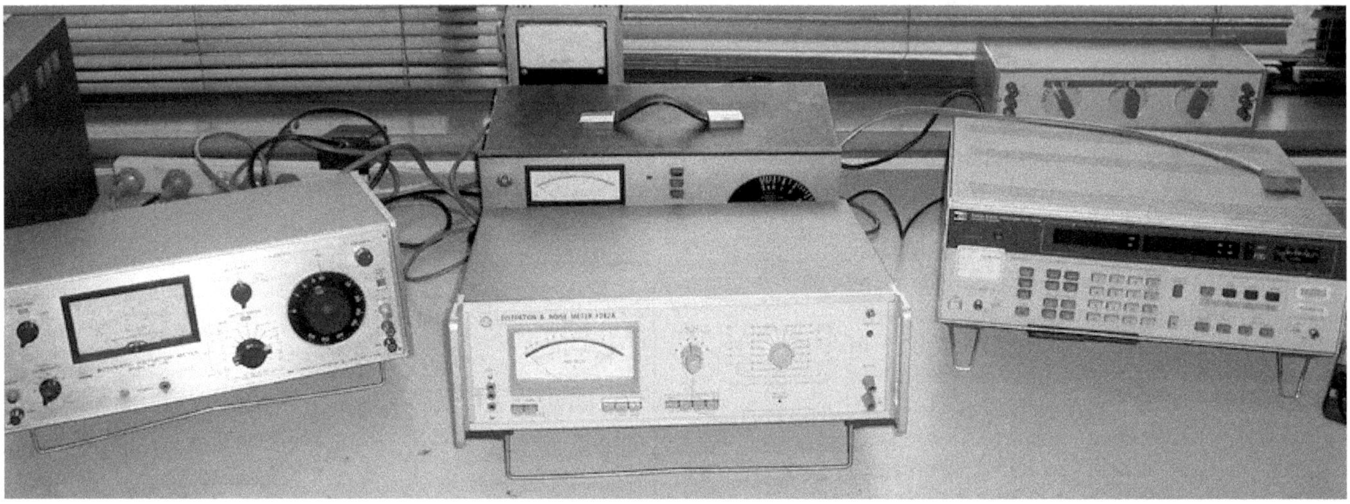

ABOVE: Three examples of THD meters on our test bench, left-to-right: Japanese-made NF analog meter, Australian made AWA F242A analog meter, both dating from mid-to-late 1970s and USA-made digital HP analyzer HP8903A, which dates from the early 1980s and includes a low distortion programmable signal source and microprocessor control. Behind the AWA unit is Radford low distortion oscillator.

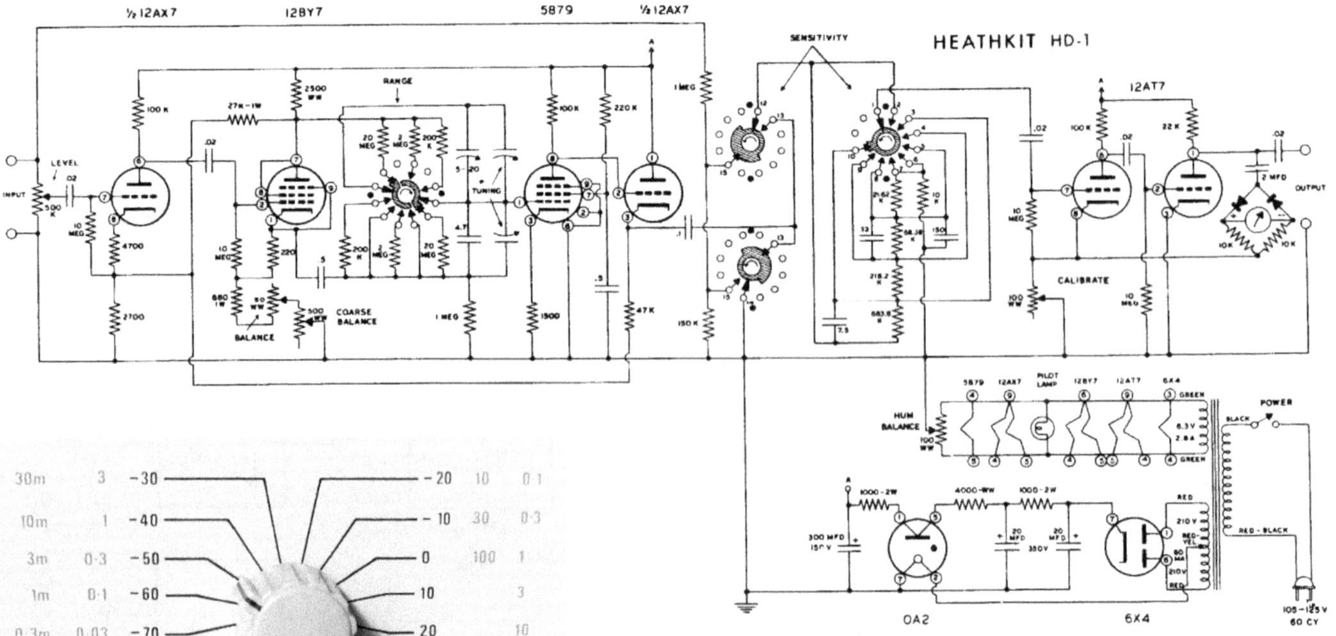

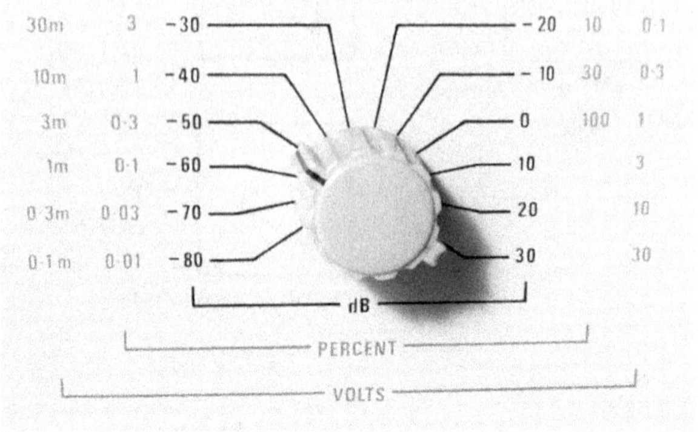

ABOVE: Circuit diagram of Heathkit HD1 and IM-12 THD meters © Heathkit

LEFT: The range switch of AWA (Australian Wireless Amalgamated) F242A. Notice the lowest mV range of only 0.1mV or 100 microVolts full scale. This is one very sensitive AC voltmeter, suitable for audio tests on moving coil step-up transformers and phono preamplifiers.

Heathkit THD meters are the most common of all brands and are much simpler, so they can be repaired quickly and easily. HD-1 was their first harmonic distortion meter, followed by IM-12 and IM-58, which were almost identical, except for the exterior looks. All are tube-based. The last model, IM-5258, was a different beast, a more advanced solid-state unit with auto-nulling, optical isolation, and a meter-amplifier combination.

How to use a typical manual analog THD meter

The Leader LDM-170 THD meter pictured has a declared accuracy of +/-5% of full scale, so it is not the most precise of instruments but is typical of its class. It is a manually-tuned THD meter, like the NF unit, HP332A, and all Heathkit THD meters. AWA F242A features automatic tuning, as do HP8903A and HP8903B.

Everything has to be manually selected and tuned, from the function (2) & (5), the input voltage and frequency ranges (6) and (7), to test frequency (10).

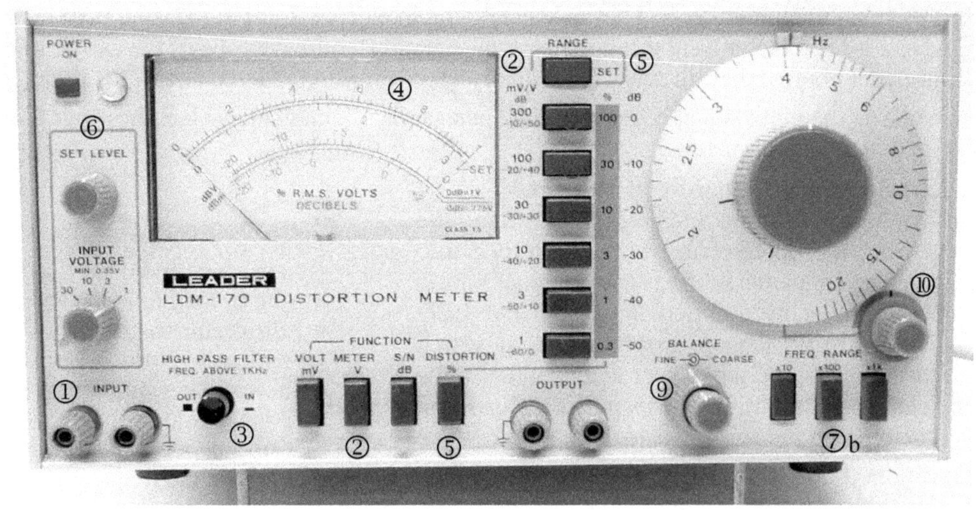

The THD measurement steps are almost identical for all manually-tuned THD meters, so here is the sequence for Leader LDM-170:

1. Connect the output of the amplifier to the input of the THD meter and power the unit up.
2. Select "V" function and a desired input voltage range (1-3-10-30V)
3. Set "HIGH PASS FILTER" to OFF or "OUT" position.
4. Measure the input voltage on the analog meter and adjust it to the required power level (by changing the amplifier's volume control or the output level of the audio oscillator feeding the amplifier's input), to say 2.83V, which is 1 Watt on an 8Ω load.
5. Press the "RANGE SET" and the "DISTORTION" switch.
6. Advance the SET LEVEL knob to get the full-scale reading on the meter (marked "SET").
7. Press the "100%" switch. Depending on the test frequency used, choose one of three frequency ranges. Say you are using a 400Hz test signal. Press the "x100" button and rotate the main "Hz" knob & dial. Then, using the more precise vernier drive, marked (10), adjust the notch frequency until you get minimum deflection.
8. Press the next range button down (30%) and repeat the adjustment until you get the lowest reading on that range.
9. Select the next lower range, 10%, and repeat the adjustment until you get a minimum reading.
10. Adjust the "BALANCE" knobs ("course" and "Fine") to get the minimum deflection of the meter.

Once you move to a lower range, for instance, the 3% range, and you cannot get any reading (the needle is off the scale to the right), that means you have to go back to a higher range, as in the previous step and read the distortion figure on that scale.

Say the needle points to 4.5 on the 0-1 scale. Since your range switch is at "10%", the distortion is 4.5%! That is why when you switched the range down to 3%, the meter was off the scale.

If your test signal is in the 2kHz - 20kHz range, switch "HIGH PASS FILTER" to the ON position. That will filter out low-frequency noise and hum.

Correction for oscillator distortion

As we have seen from the specs of two typical low distortion oscillators by Philips and Gould, even they distort ("low" is a relative term), especially at higher amplitude levels and at frequency extremes. So, you may feel that you need to correct the measured THD results by subtracting the oscillator distortion. The actual amplifier THD is $D_A = \surd(D_M^2 - D_S^2)$, where DM is the measured THD and DS is the distortion of the source (oscillator), both in [%].

For example, say you measured the oscillator's distortion as 0.4% at 100Hz and $1V_{RMS}$ output level. Then you tested an amplifier at the same frequency and using the same input voltage level, and your THD meter showed 1.3%. The distortion of the amplifier is $D_A = \surd(D_M^2 - D_S^2) = \surd(1.3^2 - 0.4^2) = \surd(1.69 - 0.16) = \surd 1.53 = 1.24$ %.

The error introduced by the oscillator distortion is $\varepsilon = (THD_M - THD_R)/THD_R = (1.3 - 1.24)/1.24 = 0.0484$ or 4.84%. The measured distortion is larger than the real (actual) distortion, so the error is positive. However, for most practical purposes, a 5% error can be ignored.

Our example was for a low frequency, where the oscillator distortion is more than ten times higher than their midrange distortion. At 1kHz these oscillators have a distortion of around 0.03%, so assuming the same figures at 1kHz test, $D_A = \sqrt{(D_M^2-D_S^2)} = \sqrt{(1.3^2-0.03^2)} = \sqrt{(1.69-0.0009)} = 1.29965$, a truly insignificant difference from the measured 1.3%!

GW GAD-201B THD meter

According to their website (yes, they are still in business), "founded in 1975, Good Will Instrument Co., Ltd. was the first professional manufacturer in Taiwan specializing in electrical test & measurement instruments". While in production, the GAD-201 THD meter went through numerous versions, ours was one of the earlier ones, but according to their specs and front panel photos, they all seem functionally identical (GAD-201B, GAD-201D, and GAD-201G).

As soon as the input signal is detected, the meter will automatically select the input voltage range (2) and display its RMS value on the analog meter (1).

The same will be done with the measured THD (4 and 3), but the "automatic" meter still needs a little help from the operator to get the auto-tune into range. The fundamental frequency manual tuning knob (6) must be adjusted until both red LED indicator arrows (7) are extinguished, at which point the auto-tune takes over and eventually the indicator needle stabilizes, meaning the distortion can be read on the analog scale (3).

During both the voltage detection and THD range tuning process, the clicking of reed relays inside the unit can be heard. If only a single test frequency is used (400Hz, 1kHz, or 10kHz), the spot PB can be pressed, and the whole manual-assisted tuning need not be done again.

GW GAD-201B, GAD-201D and GAD-201G THD meters

- Auto range and distortion measurement
- Auto or hold function (switch selectable)
- 7 distortion ranges: 0.1%-100%
- Frequency: 20Hz-20kHz, cont. adjustable in 3 ranges
- Shortcut keys for 400Hz, 1kHz, 10kHz spot frequencies
- 12 input voltage ranges: 1 mV-300V_{RMS}
- Distortion measuring range 0.1%-100% in 7 steps
- Fundamental frequency range 20Hz-20kHz
- Fundamental wave suppression > 80dB
- Frequency response as AC voltmeter: 100mV~300V range 20Hz~200kHz, ±0.5dB, 1mV~30mV range 20Hz~200kHz, ± 1dB
- Outputs to oscilloscope: X-axis 1V_{RMS} (input signal), Y-axis 0.5V_{RMS} (distortion components)
- Max. Input Voltage < 400V (DC+AC peak)

1) Input signal AC voltmeter
2) LED indication of the input voltage range (auto detected & selected by the meter itself)
3) "Distortion indicator (moving coil meter) in % and dB
4) LED indication of the distortion range (automatic)
5) Auto range or "Hold" and shortcut keys for 400Hz, 1kHz, 10kHz spot frequencies
6) Fundamental frequency manual tuning to get the auto-tune into range
7) "High" and "Low" tuning indicators
8) Binding posts for X-output (input signal) and Y-output (distortion components) to a meter or oscilloscope
9) Input terminals (binding posts)

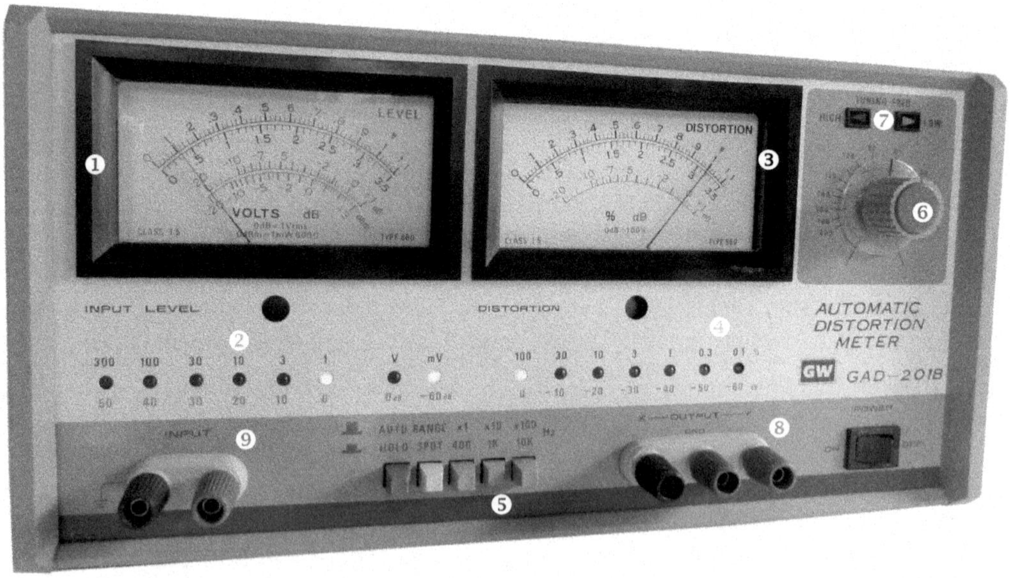

CASE STUDY: THD measurement of a push-pull amplifier on GW GAD-201B

One of our designs, this Class A push-pull amplifier (photo on the next page) uses EL-12N, great sounding, and still reasonably priced NOS power tubes. The interstage transformers provide phase inversion and coupling, so there are no coupling capacitors, one of the reasons the amp sounds incredible. No negative feedback of any kind (local or global) was used, either.

Referring to the photo of the test setup (next page), at 1 kHz and 2.83V_{RMS} output (1), meaning the output power was 1 Watt into an 8Ω dummy load, the THD was measured as 0.39% (3).

Notice the "3V" scale input signal range LED lit (2) and 1% scale THD scale LED lit (4).

The Phillips low distortion oscillator was used as a signal source (not in the photo). After pressing the 1kHz range switch (5), the manual notch filter frequency centering adjustment (6) took us a few seconds. The two red LEDs (7) went off (meaning the auto-tune was enabled), and the THD meter started its automatic fine-tuning process, which took another 20 seconds or so, and then the "Distortion" indicator (3) stabilized at 0.39%.

THD measurement of EL-12 push-pull amplifier on GW GAD-201B automatic distortion meter

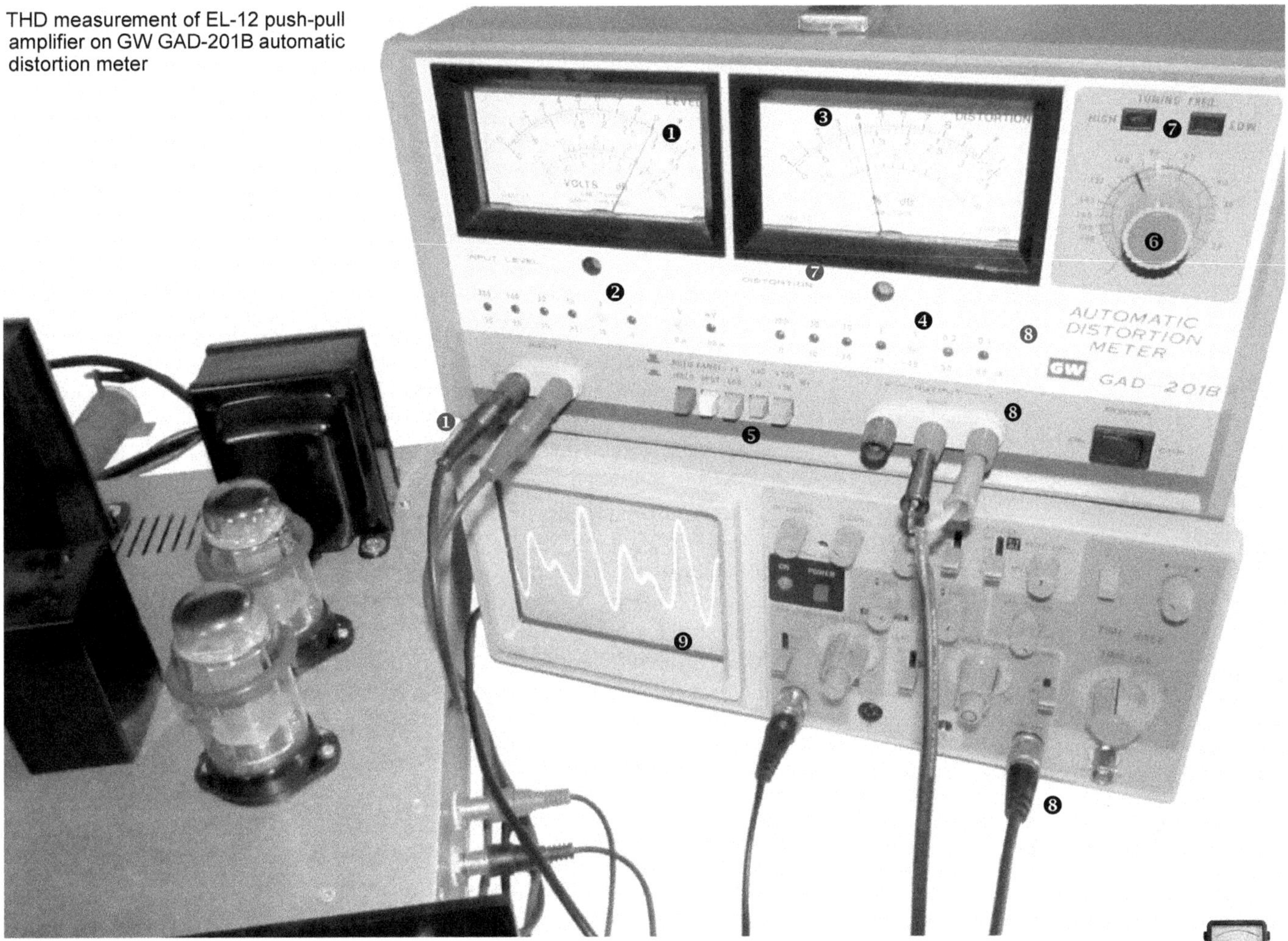

The "Y output" (8) was fed into the oscilloscope's Ch. 2 vertical input), and the waveform combining all of the distortion harmonics appeared (9). Instead of using the "X output" from the THD meter, we fed the amplifier output straight into Channel 1 of the scope.

Repeating the test at various power levels by adjusting the amplifier's output voltage amplitude and different test frequencies was very fast. The results for 1kHz and 100Hz tests are in the table (right).

OUTPUT VOLTAGE V_{RMS} on 8Ω	OUTPUT POWER (Watts)	THD @ 1 kHz	THD @ 100 Hz
2.83	1.0	0.39	0.42
10	12.5	1.10	1.35
11	15.125	1.60	2.10
12	18	2.60	3.60
13	21.125	4.20	6.00
14	24.5	6.50	8.00

SPECTRUM ANALYZERS AND FFT

Amplifier behavior in time- and frequency domains

The amplitude versus time display on an oscilloscope's screen (next page) shows a distorted periodic waveform that can be broken down into two components, the fundamental harmonic and the second harmonic of twice the fundamental frequency ($2f_1$). There is also a phase shift α (alpha) between the fundamental and the second harmonic. This is the "time domain" in which we see these three signals' amplitude and phase relationships.

If we depict the signal in a three-dimensional space with frequency as the 3rd dimension, the projection on the A-f plane will be its spectrum in the frequency domain (next page). We don't see the waveforms or phase relationships in that domain, but we see the frequencies and the absolute and relative amplitudes of all harmonics, something we don't see in the time domain. Thus, the two depictions illustrate different aspects of the same signal. Test instrument that displays amplitudes of signal harmonics in the frequency domain is called a spectrum analyzer.

Initially, spectrum analyzers were stand-alone instruments that looked like oscilloscopes (had their own CRT screen and similar controls) and were more expensive than a family car. Vintage Tektronix and HP audio (LF) and radio frequency (RF) spectrum analyzers are still priced in thousands of dollars. The more modern ones are even more expensive.

Some T&M manufacturers (such as Tektronix) even made plug-in spectrum analyzer modules for their larger "mainframe" oscilloscopes.

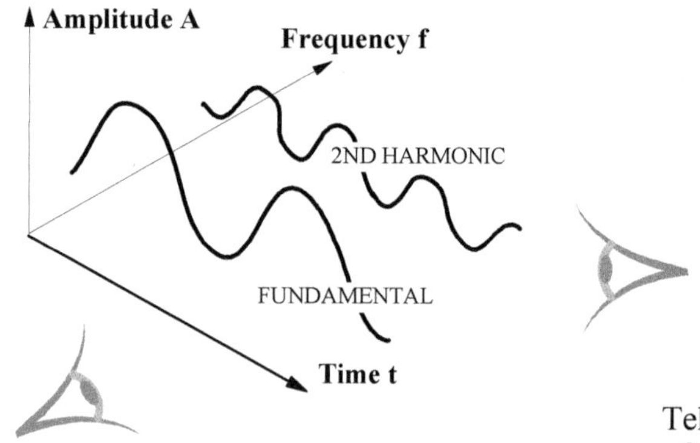

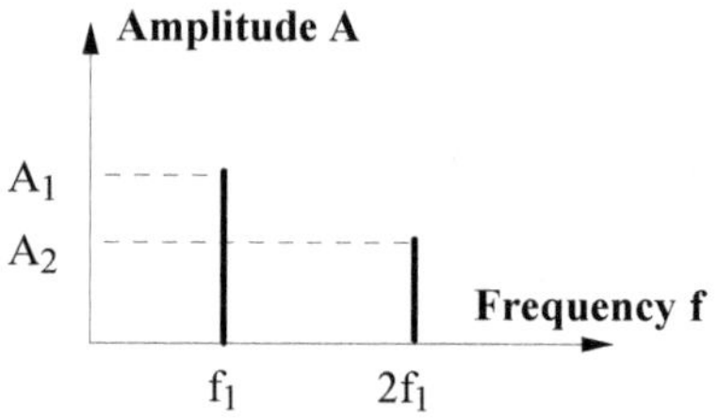

Spectrum analyzer display (frequency domain)

LEFT: The relationship between the time and frequency domains in a 3D space

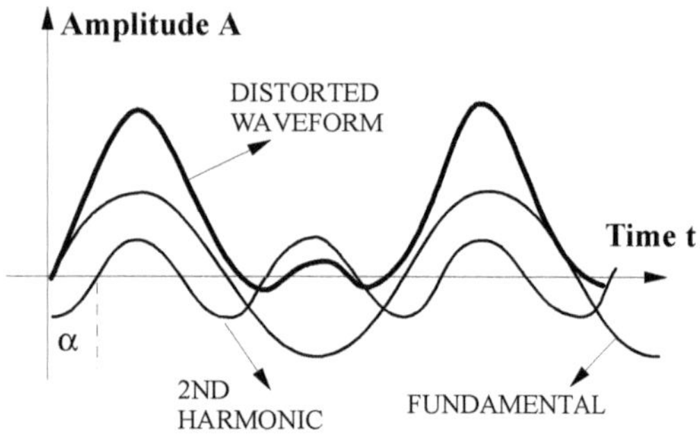

Oscilloscope display (time domain)

Tektronix even made plug-in modules in the late 1990s for their range of digital oscilloscopes, TDS210, and TDS220. TDS2MM extension module's Fast Fourier Transform (FFT) mathematically converts a time-domain signal into its frequency components so a spectral analysis can be performed and harmonic content and distortion displayed as on a spectrum analyzer.

If you can financially justify buying a spectrum analyzer or are lucky enough to pick one up for a song at garage or deceased estate sales, by all means, get one. It will give you hours of pleasure. You'll be able to compare distortion "signatures" of various amplifiers and preamplifiers, their noise floors, and even plot amplitude-frequency characteristics. You can even use them to test audio transformers and other devices, such as filters, attenuators, crossovers, signal and function generators.

How FFT spectrum analyzers work

The spectacular development of the microprocessor and A/D converter technology, their ever-increasing speed and capabilities, together with the rapid fall in the price of memory chips, enabled T&M equipment makers to implement digital signal processing solutions that would not be feasible or even possible using analog circuitry.

The Fast Fourier Transform is a methodology taught at the university level, and that mathematical technique is at the heart of FFT spectrum analyzers. FFT transforms a signal from its time domain into the frequency domain and thus computes its spectrum. The waveform must be digitized (turned into a digital form) before FFT is applied to that data array in a microprocessor. The signal's amplitude is attenuated (or amplified) to a referent level and then sampled at regular intervals. If the sample points are taken at a high enough frequency, the original waveform can be reconstructed by "filling in" the values "in-between."

If the bandwidth of a spectrum analyzer is BW, then the minimum sampling frequency is twice that value. So, for a 100kHz spectrum analyzer, the f_{SMIN}=2*100 = 200kHz or 200,000 samples per second. Practical consideration require the sample rate to be even higher than f_{SMIN}.

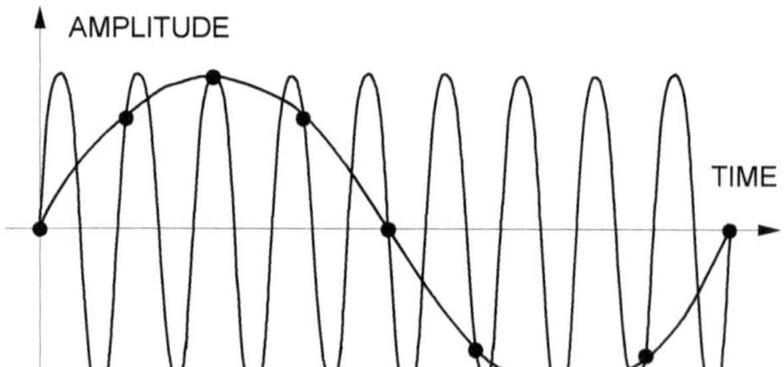

ABOVE: Aliasing is a phenomenon when, due to unfortunately chosen sampling frequency, two signals of different frequencies and even different waveforms produce identical sample points and thus identical frequency domain spectrum

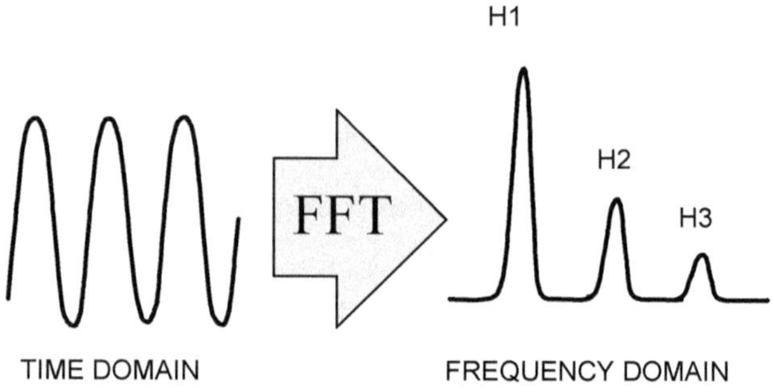

ABOVE: Fast Fourier Transform is a mathematical (computational) time-domain to frequency-domain converter

The block diagram of a typical FFT spectrum analyzer is deceptively simple (next page). The analog front-end starts with the attenuator and input amplifier, followed by a low pass filter that removes undesirable high-frequency signals that may act as an alias of the measured signal. Aliasing is a phenomenon when, due to unfortunately chosen sampling frequency, two signals of different frequencies and even different waveforms produce identical sample points and thus identical frequency domain spectrum, as illustrated above.

ABOVE: Simplified block-diagram of a typical FFT spectrum analyzer

BELOW RIGHT: HP 35670A with numerous add-on options installed

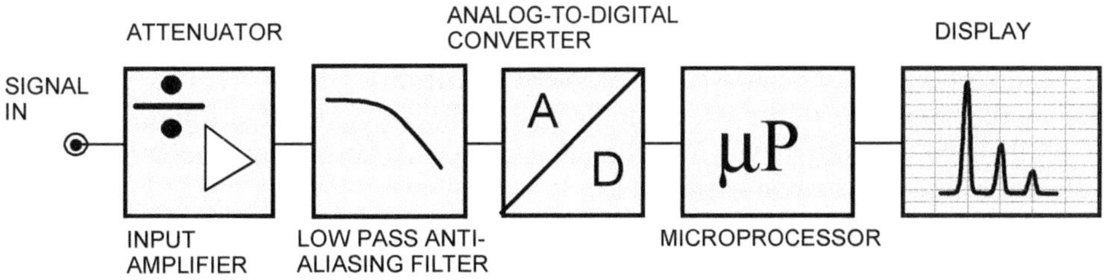

The A/D converter is followed by the microprocessor, which processes the signal, and the results of such FFT processing are displayed on a screen. The increase in the processing power and speed of microprocessors enables the bandwidth (frequency range) of FFT spectrum analyzers to reach 100MHz or even higher, but the ones suitable for audio work are the lower spec ones, with a BW of only 100kHz or so. Typical examples are HP/Agilent/Keysight models 35665A, dating from the 1980s, and its more modern successor, 35670A, discontinued in 2011. 35665A is a 2-channel Dynamic Signal Analyzer used not only in audio tests but also in vibration, acoustics, and control systems T&M. Its bandwidth is 102.4 kHz, which halves to 51.2 kHz when two channels are used.

It has an integral signal source (random, burst random, pink noise, sine, swept-sine, arbitrary, periodic chirp or burst chirp signals) and can perform various measurements such as linear spectrum, cross-spectrum, power spectral density, frequency response, time waveform, auto-correlation, cross-correlation, histogram, PDF and CDF. Using optional plugins, real-time octave, computed order tracking, swept-sine, curve fit, synthesis, and arbitrary source are also available. 100, 200, 400, or 800 lines of resolution are available (frequency resolution = frequency span/number of lines of resolution); model 35670 adds 1,600 lines of resolution and has a wider dynamic range (90dB).

Case study: Two single-ended tube amplifiers and their distortion signatures

Spectral diagrams answer the most common question in tube audio: Why do triode amplifiers usually sound better than those using pentodes? The spectra of 2A3 single-ended amplifier at half-a-watt and three-watt power outputs show a dominant second harmonic. At lower power levels, that is the only harmonic measurable. The third and fourth only appear when the amp approaches its maximum power of 3.5W. The harsh-sounding 3rd harmonic never exceeds the harmonically pleasing second harmonic and is "masked" by it.

The pentode's sonic signature is very different. Even at low power levels, the third and the fifth harmonics are present, and higher-order ones are measurable. At higher power levels, those odd harmonics increase rapidly (from -53 dB to -25 dB, in the case of the 3rd harmonic), which makes many pentode amps sound shrill and harsh compared to triodes.

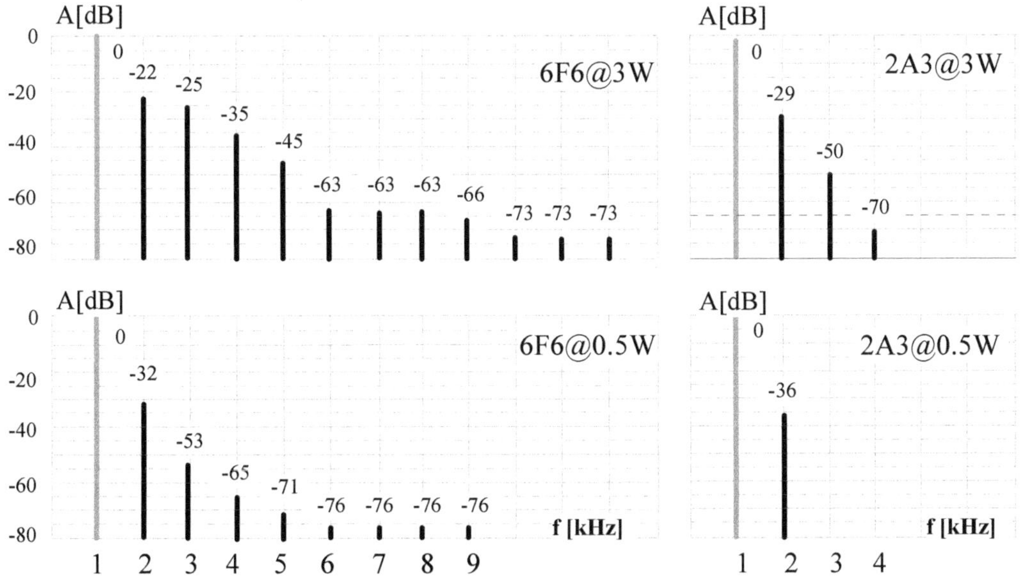

LEFT: Harmonic distortion spectra of two low powered SE amplifiers, 6F6 pentode and 2A3 triode, for two power levels, linear frequency scale

Calculating distortion figures from the harmonic distortion spectrum

How do we determine harmonic distortion in % from the spectral figures in dB, such as those just illustrated? The fundamental formula is dB = 20log(x), so "x" would be our harmonic distortion coefficient H (but not in % yet!) Remember, if the base of the logarithm is not specified, it is assumed to be 10, so log(x) means $\log_{10}(x)$. Therefore $x=10^{(dB/20)}$.

Let's say the 2nd harmonic H_2 is 40 dB below the zero level of the 1st harmonic H_1. We have $H_2 = 10^{(dB/20)} = 10^{(-40/20)} = 10^{(-2)} = 0.01$ To convert that figure into percentages we simply multiply it by 100 and get $H_2 = 1.0$ % .

One way to remember this easily is that -20 dB means 10% distortion, -40dB equals 1% distortion, -60dB corresponds to 0.1% distortion, or, in other words, for every 20 dB drop, the distortion reduces by the factor of 10! What if, for some reason, the amplitude of the 1st harmonic is not 0dB but some other figure, such as -4 dB? Well, this is where the beauty of decibels comes into play. Since the dB is a relative unit, simply subtract one figure from the other to get a relative difference between the two harmonics.

Let's say H_2 is 40 dB below the -4 dB level of H_1. Since all harmonics must be referenced to the fundamental's level we have $H_2 = H_2 - H_1 = -40-(-4) = -40+4 = -36$ dB or $H_2 = 10^{(dB/20)}*100$ [%] $= 10^{(-36/20)}*100$ [%] = 1.585 %.

Calculating THD from individual harmonics

If the amplitudes of the individual harmonics are known (in Volts), the overall THD figure can be calculated as **THD [%] = $\sqrt{}(H_2^2+H_3^2+...+H_N^2)/H_1$ *100%**

How to calculate THD from individual harmonics

THD [%] = $\sqrt{}(H_2^2+H_3^2+...+H_N^2)/H_1$ *100%

Let's calculate THD figure for our SE 2A3 amplifier at the output level of 3 Watts (from the illustration on the previous page). The fundamental H_1 is at 0dB, the 2nd harmonic H_2 is at -29dB, the 3rd H_3 at -50 and the 4th H_4 at -70 dB.

Since -29dB=20log(H_2/H_1) we get H_2/H_1=0.03548. Since -50=20log(H_3/H_1) we get H_3/H_1=0.003162, and H_4/H_1=0.00031623. Our harmonics are already expressed as percentages of H_1, so the square root of H_1^2 in the formula cancels out H_1 in the denominator and we have

THD= $\sqrt{}(H_2^2+H_3^2+...+H_N^2)$*100% if H_2 to H_N are in % of H_1. In our case THD [%] = $\sqrt{}(0.03548^2+ 0.003162^2+ 0.00031623^2)$ *100% = $\sqrt{}(3.548^2+0.3162^2+0.031623^2)$*100% = 3.562%

MEASURING INDIVIDUAL HARMONICS WITH A SELECTIVE VOLTMETER (WAVE METER)

Since there is only a discrete number of distortion harmonics whose amplitude rapidly diminishes with frequency, it is feasible to measure them individually using a selective frequency voltmeter (also called wavemeter). The same test could be performed with a separate tunable filter and a broad range electronic AC millivoltmeter, but frequency-selective voltmeters (FSVM) incorporate both functions in one compact instrument. Also, the selective voltmeters' tunable filters are generally sharper and narrower than those in general-purpose filters, so their measurement results are more accurate.

THD measurements are done at 1 Watt power levels and sometimes at the full rated output power. Except at very high power levels where an amplifier is driven into severe overload, only the first four or five harmonics are of interest.

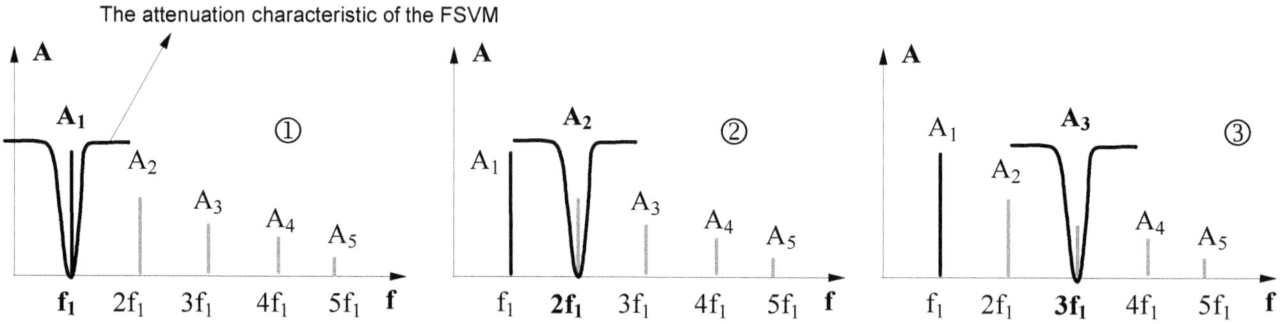

ABOVE: Test steps for measuring the amplitude of individual harmonics A_1 to A_5 (only the first three steps are shown)

Rycom 3111A selective voltmeter

Rycom or Railway Communications, Inc. was one of the smaller manufacturers of communication T&M equipment. Model 3111A is a relatively low frequency-selective voltmeter for signals between 200Hz and 150kHz. This solid-state instrument (bipolar transistors) is based on a single conversion superheterodyne circuit with a fixed bandpass filter of very narrow 25Hz bandwidth. It is particularly useful for analyzing closely spaced signals as in harmonic and IM distortion tests. Most other selective voltmeters were designed for high-frequency signals (in MHz) and aren't suitable for audio tests since their filters' bandwidth is much wider, say 500Hz or even 2.5kHz.

The input signal is attenuated to the proper value by the attenuator's 10dB steps to control the instrument's sensitivity from -80 to +20 dBm (100dB range).

The signal is then heterodyned with the local oscillator signal in a mixer to produce a difference frequency. The IF amplifier, tuned to that frequency, amplifies the mixer output. The signal is demodulated to 250 kHz by a shunt type demodulator and variable oscillator, detected (rectified), filtered in a meter detector circuit, and fed into a 200mA moving coil meter.

The meter has one dBm and two voltage scales (0-3V_{RMS} and 0-10V_{RMS}). An internal oscillator and calibrating circuit are provided so calibration can be made or checked as desired.

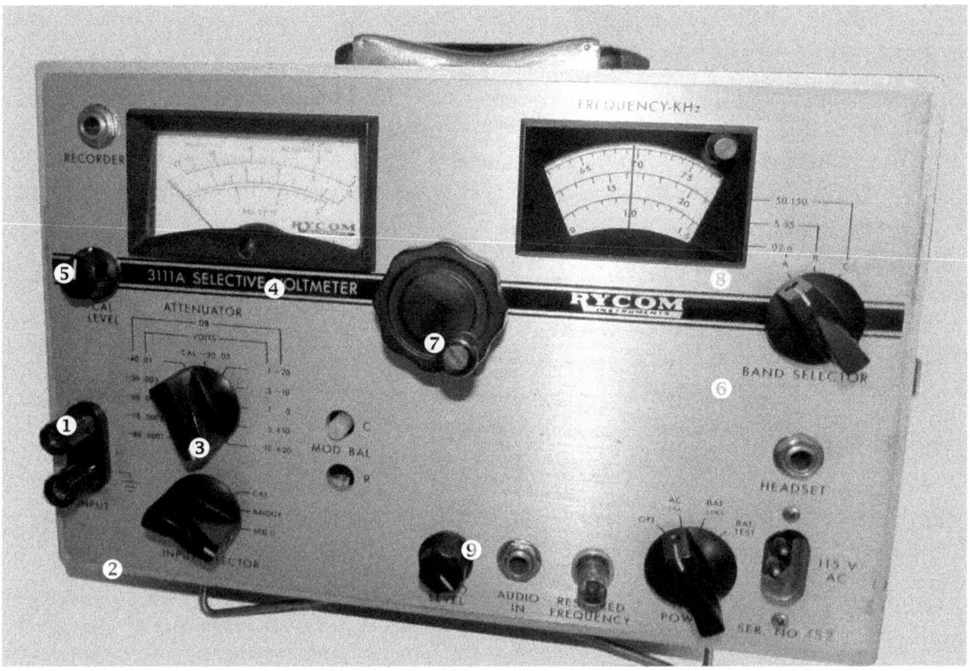

LEFT: The controls of Rycom 3111A Selective Voltmeter

1) Input terminals (binding posts)
2) CAL-Bridge-600Ω input selector
3) Input signal attenuator
4) Analog voltmeter
5) Calibration level adjustment
6) Frequency band selector (selected range 200Hz - 6 kHz)
7) Frequency tuning knob
8) Tuning dial (frequency indicator)
9) Audio volume level for the speaker or headset (if plugged in)

To compare the results of the THD measurement (page 109), the individual harmonics of the same amplifier and at the same power level (1 Watt) were measured using Rycom 3111A with following results: A_1=1.3V, A_2=5.3mV, A_3=0.9mV, and A_4=0.08mV. Thus, using milliVolts, THD [%] = $\sqrt{(A_2^2+A_3^2+ ...+A_N^2)}/A_1*100 = \sqrt{(5.3^2+0.9^2+0.08^2)}/1,300*100 = \sqrt{28.9}/1,300*100$ = 5.38/1,300*100 = 0.414%, pretty close to the 0.39% figure obtained by the THD meter.

Since the 3rd and 4th harmonics' amplitudes, 0.9mV and 0.08mV, contribute very little to the overall THD calculation, the main error here comes from the first two readouts on the analog voltmeter's scale. The main aim of this exercise was to demonstrate the validity of this approach and the value of such a 40+ years old instrument. In the 1970s & 80s, these voltmeters cost thousands of dollars but now are priced in hundreds, and often even lower, which is incredible considering their quality construction and first class components used.

The use of dBm units is very simple. In the same amplifier test, the 2nd harmonic A_2 was detected by turning the tuning knob to its frequency (2kHz, since the test signal frequency was 1 kHz). With the attenuator in the -40dB position, the meter indicated -3.3dBm, so by simply adding the we get A_2= -40 dB + (-3.3 dB) = -43.3 dB! Since the fundamental was 1.3V or -3.5dB on 0-3V scale (+10dB range), A_2 was 43.3+10-3.5= 49.8 dB below the fundamental A_1!

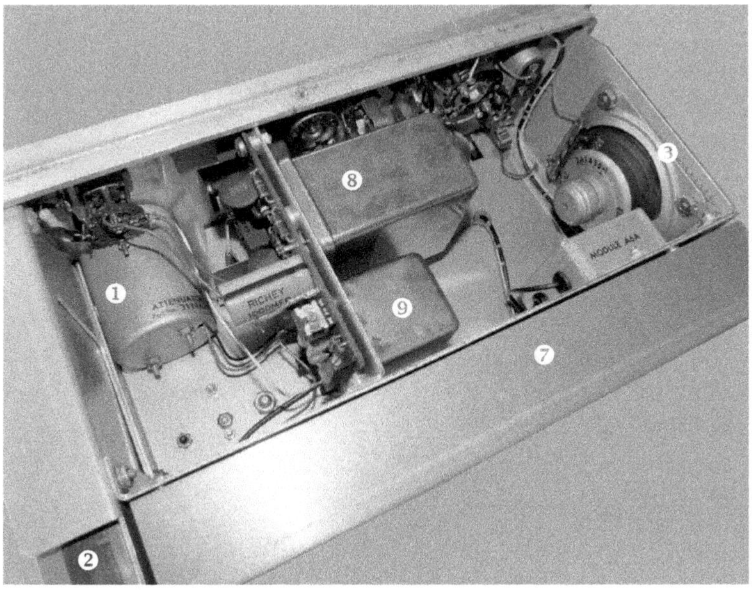

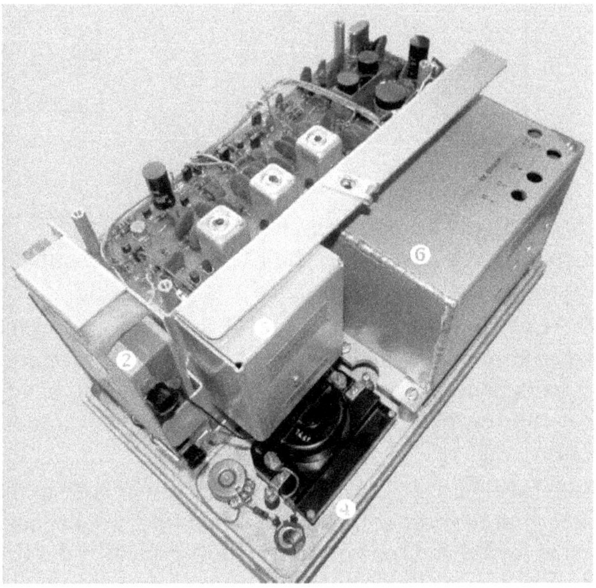

ABOVE: The internal views of Rycom 3111A Selective Voltmeter

1) Input signal attenuator
2) Rechargeable Ni-Cad battery 12V 1A
3) Loudspeaker (module A6A next to it)
4) Analog voltmeter
5) Power supply (module A5)
6) A3 module (local oscillator & modulator)
7) IF amplifier (cover removed on the second photo)
8) Low pass filter
9) Narrow bandpass filter (f_C=250kHz, BW=25Hz)

MEASURING INTERMODULATION DISTORTION

IM (Intermodulation) distortion meters (analyzers) - principle of operation

Intermodulation meters or analyzers mix two audio signals of different frequencies and feed that mix to the amplifier under test (the output of the IM analyzer). The amplifier's output voltage (across the dummy load) is then fed back into the analyzer's input and "analyzed." If the two signals interact and modulation happens, the meter indicates the IM distortion on a calibrated meter. Modulation means that two signals create other unwanted or distortion signals or harmonics at the amp's output that were not present at its input.

The SMPTE method uses a moderately high HF signal, usually 6 or 7kHz, while the LF signal is simply the mains voltage of 50 or 60Hz and the required amplitude (four times larger than the HF signal), usually taken from a secondary winding of the mains transformer (to save money on the second oscillator that would otherwise be required). German DIN standard uses slightly different frequencies, 250Hz, and 8kHz, for which two oscillators are needed.

The European way (which later became the international CCIF standard) is to mix two signals of the same amplitude and close frequencies, 14 and 15 kHz. The lower sideband f_2-f_1=1kHz then falls in the center of the amplifier's midrange, while the second sideband $2f_1$-f_2 = 13kHz is in its treble range.

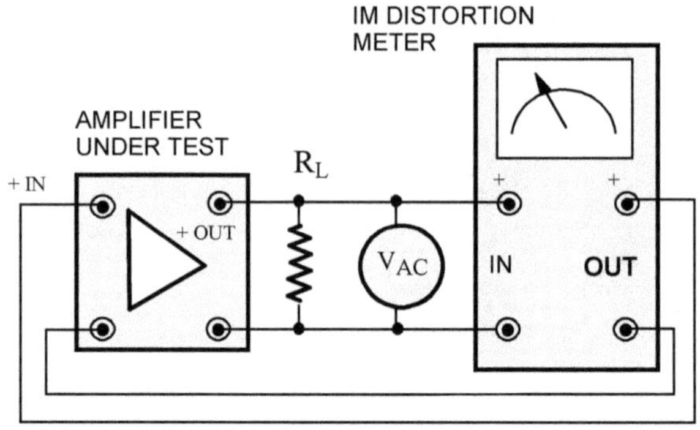

ABOVE: The IM test setup is very simple, the AC voltmeter to measure the output voltage confirms at what output power was the measurement taken.

BELOW: Two different methods of IM measurements, SMPTE (and DIN) versus CCIF

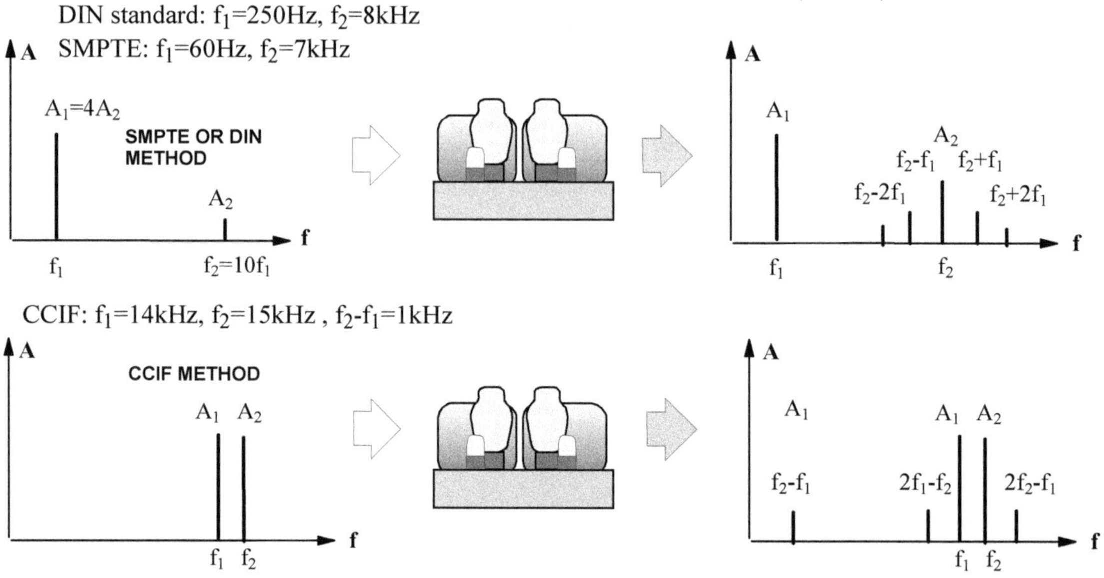

The IM distortion measured by the SMPTE method is encountered primarily by the LF signal because its amplitude is much higher than the HF signal.

The CCIF method measures IM distortion in the treble region of the audio band, and its results will be different from the low-frequency-focused SMPTE method. If you have equipment that can use both methods, it would be useful to compare their results.

Both IM figures (CCIF and SMPTE) seem related to the harmonic distortion figures. For instance, when single-ended amplifiers are tested for IM and harmonic distortion, typically, the IM figures by the SMPTE method are 3.2 times higher than the THD figure, while the CCIT test results in IM distortion percentage about half of the THD figure.

If, for instance, THD=1%, then IM_{SMPTE}=3.2% and IM_{CCIT}=0.5%! However, for some reason, these relationships do *not* apply to push-pull amplifiers!

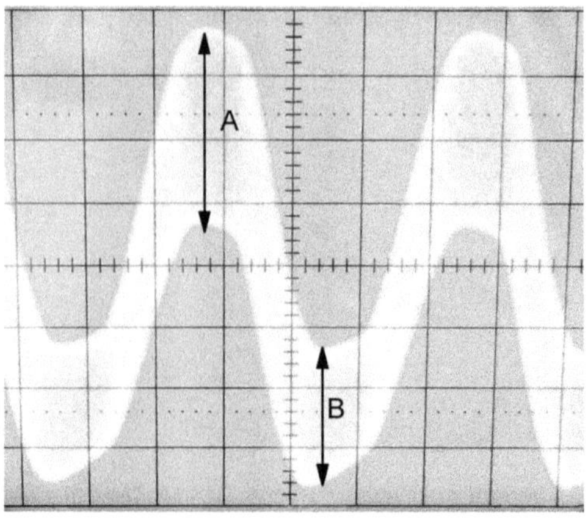

ABOVE: Distorted modulated signal at amplifier's output observed on an oscilloscope (SMPTE method). Notice how at points A and B the modulated wave thickened or "stretched". At the input of the amplifier, the amplitude was constant and equal, meaning there was no IM distortion.

How IM distortion meters work (a more detailed look at the SMPTE method)

The low- and high-frequency signals from the output of the IM analyzer are fed to LF-HF potentiometer control and mixed. The HF source has its level control, so the required 4:1 ratio of LF to HF signal amplitudes can be adjusted. The mix is then fed into the input of the amplifier under test. The amplified output is fed back into the IM analyzer. Firstly, the low-frequency component is removed, but the amplitude variations caused by intermodulation distortion remain (point A). As illustrated in point B, a diode rectifies this signal (removes the negative or bottom half). This rectified signal is fed into a low-pass filter which removes all high-frequency components, as in point C.

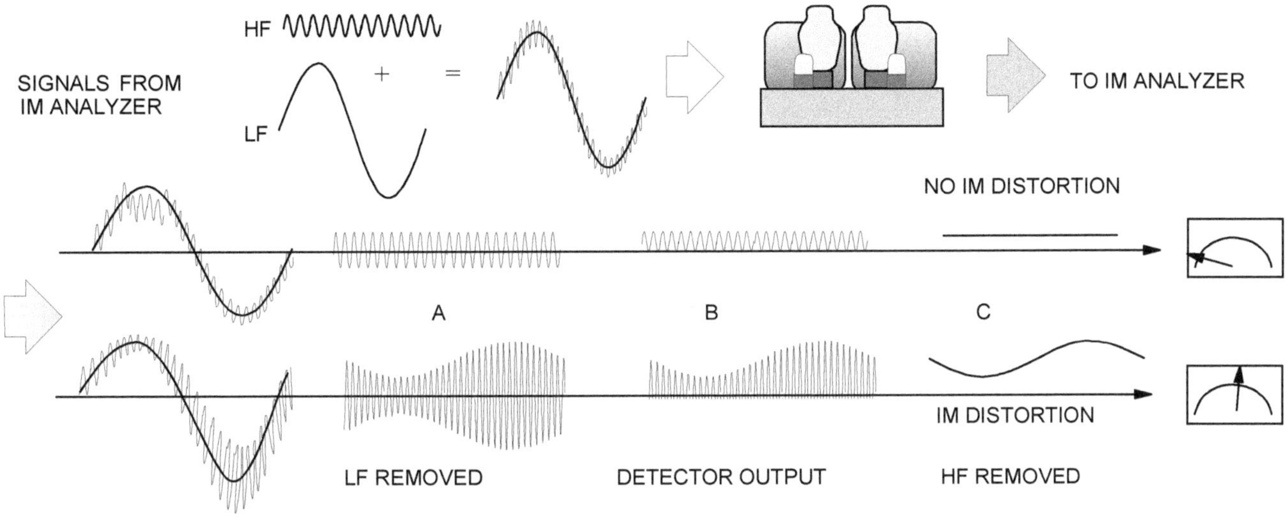

ABOVE: Waveforms in significant points of the IM distortion analyzer (Heathkit AA-1, SMPTE method)

BELOW: Functional block diagram of Heathkit AA-1 audio analyzer

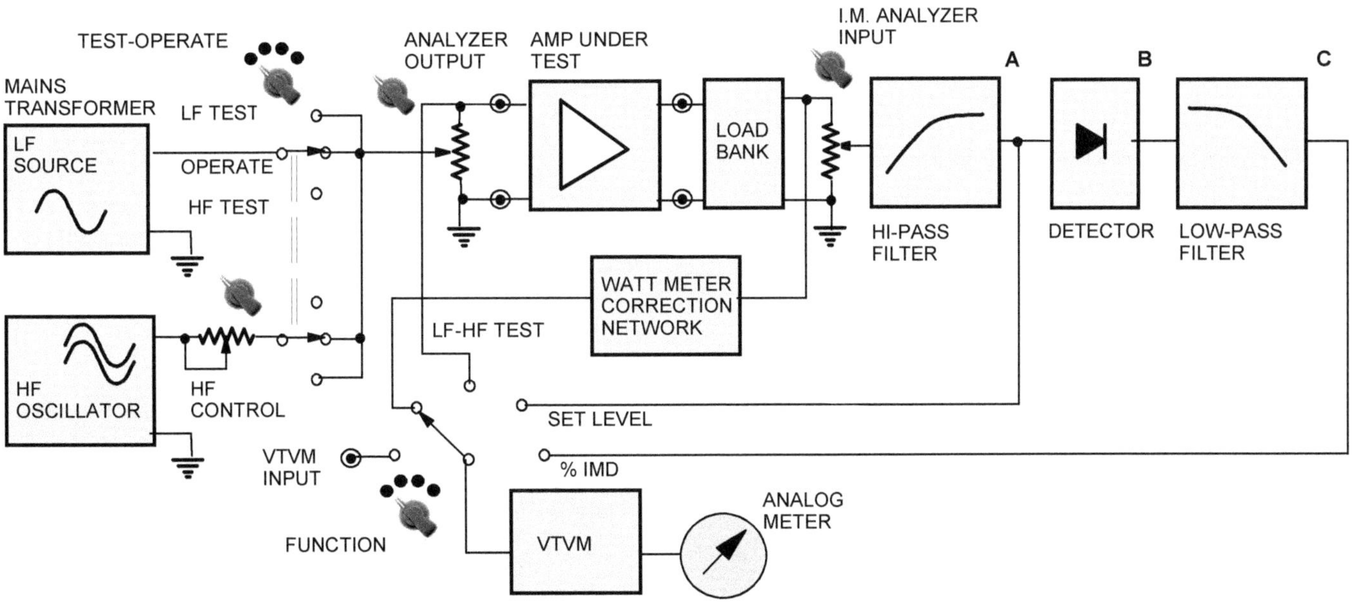

This signal is fed into the AC VTVM (vacuum-tube voltmeter) and measured. Without IM distortion, the signal in point C would be a steady DC level without any AC component so that the VTVM would show zero % IM distortion. If IM distortion artifacts are present, the VTVM would amplify them and drive the analog meter. The IM scale is marked in % of the high-frequency signal present at the output of the high pass filter, point A.

Heathkit AA-1audio analyzer

Heathkit AA-1 audio analyzer is a dummy load (4, 8, and 16Ω ceramic resistors), a vacuum-tube AC voltmeter, a power meter (albeit a primitive one, which does not include the cosϕ factor of the load, so it displays VA or apparent power, not Watts!) and an intermodulation distortion meter. The later version is identical inside, but the name was changed to IM-48. There are other similar Heathkit instruments; IM-1 and IM-22 are IM analyzers, and IM-58 is a THD meter.

The 6 kHz test signal is generated by its Hartley tube oscillator. Since the lower test frequency is the mains frequency that changes with location, you need to calibrate this instrument after purchasing it from a country with a different mains frequency (60Hz in USA and 50Hz in Australia, for example).

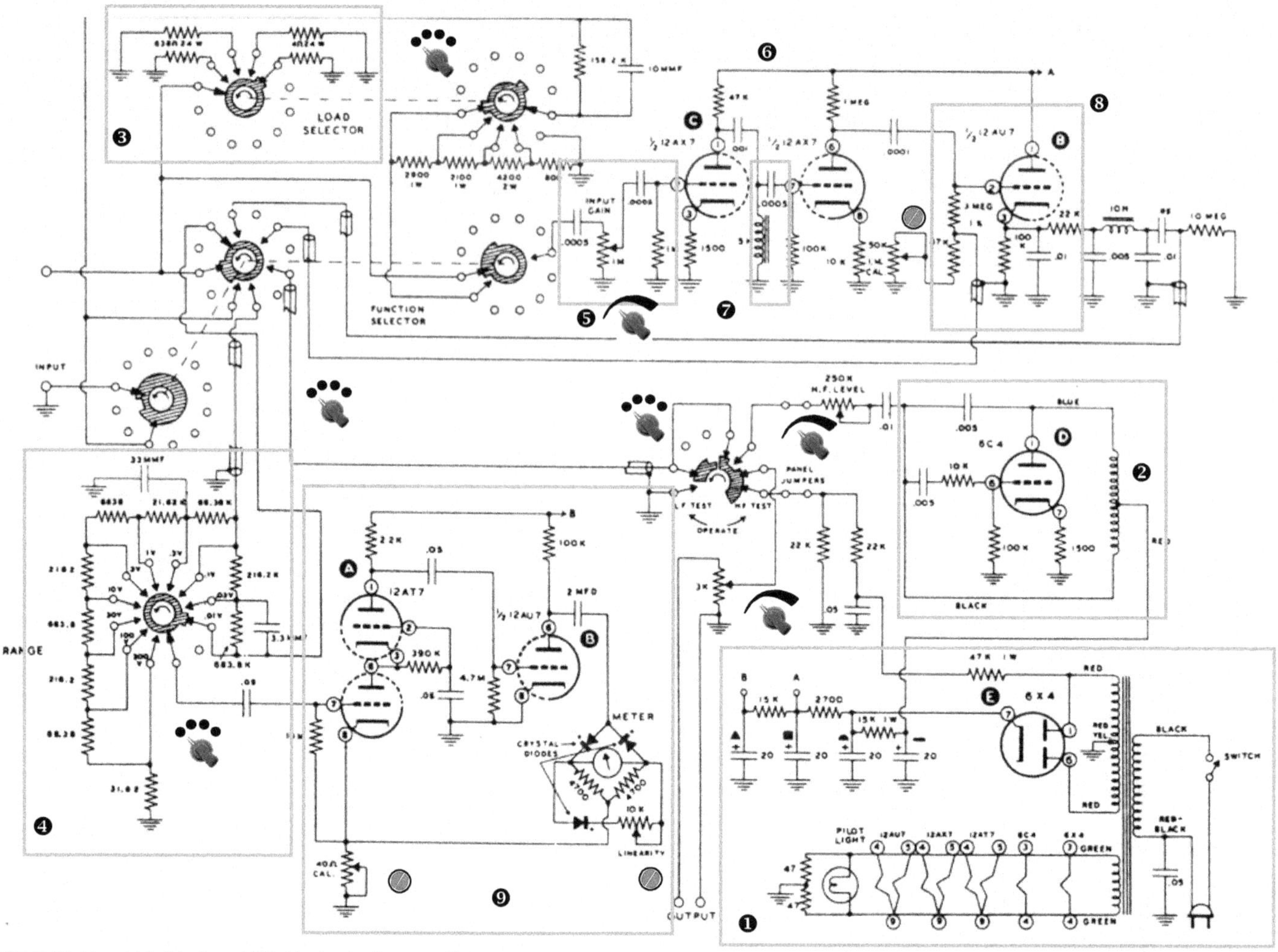

ABOVE: Heathkit AA-1 and IM-48 circuit diagram © Heathkit

The power supply (1) provides all DC voltages and heater AC supply. The Hartley HF oscillator (2) is easily identifiable on the diagram by its center-tapped coil. The internal load bank (3) has four ceramic disc resistors and a selector switch that selects one of them.

The input signal first passes through the "Range" voltage divider network (4) and the first passive high pass filter (5), a CRCR network. The first stage of the 12AX7 amplifier follows, with the 2nd high pass filter (7) between the two stages.

The detector or half-wave rectifier is a 12AU7 cathode follower (8), whose cathode output feeds the low pass CLC filter. The AC signal proportional to IM distortion % is then amplified by the internal vacuum tube voltmeter (9).

VTVM's first stage is a 12AT7 cascode, and the second stage drives the meter. The germanium diode bridge rectifier is placed in a negative feedback (NFB) loop back to the cathode of the cascode stage. This improves the linearity and accuracy of the instrument.

All in all, a very cleverly designed tube-based instrument that is easy to repair and refurbish. It can be used as an ordinary wideband vacuum tube voltmeter and an intermodulation distortion meter.

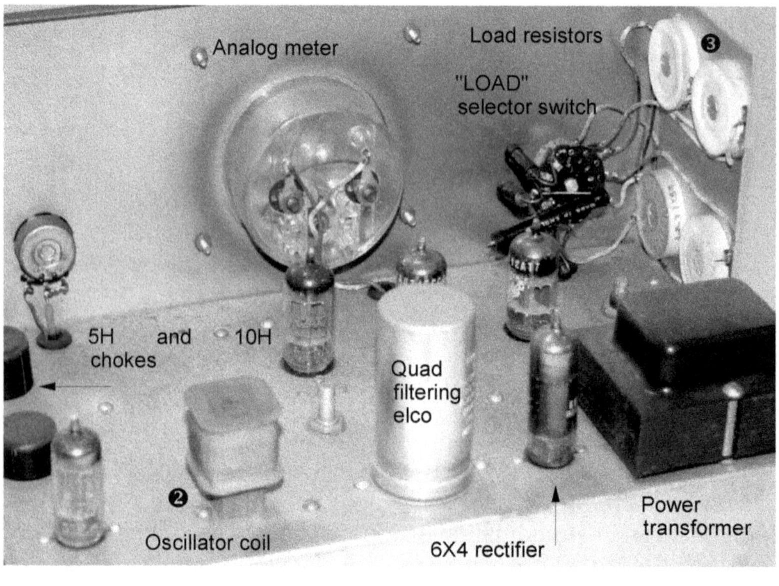

ABOVE: Top inside view of Heathkit AA-1. Most resistors and capacitors are under the chassis.

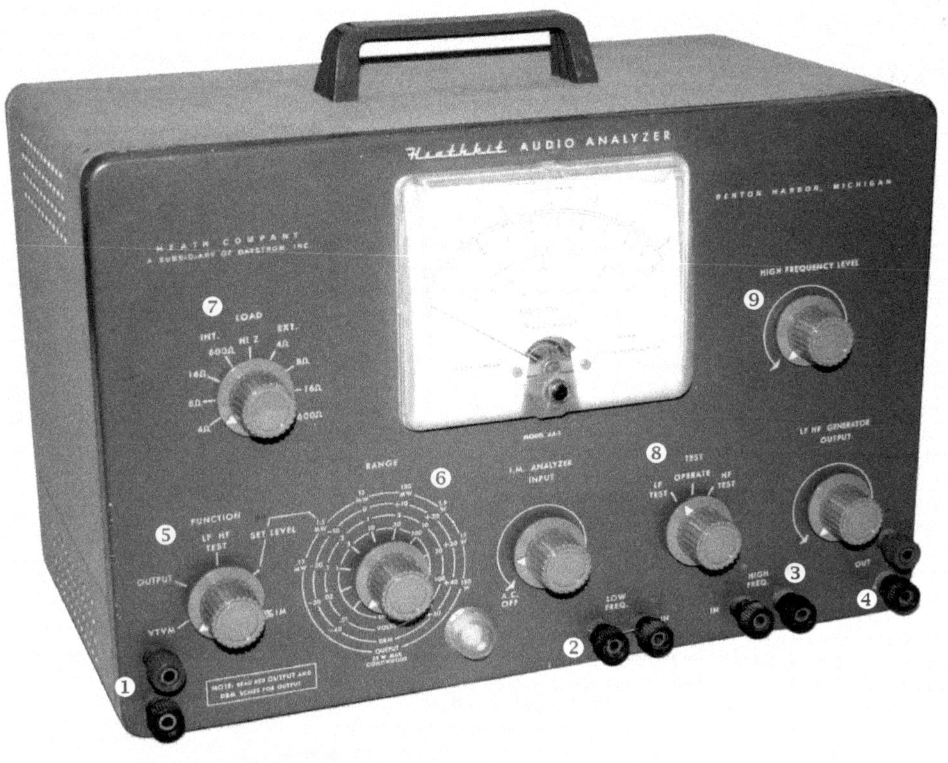

LEFT: Heathkit AA-1 audio analyzer looks ancient but is a very useful instrument, by no means obsolete.

1) Input (output signal from an amplifier under test)
2) External low frequency input
3) External high frequency input
4) LF & HF output
5) Function switch
6) Range selector
7) Load selector
8) Test switch (LF Test - Operate -HF Test)
9) HF amplitude adjustment

Eico 902

Released in 1963, EICO 902 can measure both THD and IM distortion or be used as a sensitive AC VTVM and dB meter. Due to its complexity and the required calibration, the 902 analyzer was only sold as factory-wired units. It cost a whopping US$250, a small fortune at that time!

Eico 902 uses a continuously adjustable 20Hz-20kHz Wien Bridge rejection filter for THD measurements. The overall accuracy is 5% on distortion measurements and 4% as a vacuum tube voltmeter.

For IM distortion measurement Eico 902 uses a 7kHz oscillator as the HF source and filtered line frequency signal as LF source. Front panel switches select ratios of LF to HF of 4:1 or 1:1.

For more precise measurements, external low-frequency sources up to 400Hz and high-frequency sources above 2kHz can be connected to the mixing bridge through the front panel binding posts. The residual distortion of the instrument is around 0.05%.

As an AC VTVM, the 902 has ranges from 10mV to 300V. The tube complement includes 12DW7, 12BY7, 2xECF80, EF86, 6D10, 6C4, OA2 and 6X4. The calibration is complex and tedious, so refurbishing and bringing the non-working unit up to acceptable accuracy is not for fainthearted or for beginners.

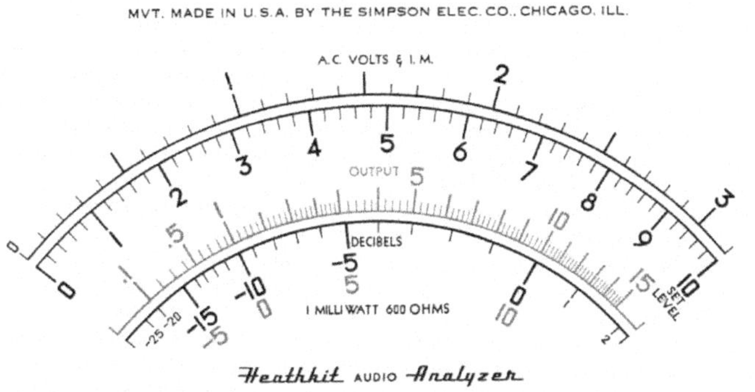

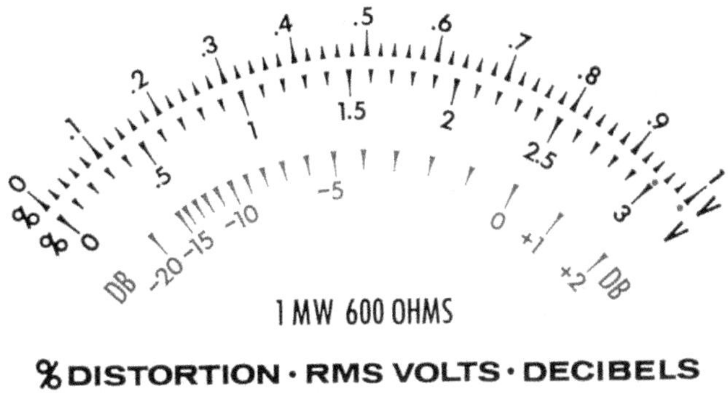

ABOVE: The scales of Heathkit AA-1 (LEFT) and Eico 902 (RIGHT) audio analyzers

How to measure intermodulation distortion without an IM analyzer

You can measure the IM distortion without a specialized instrument such as an AA-1 audio analyzer. You'll need two AC signal sources, say two function generators, and a mixing network (a resistive bridge as pictured). Instead of a second RC oscillator or function generator, you can use a small mains transformer with low voltage secondary (a few Volts) as the LF (low frequency) source. Notice that the LF source's output must not be grounded; it must be fully floating.

A high pass filter is also needed to filter out the LF from the tested amplifier's output, which is displayed on the oscilloscope. You can calculate the IM distortion from the typical waveform of a modulated signal.

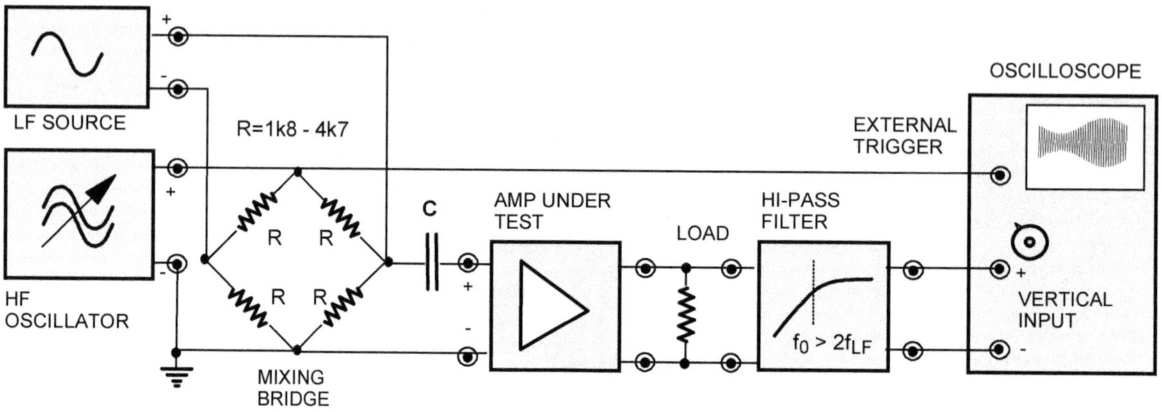

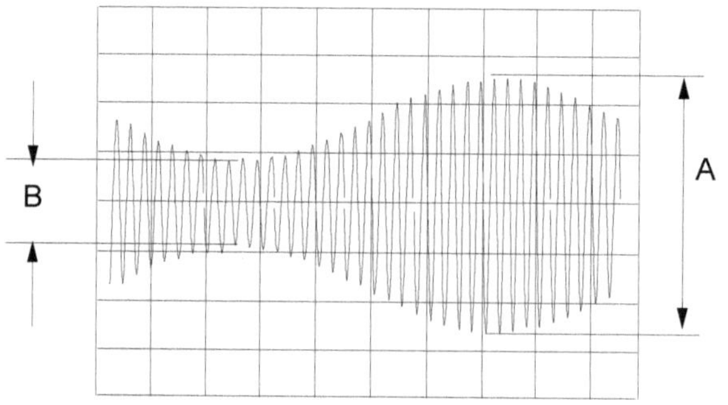

IMD = (A-B)/(A+B) * 100 [%]

ABOVE: IM distortion setup using two signal sources, a mixing bridge, high-pass filter and an oscilloscope. The LF source could be a mains (power) transformer, in which case only one function generator is needed (HF oscillator).

LEFT: How to determine IM distortion from the AM (Amplitude Modulated) signal on oscilloscope's screen, as per the test setup above.

8 TRANSFORMER TESTS & MEASUREMENTS

- IDENTIFYING AND TESTING UNKNOWN POWER AND AUDIO TRANSFORMERS
- TESTING MAINS (POWER) TRANSFORMERS
- TESTING POWER SUPPLY CHOKES (INDUCTORS)
- TESTING AUDIO (INPUT, OUTPUT AND INTERSTAGE) TRANSFORMERS
- USING AN OSCILLOSCOPE TO EVALUATE THE FIDELITY OF AUDIO TRANSFORMERS
- EVALUATING COMMERCIAL OUTPUT TRANSFORMERS BASED ON THEIR PUBLISHED OR MEASURED SPECIFICATIONS

FURTHER READING

For a more detailed, in-depth look at the operation, design and construction of electronic transformers, please consult my book "TRANSFORMERS FOR TUBE AMPLIFIERS - HOW TO DESIGN, CONSTRUCT & USE POWER, OUTPUT & INTERSTAGE TRANSFORMERS AND CHOKES IN AUDIOPHILE AND GUITAR TUBE AMPLIFIERS"

ISBN 978-0-9806223-8-6

It's available on Amazon, Book Depository, Barnes & Noble, and all good online bookstores.

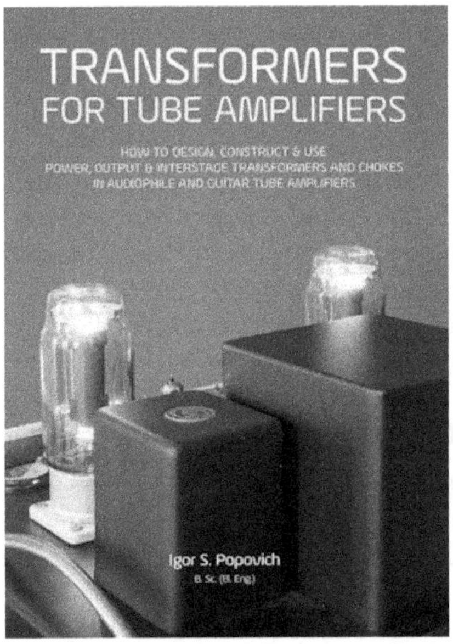

IDENTIFYING AND TESTING UNKNOWN POWER AND AUDIO TRANSFORMERS

Power transformers can be considered a special case of audio transformers that operate on a single frequency - the mains frequency, which is either 50Hz or 60Hz. Since tube amplifiers' output transformers are always of a step-down type, their primary voltages are always high, in the order of magnitude of mains voltages (100V to 240V, depending on a country).

It is important to know how a transformer (both audio and power types) will behave with an increasing load (the output power of a class AB amplifier raising, for instance) and how hot it gets in normal operation. A transformer must be properly sized for the load, both in magnetic (the core size) and electric terms (the wire sizes). Efficiency is not of special interest in hi-fi unless power losses are extremely high, which points to other problems in the design.

How to identify windings of an unknown power transformer

A power transformer for a tube amplifier or preamplifier will have at least three separate windings, primary, high voltage secondary and low voltage heater winding. The rule for identifying transformer windings is: the lower the voltage, the higher the current, the thicker the wire, the lower the resistance of the winding. The heater winding will have the thickest wire and the lowest winding resistance (say 0.3-3Ω). The primary winding will have an order of magnitude higher resistance (say 10Ω), and the high voltage winding order of magnitude higher again (1:10:100Ω).

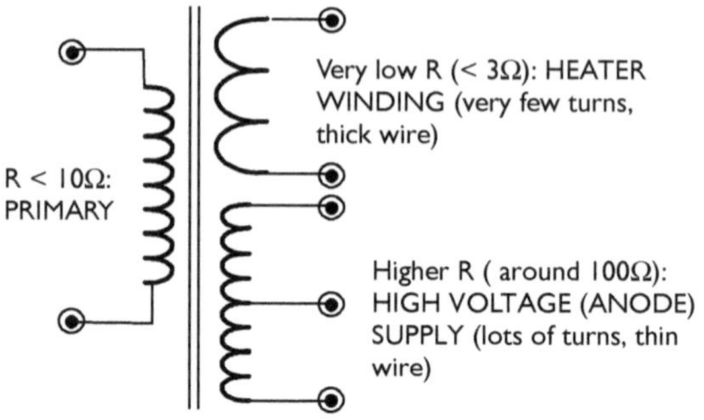

ABOVE: A typical power transformer for a tube amplifier or preamplifier

How to determine the turns & impedance ratios of mains and audio transformers

You have a salvaged or unmarked output transformer and want to determine its turns ratio(s). First, you must measure DC resistance between wires and determine the transformer's wiring diagram. Next, connect its primary to the mains outlet and measure the primary and secondary voltages with an AC voltmeter or multimeter on AC volts. Do not touch any wires!

In step three, calculate voltage ratios and square them to get impedance ratios (for audio transformers only). Finally, multiply the nominal secondary loads (4, 8, and 16 Ω) by their impedance ratios to get their reflected primary impedances.

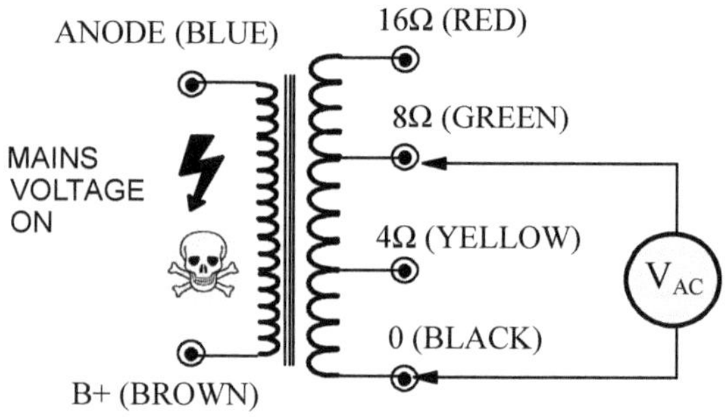

ABOVE RIGHT: Using the mains voltage on output transformer's primary to measure its secondary voltages. Note: The most commonly used color-coded wire insulation scheme is indicated, but that is not always the case, the wire colors may be different on a particular transformer!

EXPERIMENT: Turns & impedance ratios through windings' inductances versus the AC voltage test

The user manual of a certain LCR meter claims that an impedance ratio of a transformer can be determined (under no-load conditions) by measuring primary inductance L1 and secondary inductance L2 and then dividing L1 by L2. We performed this test on various transformers. For example, at the 120Hz LCR meter test of one output transformer, we had L1=9.72H and L2=23.88mH, so according to that formula, IR should be 407!

However, using a live AC test with 250.8V mains voltage (50Hz) on the primary, there was 10.10V on the secondary, so the turns (and voltage) ratio was VR = TR = 24.83, and the impedance ratio IR = TR^2 = 616! Since that was a brand new transformer sold as 2k5 to 4Ω impedance (IR=626), the AC voltage test method was the accurate one, while the inductance method resulted in an error of ε =(407-616)/616 = 0.3393 or -33.93%! This assumes our result of IR=616 was accurate, not IR=626 declared by the manufacturer. We've had many cases where the actual IR varied significantly from that declared by the transformer maker.

Determining TPV (Turns-Per-Volt) visually

If a power transformer had its heater secondary wound last (topmost), which has been the most common coil arrangement over the years, it is possible to carefully open the final insulation layer (always paper for older and plastic for newer transformers) and count the number of windings. This makes it easy to determine TPV used by its designer simply by visual means.

This vintage German power transformer from a small tube amplifier didn't even have such a top insulation layer, making the heater winding fully visible. There are 36 turns for 6.8V (measured unloaded heater voltage), so its TPV figure is TPV = 36/6.8 = 5.3 Turns-Per-Volt.

Determining TPV (Turns-Per-Volt) experimentally

Some transformers are encapsulated; others have been wound so tight that there is no clearance between the coil and the core. Neither the visual method nor this experimental test will work in those cases. Assuming there is a gap of 1mm or more between the coil and the laminations (1), feed through a piece of winding (magnet) wire, enough for 3-5 turns (four turns are illustrated below right). The wire can be any diameter, whatever you have on hand, but it must be insulated (lacquered). Connect an AC voltmeter or multimeter on the "AC Volts" function to the ends of this temporary winding.

If it's an output transformer of a step-down kind or a mains transformer, connect its primary to the mains voltage and note the reading of the voltmeter. Turns-per-volt used are easily calculated as $TPV=N/V_{AC}$, where N is the number of turns you have used for this test (in this example N=4), and V_{AC} is the measured voltage. Say you measured $1.25V_{AC}$; then TPV=4/1.25 = 3.2.

Alternatively, use a function generator (on sine voltage setting, 50-100Hz frequency) on one of the transformer's windings.

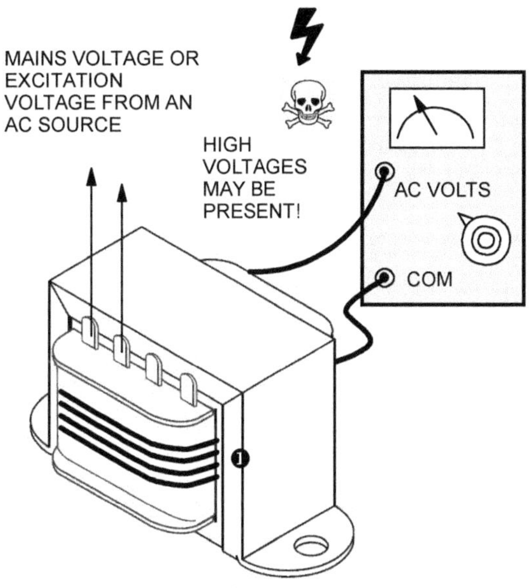

DANGER - TRANSFORMER TURNS-PER-VOLT TESTS INVOLVE LETHAL VOLTAGES!

Perform this test under extreme caution since high voltages will be present on some windings. Be careful where you connect the "excitation" voltage; if you connect, say $8V_{AC}$ from a function generator onto a 6.3V heater winding, you will get 240*8/6.3 = $305V_{AC}$ on the 240V primary!

TESTING MAINS (POWER) TRANSFORMERS

Measuring transformer's no-load current and power losses

"Bob's Black Box" (see page 25), a simple VA-meter, is not useful just for powering up and testing amplifiers and preamplifiers; by connecting unloaded mains or even output transformers to it, we can measure their no-load or magnetizing currents.

EXAMPLE: What is the power consumption in VA of this mains transformer for 300B amplifier without load (open secondaries) if the measured voltage drop on BBB is 0.527 V_{AC} and the measured mains voltage is 247 V_{AC}?

The magnetizing current is $I_M = 0.527V/10\Omega = 0.0527\ A_{AC}$ or 52.7 mA_{AC}.

The power draw, representing transformer losses in idle mode (with no load), is P=V*I = 247V*0.0527A = 13 VA!

If you are winding your own power transformer, connect it to BBB (without any load on the secondary) once it's finished If the magnetizing current is between 40 and 100mA (depending on the lamination quality and transformer's size) and there is no buzzing, leave it powered up overnight. If there is any problem such as insulation breakdown or overheating, the fuse in BBB will eventually blow. If the transformer is still idling happily the next morning, it is ready for installation into an amplifier!

Estimating the current capacity of the high voltage winding - Method #1

This quick & rough method works quite well for currents up to 200 mA and does not require any transformer loading.

In Step 1, with the primary disconnected from the mains (transformer de-energized), measure the DC resistance of the half of the secondary (assuming an HV secondary with CT).

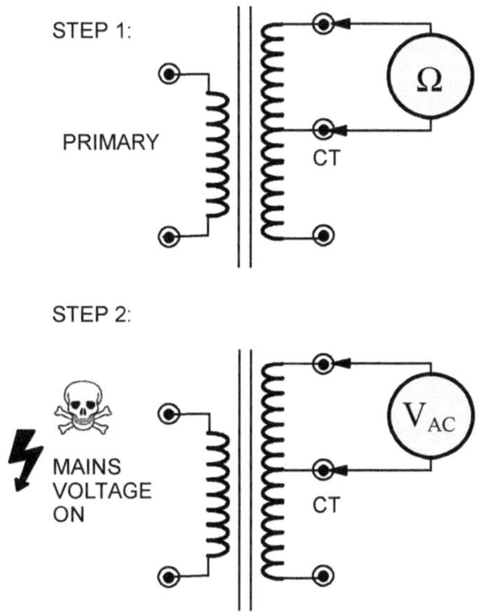

With one power transformer, we got 82.2Ω and 76.4Ω. Unless they are bifilar-wound, the two halves of the HV winding will not have equal DC resistances, but that's normal.

In step 2, energize the transformer from the mains and measure the AC voltage across the same half of the secondary. We measured $386V_{AC}$. Then, using the rough "rule-of-thumb" formula, $I_{MAX} = 25*V/R$ [mA] = 25*386/82.2 = 117 mA.

The method is also valid for tube amplifier power transformers without a center-tapped high voltage secondary winding.

Estimating the current capacity of the high voltage winding - Method #2

Once you provisionally estimate (from the diameter of the winding wire used) the rated secondary current, for example, 100mA, calculate the value of the load. You know the voltage output of the winding in question, 386V in this case, so $R_L=V_0/I_{EST}$ = 386/0.1 = 3,860Ω. Calculate the power dissipation on the load as $P=V_0^2/R_L = 386^2/3{,}860 = 38.6W$. We would need a 50W-rated 5kΩ rheostat R_X.

Lower the rheostat's resistance until the voltage under load drops 10% from the unloaded voltage level, in this case, down to $V_{90\%}=0.9*386$ = 348V. Power down the circuit, disconnect the rheostat and measure its resistance R_X (between the slider and the end B). The maximum current that winding can supply is then $I_{MAX}=V_{90\%}/R_X$. Say R_X = 3,520Ω, then I_{MAX}=348/3,020 = 115 mA. If you don't have a high-power rheostat, hook up 5-20W rated fixed resistors in series-parallel combinations until you get a 7-10% voltage drop.

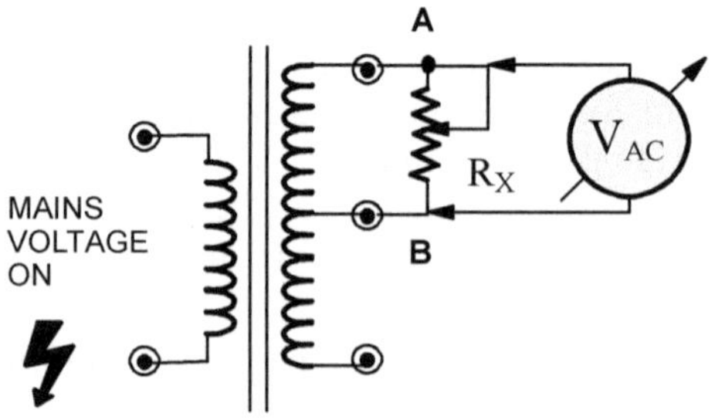

Transformer phasing check

In most applications, it is crucial to connect two or more secondaries of a transformer the right way so that their voltages are in phase. This is one way of determining the correct phasing.

Measure the two secondary voltages, V12 and V34. They may or may not be equal. Connect a jumper lead between points 1 and 3 and measure V24. If V24 is around zero (providing V12=V34), then points 1 and 3 are in phase. If V24 is approximately twice the voltages V13 and V24, then points 1 and 3 are out of phase, meaning that points 1 and 4 are in phase, and points 2 and 3 are in phase.

If the two voltages V12 and V324 are not equal, the measurements will not be zero and double, but the principle still applies. In the first measurement, instead of zero, we would get the difference between the higher voltage and lower voltage (a lower reading) and the sum of the two voltages in the second test (a higher reading). Although a mains transformer is shown, this method also applies to audio transformers. Instead of the mains, feed the primary audio frequency signal from a function generator.

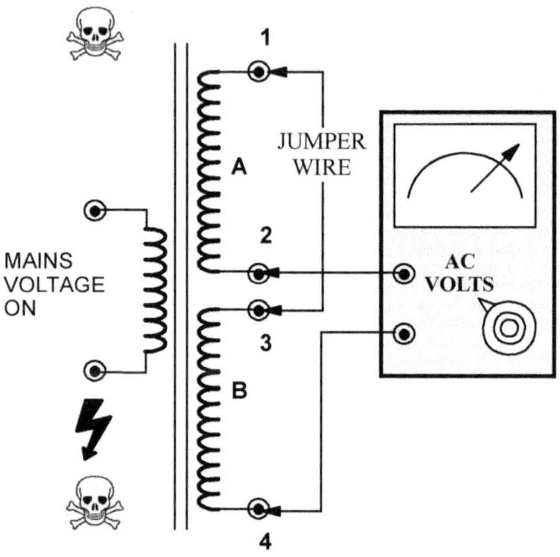

TESTING POWER SUPPLY CHOKES (INDUCTORS)

The inductance of a power supply filtering choke (or any choke designed to operate with a DC current flowing through it) is called an incremental inductance. Due to the presence of DC current, the permeability of the choke lamination stack is reduced, and so is its inductance. Thus, the actual inductance such a choke will have in a rectifier filter will be lower than the inductance measured by a digital LCR meter, which is inductance without any DC current.

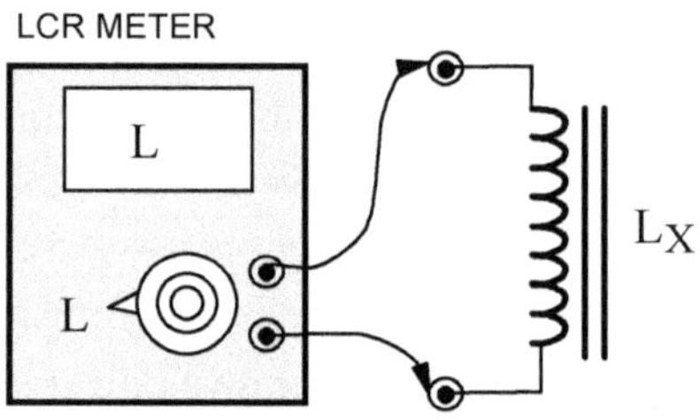

ABOVE: The inductance of a choke measured by a digital LCR meter

THE INCREMENTAL INDUCTANCE TESTER **DIY PROJECT**

Instead of using a complex and fiddly bridge test setup, this simple circuit determines incremental inductance by comparison of the AC voltage drop across the choke and the DC voltage drop across a resistor of a known resistance R. Another advantage of this circuit is that it is almost identical to the actual power supply a choke would be used in an amplifier. In this case half-wave rectification is used instead of full wave, but you can replace the single diode here with a diode bridge.

The same current flows through LX and R, so the voltage drops are proportionate to their impedances: $V_L/V_R=X_L/R$ and since $X_L=\omega L$ we have $L=V_L/V_R * R/\omega$!

Normally R would be a power rheostat (10-20W), used to adjust $V_L=V_R$, whose scale would be calibrated to read L directly. However, such rheostats are rare and expensive, so we used a trick, making $R=\omega$ or $R/\omega = 1$, thus simplifying the ratio to $L_X=V_L/V_R$!

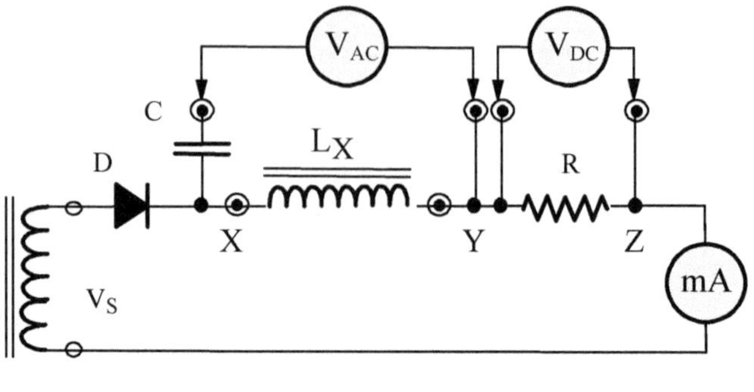

If an AC voltmeter is connected directly across the choke, it will measure a voltage drop proportionate to the choke's total impedance (including its ohmic resistance), not just its inductive reactance, introducing an error of a few %. A series capacitor is used to eliminate this error, which also separates the AC voltmeter from the DC circuit. The frequency of AC ripple for half-wave rectifiers is the same as the mains frequency f=50Hz, so $\omega=2\pi f=314$ and R=314Ω. For 60Hz countries R=377Ω! For full-wave rectification, R=ω= 628Ω, or for 60Hz countries, R=754Ω !

To determine the required secondary voltage, we must consider the fact that power supply filtering chokes have L of up to 20H, and DC resistance of between 10 and 200Ω, while their DC current ratings are usually below 200-250 mA. This means that the transformer's nominal secondary voltage VS. needs to be around 200 V_{AC} and have 200-250mA current capability. C should be 2.2μF/250V_{AC}, and the resistor R should have a power rating of 10 Watts.

Should you wish to make the secondary voltage adjustable, power the tester from a Variac. To illustrate typical test results, a choke with L=5H, tested on an LCR meter and a DC resistance of 85Ω, which was probably good for up to 150mA, was tested at three secondary voltages 43V, 65V, and 190V. The higher the secondary voltage, the higher the DC current as well.

The tabulated results show a drop in inductance from 4.7H at around 20mA to 3.7H at 110mA!

V_S [V_{AC}]	V_L [V_{AC}]	V_R [V_{DC}]	L [H]	I_{DC} [mA]
43	32	6.8	4.70	21.7
65	42	9.6	4.38	30.6
190	130	35	3.71	111.5

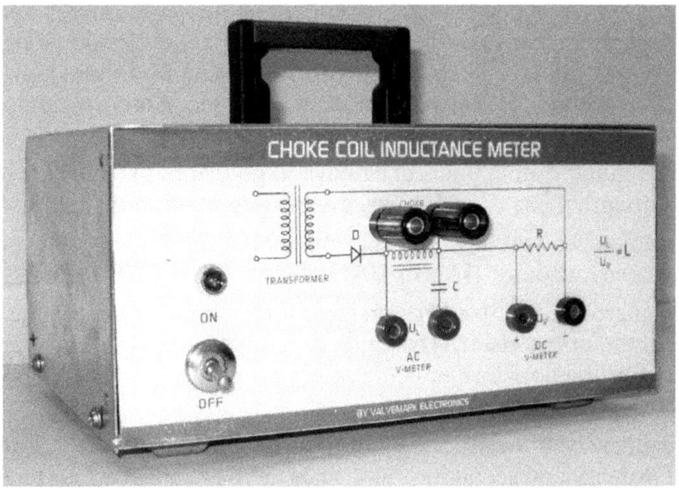

ABOVE: Test results of a 5H - 85Ω choke
RIGHT: Our prototype incremental inductance tester

TESTING AUDIO (INPUT, OUTPUT AND INTERSTAGE) TRANSFORMERS

Measuring transformer's primary impedance and DC resistance

When set on the "R" or resistance range, digital LCR meters measure AC resistance, not DC resistance. The AC resistance measured this way seems to be closer to overall impedance and will be very different from the DC resistance measured by a DMM.

So, if you terminate a transformer with its nominal load R_L (for output transformers usually 4, 8, or 16Ω) using a dummy (resistive) load, the "R" reading will be its reflected impedance onto the primary side.

Perform this test on both 120 and 1kHz frequencies of a typical LCR meter to see how much the two impedances differ due to the transformer's complex behavior and frequency-dependent nature of its parameters!

The measurement of primary and secondary DC resistances is trivial, using an analog or digital multimeter set on "Ohms." The primary DC resistance is of no particular importance, except to calculate the DC voltage drop across the primary inside an amplifier and the actual anode-to-cathode voltage of the power tube, to determine the DC operating conditions and the required bias voltage.

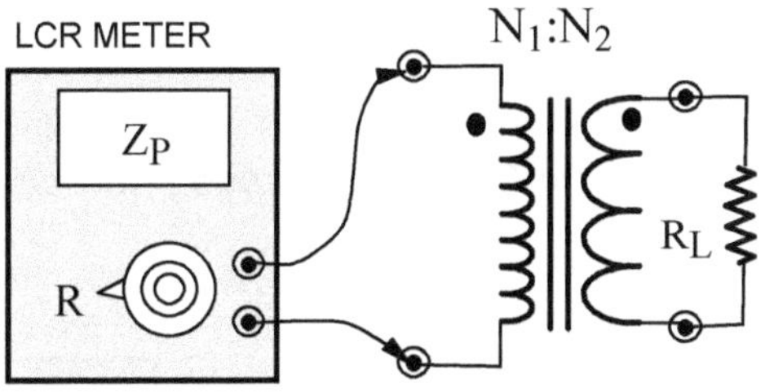

ABOVE: The primary impedance is measured with secondary loaded by its nominal load

BELOW: measuring primary DC resistance with an ohmmeter

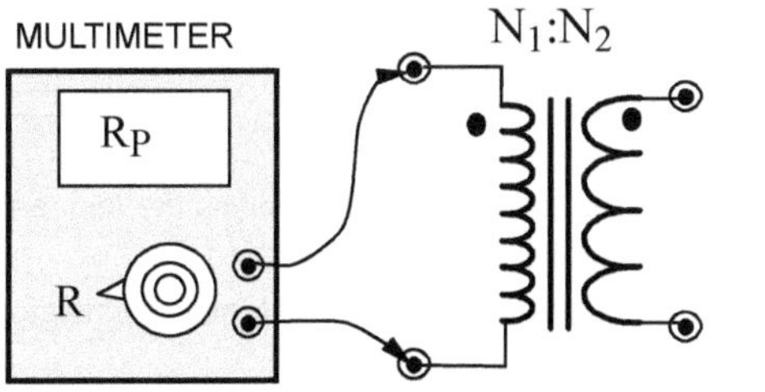

Measuring transformer's primary and leakage inductance

In the first approximation, the easiest and fastest way to measure both the primary inductance and the leakage inductance is with a digital LCR meter. Since inductance changes if a DC current flows through the primary (as in single-ended output transformers), a more precise way to measure these parameters would be by using an LCR bridge.

Better quality ones have the capability of adjusting the DC current through the measured coil. For estimation purposes the results without any DC flowing are close enough.

BELOW: The primary inductance is measured with open secondary winding(s).

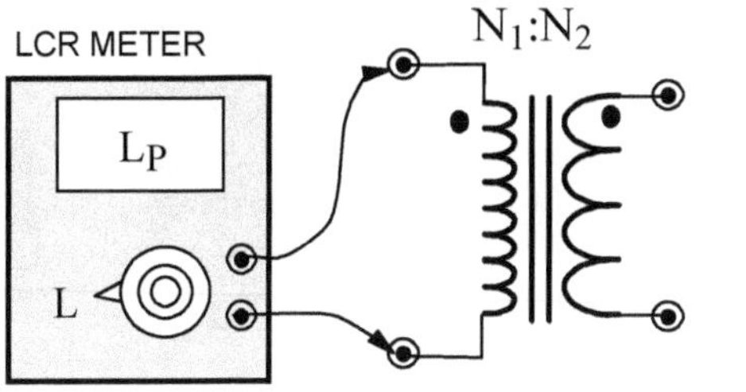

The primary inductance should be measured at a lower frequency such as 120 Hz since its importance is only to the bass frequencies in the audio range. This measurement is performed without any load on the secondary (open secondary).

The leakage inductance should be measured at a frequency as high as possible, which is only 1kHz in most LCR meters, but some models have a ten and 100kHz measurement option as well. Secondary terminals are shorted for this measurement.

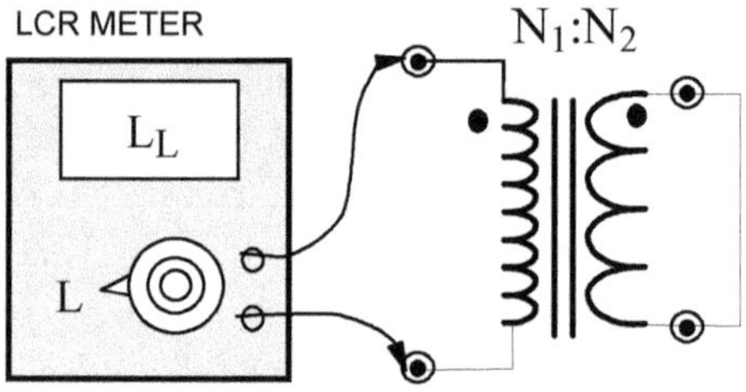

ABOVE: The leakage inductance is measured with shorted secondary winding(s).

Measuring interwinding capacitances

Transformer capacitances are important for hi-fi output and interstage transformers because these need to have an extended upper-frequency limit, way above 20 kHz, and parasitic capacitances shunt high frequencies to the ground or between primary and secondary windings, reducing transformers' upper-frequency limits.

Since guitar output amps don't have to go that high (8 kHz is considered a minimum), these capacitances are less important in that application.

Capacitances should also be measured at the highest possible test frequency, usually 1kHz or 10kHz for handheld LCR meters.

Using crocodile clips, perform these tests without touching the test leads or the transformer itself. The capacitance between your body and the LCR meter or the transformer could affect the results.

The capacitance of the primary is measured between its ends, as is the capacitance of the secondary. Of more interest is the capacitance between the primary and secondary, and since there are at least four terminals, there are four possible measurements. The results are very close in most cases, so you don't have to perform all four tests.

The capacitance between the primary/secondary winding and the magnetic core or the metal frame or case (considered "ground" since they are usually earthed) is measured similarly. The two ends of a winding may have different capacitance to the ground, so you may need to measure both.

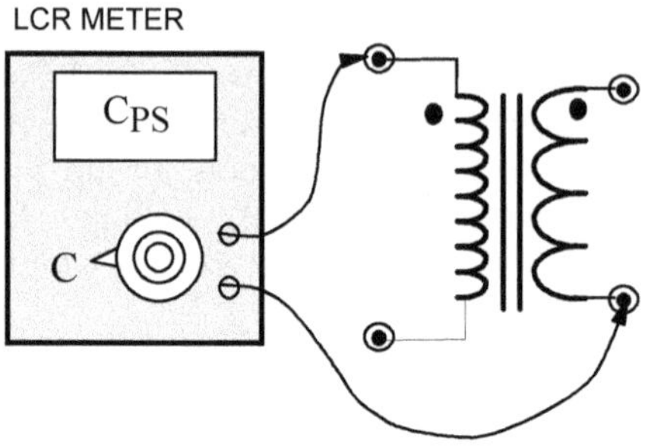

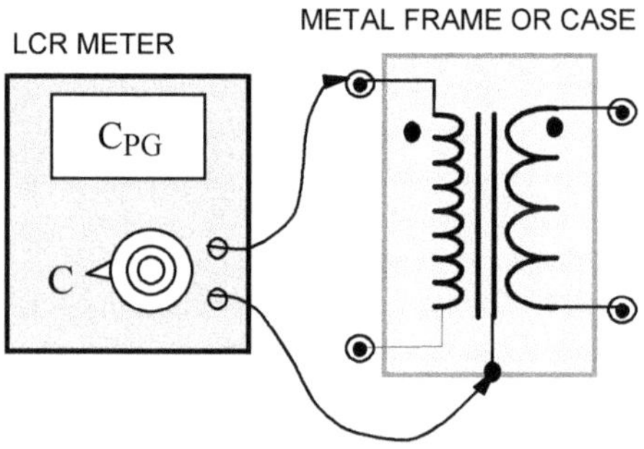

ABOVE: The primary-to-secondary and primary-to-ground (metal case or core, often bound together) capacitance measurements

Testing the frequency range and voltage ratio of MC (moving coil) input transformers

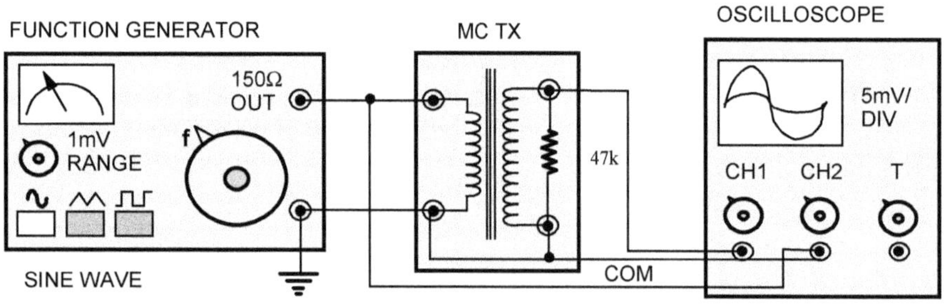

ABOVE: The test setup for measuring the frequency range and voltage ratio of MC (moving coil) input transformers

BELOW: The four microphone/MC step-up transformers used in the experiment

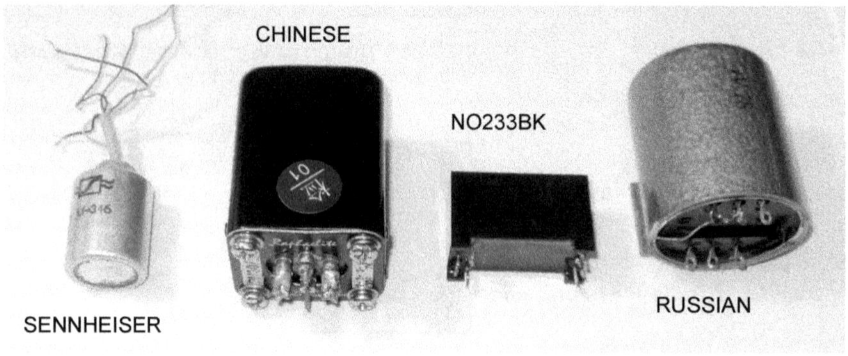

In this experiment, we tested four input transformers (for microphones or moving coil phono cartridges). Notice the apparent relative size difference. The rectangular-shaped transformer (NO233BK) was not shielded, so it is most likely the same size as the Chinese and Russian transformers, which are shielded in large cans.

Most oscilloscopes' vertical sensitivity goes down to only 5 mV/division, and if a 0.1mV signal is used for this test, it will not be possible to measure the signal amplitudes on the screen (the amplitude will be too small).

So, although MC cartridges produce 0.1 mV order of magnitude signals, it is more practical to perform this measurement at 1 mV input level (10x higher).

Since the Russian transformer had a CT primary, the measurements were taken between terminals 3 and 6, so they refer to the whole primary winding.

Columns C_{PS} A and B refer to measurements of the capacitance between the primary and secondary windings taken between different ends of the primary and one end of the secondary winding.

The primary-to-secondary parasitic capacitance of the Russian transformers is roughly 1/2 of the Chinese ones. The leakage inductance of the Russian TXs is 9X lower! The primary inductance at 120Hz (important for bass reproduction) of the Russian TXs is roughly 15X higher! Despite all that, which would predict that the Russian transformer would have a much wider frequency range than the Chinese one, that was not the case; the results were very close.

NO233BK had by far the highest primary inductance of 35H at 120Hz, yet it had the worst f_L (40 Hz)! It had the lowest parasitic primary-secondary capacitance and, as a result, had the highest fU of all, by far, a whopping 90 kHz!

These measurement results are highly dependent on the output impedance of the signal source. Although the 150Ω Z_{OUT} of the Radford precision oscillator we used is quite low compared to other generators (typically 600Ω), it is relatively high compared to a very low impedance of moving coil cartridges, which contributed to the treble roll-off.

You can insert a voltage divider at the function generator's output, say 5kΩ and 5Ω, which will ensure that the transformers' primaries see a very low source impedance of around 5Ω. With a ratio of 1,000:1, the divider will give you 1mV out for 1V input. Our measurements were mostly concerned with relative comparisons between the tested four transformers and not absolute accuracy.

TESTING MC STEP-UP TRANSFORMERS

	C_{PS} A [pF] (at 1 kHz)	C_{PS} B [pF] (at 1 kHz)	L_P 120Hz [H]	L_P 1 kHz [H]	L_L 120Hz [mH]	L_L 1 kHz [mH]	f_L (-3dB) [Hz]	f_U (-3dB) [kHz]
Russian	119	395	15	2.2	4.5	4.2	9	22
Chinese	258	660	1.23	0.2	31	35	12	20
Sennheiser	27.8	27.8	0.46	0.1	0.23	0.24	30	26
NO233BK	16.5	17.2	35	9.8	47.3	16.8	40	90

Quick power transformer lamination quality test using an LCR meter

Finding transformer laminations in many parts of the world is almost impossible. However, EI power transformers with 6-24V secondaries are widely available. Most use low-grade laminations (cheap 3% non-oriented silicon steel), but some use GOSS materials, and you can use those to wind audio transformers. How can you tell them apart?

First, learn to recognize 0.5mm lamination thickness from 0.35mm. The 0.5mm laminations are never grain-oriented, so forget those. Some 0.35mm laminations are GOSS, but not all. Use this quick test to find out which is which. Using a digital LCR meter, measure the primary inductance L_P of the mains transformer at 120 Hz, then measure it at 1kHz. The closer the two values are to each other, the better suitable are the laminations for audio use!

Say you measure the primary inductance as 9.2H at 120Hz and only 2.1H at 1kHz. These laminations would not perform well in an audio transformer; the inductance drops too much with the frequency. Another transformer may have an L_P of 6.7H at 120Hz and 5.6H at 1kHz. These laminations could be suitable for use in output or interstage transformers.

A simple way to plot the impedance curve of an audio transformer

The transformer under test is terminated by its nominal impedance, in this case, 8Ω. DC voltage source V_{DC} supplies the required DC current to the primary winding (for single-ended transformers only, of course), 100mA (as measured by the DC ammeter), while the AC signal source provides the test signal of a suitable amplitude.

The frequency of such source must be continuously variable from 10Hz to 80+ kHz!

The same current flows through the series resistor R and the primary impedance Z_L and the voltage drops across them, measured by AC voltmeter, are proportionate to their impedances: $V_L/V_R=Z_L/R$ so $Z_L=V_L/V_R* R$.

Instead of one switchable voltmeter, a more elegant method is to use two voltmeters, one across the primary winding, the other across resistor R. The voltmeter(s) must be capable of operation up to at least 80 kHz, so cheap multimeters are out of the question.

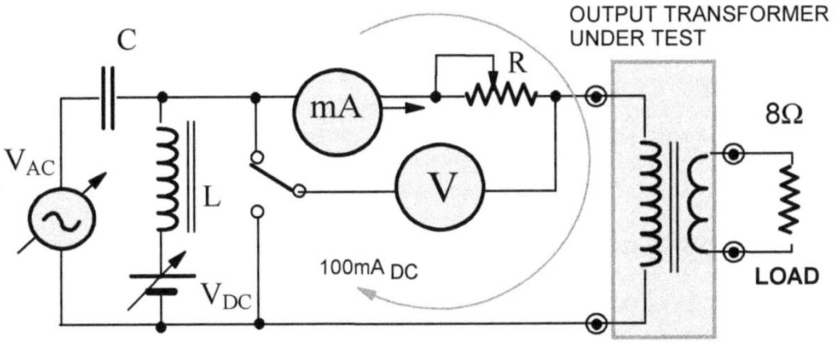

ABOVE: Test setup for manual (point-by-point) plotting of the impedance versus frequency curve of a SE output transformer. For push-pull transformers the DC source and the choke are not needed (no DC current in the primary winding).

USING AN OSCILLOSCOPE TO EVALUATE THE FIDELITY OF AUDIO TRANSFORMERS

The oscillographic methods outlined below are more precise and do what the LCR method cannot do, and that is to evaluate the linearity and losses of the transformer's laminations, which can be deduced from the shape of its hysteresis curve and the magnitude and waveform of the magnetizing current. The test setup for the two measurements is very similar. While identical on the primary side, the main difference is the way an oscilloscope is hooked up to the secondary.

Observing transformer's magnetizing current on an oscilloscope

Once the primary winding is energized, a sinusoidal primary voltage (either the mains voltage in power transformers or a sine test signal for audio transformers) establishes a sinusoidal magnetic flux through the magnetic core. The flux lags behind the voltage by 90 degrees, so the flux peaks when the primary voltage drops to zero (point A).

If point-by-point construction for a specific hysteresis loop is carried out, a waveform of the primary (magnetizing) current can be drawn. Due to the nonlinear dependence between the flux and the said current, its waveform will be significantly distorted. If displaying the hysteresis curve (the next experiment) is too complex or time-consuming, this simpler test setup that displays the magnetizing current can also be used as a proxy, and meaningful comparisons between different transformers can still be made.

The wider and more tilted the hysteresis curve, the more distorted the magnetizing current and the higher the insertion losses and harmonic distortion of the audio transformer.

A variable autotransformer (popularly called Variac®) is used to adjust the amplitude of the mains signal used for this test. Its output is fed into an isolation (1:1) transformer for safety reasons. A small value (1- 10 ohms) resistor is connected in series with the primary winding. The voltage drop across it is proportional to the exciting current I_1, and this voltage is observed on the oscilloscope. The transformer under test is left unloaded (open secondary).

The mains voltage is used as a signal source for mains (power) and tube interstage and output transformers, which operate with similar primary voltage amplitudes (100-400V). A function generator needs to be used for low-level audio transformers such as MC step-up and input types since very low amplitudes (under 1 Volt) are needed. In that case, neither the variac nor the isolation transformer is required.

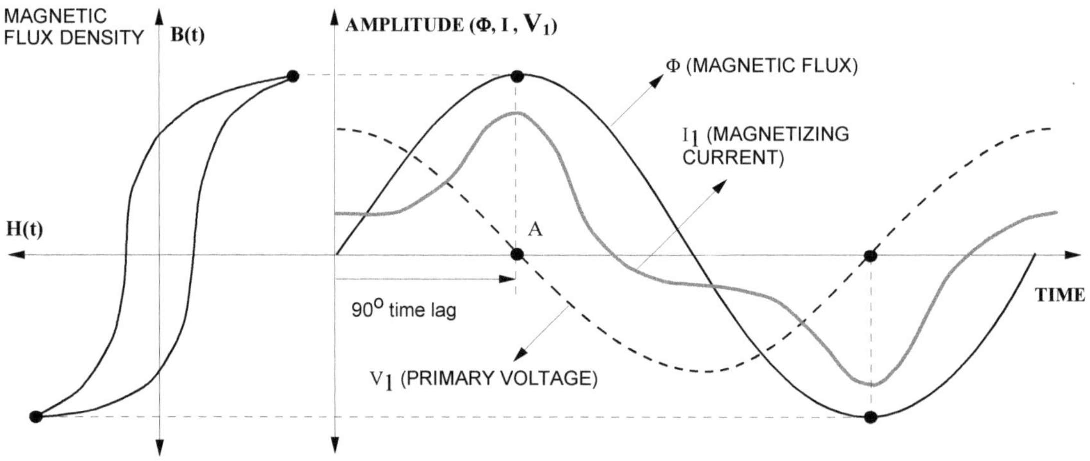

ABOVE: While the primary voltage and the magnetic flux it produces are undistorted sine waves, the primary magnetizing current isn't. Due to the hysteresis curve, it has the waveform illustrated.

RIGHT: The test setup to display the waveform of the magnetizing current on an oscilloscope

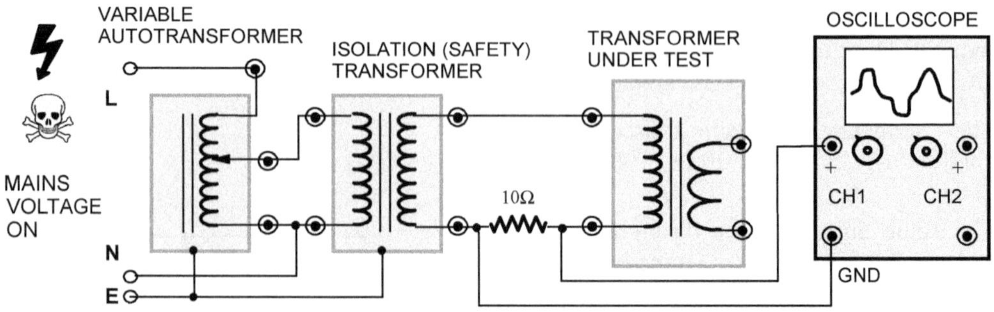

Estimating transformer's laminations quality by observing its hysteresis loop on an oscilloscope

The hysteresis curve of a particular transformer's core (lamination stack) has H, magnetic field strength or "magnetomotive force", on the X (horizontal) axis and B, magnetic flux density as a dependent variable on the vertical (Y) axis.

Since H is a direct function of the excitation current (primary current of the transformer without any load), again, if we connect a small value (1 to 10 ohm) resistor in series with the primary winding, the voltage drop across such resistor will be proportional to the exciting current I_1 and thus to H (just as in the previous test).

This voltage should be applied to the horizontal input of the oscilloscope, which must be in X-Y mode (its time base generator disconnected).

Since the voltage induced in the secondary winding is proportional to the rate of change of magnetic flux Φ, this is expressed mathematically as a derivative ($V_2 = N_2 * d\Phi/dt$), where N_2, the number of secondary turns is a constant. This means the faster the rate of magnetic flux's change and the higher the number of secondary turns, the higher the induced secondary voltage.

B is obviously proportional to magnetic flux Φ. To get B, we must perform the opposite mathematical operation on V_2. The opposite of differentiation is integration, so we need to integrate the secondary voltage V_2. A simple RC network can perform that task under the provision that its time constant $\tau = RC$ is much larger than the period of the test signal.

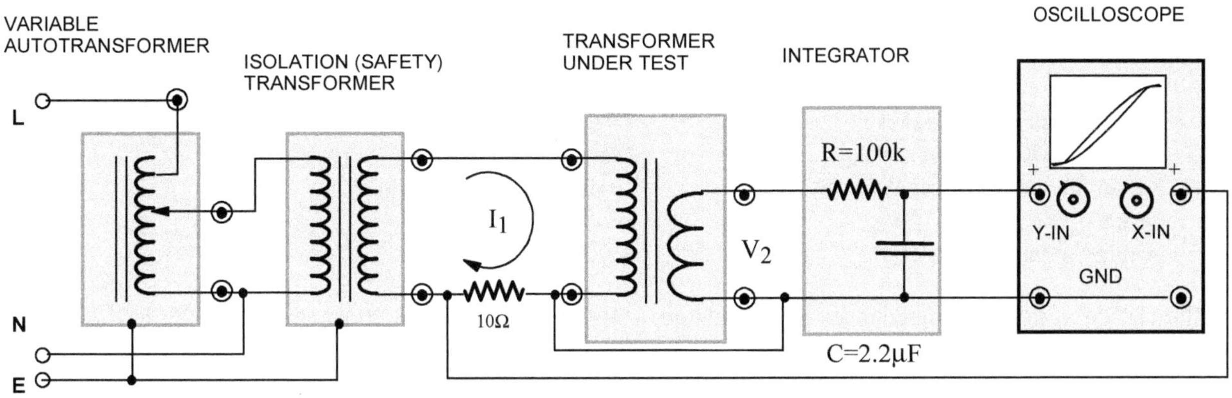

ABOVE: The test setup for displaying the hysteresis curve of transformer's magnetic material on an oscilloscope

In other words, we must ensure that $\tau = RC >> 1/\omega$, or that $\omega RC >> 1$! (>> means "much bigger than"). Angular frequency ω (omega) is ω=2pf! If this condition is not satisfied, the display will not be a true representation of the hysteresis curve; the curve will be "folded" or warped as in the illustration below (far right).

If the mains voltage is used for this test, a frequency of 50 Hz (or 60 Hz in the USA and some other countries) needs to be used in the formulas above, so let's see what we get with our chosen values of R=100kΩ and C=2.2mF

$\omega RC = 2\pi fRC = 2*\pi*50*10^5*2.2*10^{-6} = 10*\pi*2.2 = 69$, which can be considered much larger than 1 (69>>1), so the hysteresis curve will be displayed properly.

Going back to the illustrative examples below, even the far left example isn't bad at all; the curve is fairly narrow (low losses) and fairly straight (low distortion). The example in the middle is even better; only the highest fidelity audio transformers using EI laminations or C-cores of superior quality will have such a linear hysteresis curve.

Here are a few pointers regarding the test circuit. The capacitor used in the integrator network should be a low loss film type, polyester, or polypropylene. The test should start with the Variac in zero output position. Its dial should be slowly raised until the nominal primary voltage is reached.

The vertical sensitivity control of the oscilloscope and the time base will need to be adjusted until a display as illustrated is obtained. Due to the polarities of the signals involved, the display will be a mirror image unless the invert button on the scope is pressed (activated), which flips the signal brought to the scope's X-input.

If the scope's sensitivity cannot be lowered enough to display a large enough curve (budget scopes of low sensitivity or very low magnetizing current amplitudes), the 10-ohm series resistance will need to be increased.

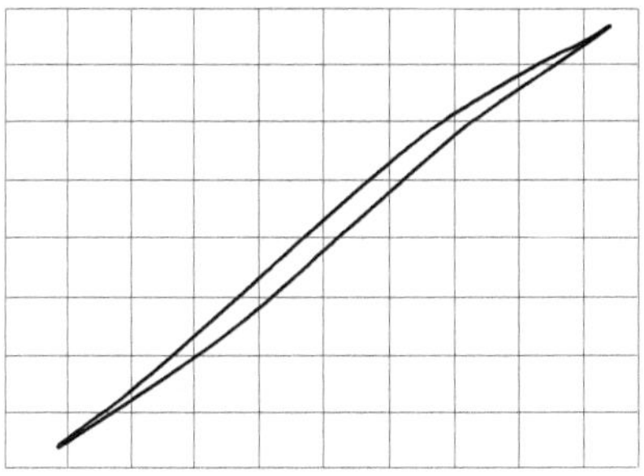

Hysteresis loop of a good quality output transformer with GOSS laminations

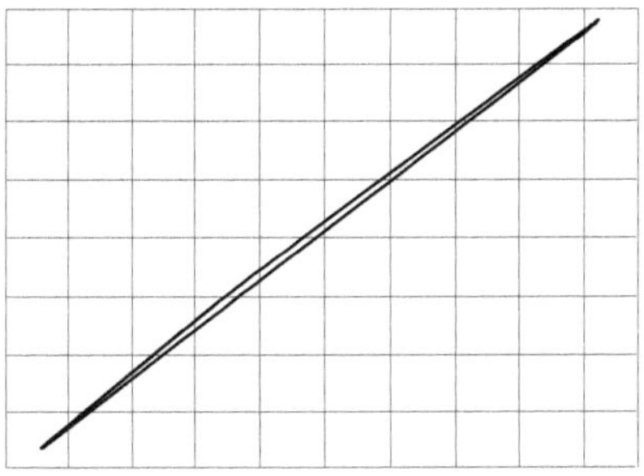

Hysteresis loop of a top quality output transformer with extremely small losses and very low distortion

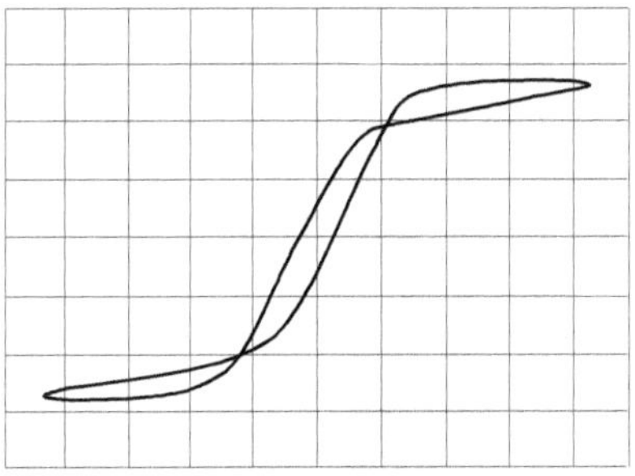

Distorted oscillogram of a hysteresis loop due to improperly selected RC integrator (ωRC product is too small).

EVALUATING COMMERCIAL OUTPUT TRANSFORMERS BASED ON THEIR PUBLISHED OR MEASURED SPECIFICATIONS

How to scrutinize transformer data and manufacturers' claims

For this exercise, we selected three output transformers from the same manufacturer (One Electron), all of the same physical size but different primary impedances. The same physical size was an important consideration since their center-leg cross-section is the same, meaning their power handling capabilities are the same. Only physical dimensions were specified in the datasheet (except stack thickness), so how did we determine the lamination size?

First, find the transformer's largest dimension (laminations only, without end bells), 86 mm in this case. Then, find the laminations' width, in this case, 72 mm. Finally, to determine if the laminations are of the scrapless (wasteless) size (the ratio of width to length of 5/6 = 0.8), divide 72/86 = 0.837. Since the manufacturer is from the USA, it is possible that an "imperial" size EI85.73 (L112) laminations were used, 85.73mm height *5/6 = 71.44 mm width, which is close to the possibly rounded specified figure of 72 mm! The stack thickness is 5cm.

MODEL	UBT-1	UBT-3	UBT-2
Primary impedance Z_P [Ω]	1k6	3k0	4k8
Maximum DC plate current I_{PMAX} [mA]	160	110	110
Maximum output power P_{MAX} [W]	13-18	13-18	13-18
Primary resistance R_1 [Ω]	165	286	432
Primary inductance L_1 [H]	8	17	29

Notice that the choice of primary impedances seems related: 1,600Ω, 2*1,600Ω= 3,200Ω, and 3*1,600Ω= 4,800Ω!The specified impedance was 3,000Ω but 3,200Ω seems suspiciously close.

The gross center-leg cross-sectional area for EI85.73 (L112) laminations with a 5cm stack is A=2*1.43*5=14.3 cm^2, so the effective area of the central leg A_{EF} is around 13.7 cm^2. The empirical formula A_{EF}=k/(P/f_L) gives us k=17 for P=13W and f_L=20Hz and k=14.4 for P=18W and f_L=20Hz. Since k should be 20-30, it seems that the available power transferred to the load at 20Hz would be below the specified 13-18 W. Although the manufacturer declared the maximum output power as a range (13-18 Watts), the center leg cross-section seems far too small for such power levels! This means the transformer's distortion at low frequencies will be significant.

One thing is immediately obvious. The higher the primary impedance Z_P, the higher the primary resistance and the primary inductance L_1. This is easily explained. To get a higher impedance ratio, a higher number of primary windings is needed (N_1); therefore, primary inductance goes up since it is proportional to N_1. More primary windings also mean that the length of the primary winding increases, so its DC resistance must also go up.

EXPERIMENT: Comparing output transformers

We took two output transformers for single-ended 300B tube amplifiers, one made by One Electron in USA, the other wound by us, and measured their primary and leakage inductances.

SIDE-BY-SIDE: OUTPUT TRANSFORMERS

	Ours	One Electron UBT-1
Primary inductance L_1 @ 120Hz and 1,000Hz	18.5 H, 15.5H	5.8 H, 4.5 H
Leakage inductance L_L @ 1,000Hz	23 mH	4.1 mH
L_L as % of L_1 @ 1,000Hz	0.124	0.091

Our transformer had a much higher primary inductance, which means its bass performance will be superior. However, One Electron has a lower leakage inductance (4.1 mH versus 23 mH). The explanation lies mostly in the size of the laminations used, EI114 in our case versus EI85.73 (L112) used by One Electron. Larger laminations mean more primary turns and higher primary inductance, but due to the larger physical size of the finished transformer, larger leakage inductance.

L_L divided by L_P is the relative ratio used as an indicator of the overall frequency range of a transformer, 0.124% for ours and 0.091% for One Electron, meaning that the frequency range of the One Electron transformer will be 124/91 = 36% wider.

Remember that One Electron is a specialist transformer maker, and we are not. They have access to better quality laminations, superior wire, and professional winding machines, while we wind everything by hand on the simplest winder possible. It would be a sad indictment of their product if our transformer measured better than theirs.

We should have used One Electron's UBT-3 model for this comparison, which is much closer in terms of primary impedance, but we purchased a pair of UBT-1 transformers, and the aim of this exercise was educational, to see how various parameters of output transformers impact their measured values, not to determine which design is superior, or which transformer sounds better in an amplifier.

9 LOUDSPEAKER TESTS & MEASUREMENTS

- ELECTRODYNAMIC (FIELD COIL) SPEAKERS
- PLOTTING SPEAKER'S IMPEDANCE-VS. FREQUENCY CURVE
- QUICK AND SIMPLE SPEAKER CHECKS
- SPL (SOUND PRESSURE LEVEL) METERS

> "Testing leads to failure, and failure leads to understanding."
>
> Burt Rutan

ELECTRODYNAMIC (FIELD COIL) SPEAKERS

A permanent magnet loudspeaker is an electromechanical transducer, transforming the electrical energy of the signal current flowing through its voice coil ("spool") into a mechanical sound wave (air pressure). A permanent magnet produces a steady uniform magnetic field where the voice coil is placed, surrounded by a small air gap.

With no audio signal, the coil is at rest. An AC signal current flowing in the coil (from the output of an amp) produces its pulsating magnetic field, which interacts with the permanent magnet field and produces a mechanical force transferred onto the speaker's cone.

The cone's movement follows the audio signal and produces sound waves whose frequency and amplitude follow the audio signal's frequency and amplitude.

The first permanent magnet speakers used Alnico (an acronym for Aluminium-Nickel-Cobalt mixture), an alloy whose chemical composition is 8–12% Al, 15–26% Ni, 5–24% Co, up to 6% Cu (copper), up to 1% Titanium, and the rest is iron.

In the 1970s, "ceramic" or "ferrite" magnets were developed, which replaced Alnico almost completely.

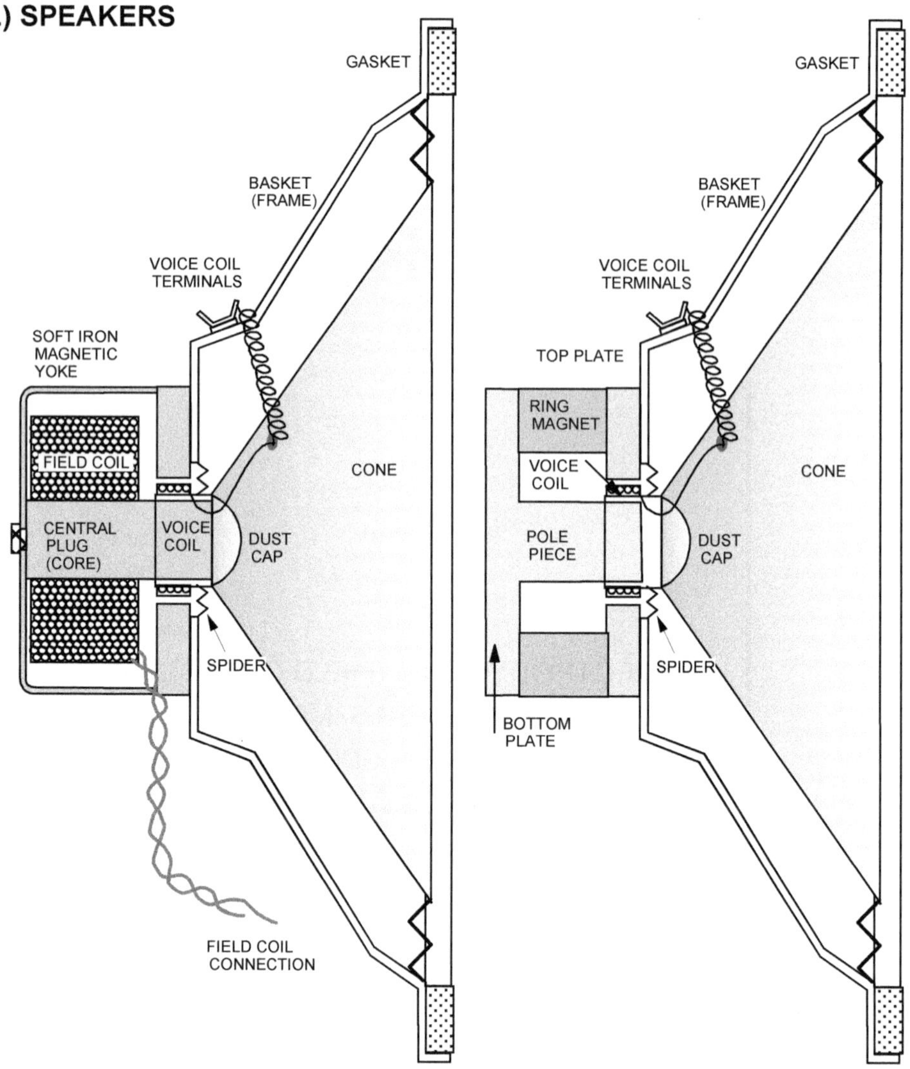

A cross-sectional view of a permanent magnet (RIGHT) and electrodynamic loudspeaker (LEFT). Only one VC terminal is shown.

Electrodynamic (field coil) speakers

Alnico and ceramic magnet permanent speakers are relatively recent inventions. In the first half of the 20th-century, speakers did not have permanent magnets. Instead, a magnetic yoke with a center insert was used, and the magnetization was provided by the flow of DC current through the "field coil."

The field coil had a few thousand turns N_F of copper wire, wrapped around a spool or cylindrical bobbin, which would snugly fit over the centerpiece (pole). The magnetomotive force (MMF) of the field coil or an electromagnet is $MMF=1.257*N_F*I_F$ [AmpereTurns]. It is also called the "total field." The constant 1.27 is $4\pi/10$, N_F is the number of turns in the field coil, and I_F is the DC field or "excitation" current. Magnetomotive force should not be confused with magnetizing force H, but the two are related by H = MMF/LMP, where LMP is the "Length of the Magnetic Path," or how long the average magnetic line of force is.

Finally, the magnetic flux density B is the number of magnetic fields lines through one unit of cross-sectional area: B=F/A $= m*H = mN_F*I_F/LMP$.

Greek letter m symbolizes the magnetic permeability of the ferromagnetic material used for the centerpiece. When signal current flows through the voice coil VC placed in the uniform magnetic field, the force acting on the voice coil is $F=B*I_S*l_S*\sin\theta$.

Since the magnetic lines of force are perpendicular to voice coil windings, the angle between them is $\theta=90^o$, and $\sin\theta=1$, so $F=B*I_S*l_S$, where B is the magnetic induction or the flux density, I_S is the signal current and l_S is the total length of the coil wire (length of one turn multiplied by the number of turns N_V).

Since l_S is constant for each voice coil, the force is directly proportional to magnetic flux density B! If B is reduced, the force on the voice coil is also reduced, which results in smaller movement (excursions) of the cone. Smaller movement means reduced air pressure created by the cone and lower loudness.

With permanent magnet speakers B depends on the type and size of the magnet ring and is thus fixed, and so is their efficiency and loudness.

Once the magnetic material is chosen, its permeability is fixed, as is its size, shape, and thus LMP, so $B=k*N_F*I_F$ (where $k=\mu/LMP$). The "loudness" depends on B, and it can be changed by varying the number of turns N_F or the excitation (field) current I_F.

Once we have the actual speaker, we cannot vary the number of turns, that is a given, so the only remaining practical way of changing the loudness of a field coil speaker is by varying its field current.

Most vintage European electrodynamic speakers have at least a couple of parameters printed on the field coil. For instance, a Telefunken speaker is marked "Tel. Bv. 665 a" and "Feldspule 100V/67 mA". *Feld* is German for field, and *spule* means coil, whose resistance can be easily calculated as $R=V/I=100/0.067=1{,}493\Omega$.

The field coil of another speaker was marked 75mA/6Ω. Since the power dissipated in the coil is $P=I^2*R$ we can calculate its resistance as $R=P/I^2=6/(0.075^2)= 1{,}067\Omega$ and the required field voltage as $V_F=I*R=0.075*1{,}067=80V_{DC}$.

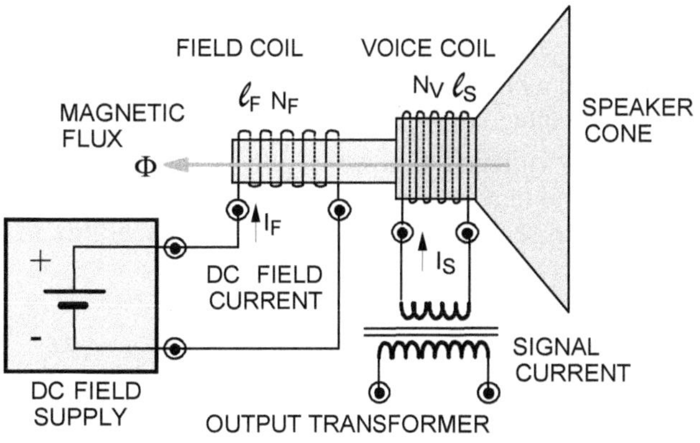

ABOVE RIGHT: The operating principle of electrodynamic speakers. For clarity, only the center slug is shown without the surrounding soft iron magnetic yoke needed to close the path of the magnetic flux.

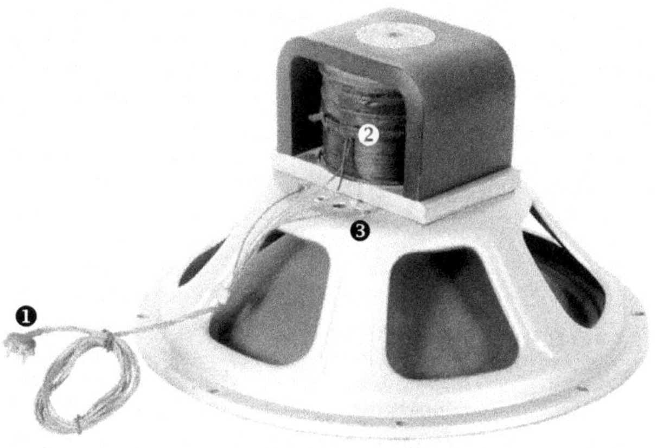

RIGHT: Electrodynamic speakers usually have a four-core cable terminated by a 4-pin plug (1), two for the field coil (2) and two wired to the voice coil terminals (3).

EXPERIMENT: Varying the SPL of an electromagnetic speaker by changing the field strength

You can either use a variable voltage DC power supply of suitable voltage and current rating or a fixed DC power supply and a rheostat for this test. If you don't know the required voltage and current for the field coil of your speaker, it is safer to use a variable DC power supply; otherwise, you may burn the rheostat out. Start with zero volts and gradually ramp the voltage up while monitoring the current. Many benchtop power supplies have meters for both voltage and current.

The required signal voltage at the amp's output will depend on the speaker's impedance. The first step is to use a digital LCR meter set on the "R" range (which is AC resistance and can be very different from DC resistance measured with a multimeter) and on 1kHz test frequency, standard on all such LCR meters. Then, adjust the amp's volume control and thus the input voltage to the speaker to get exactly 1 Watt input power. Say you measured AC resistance of R=9.24 Ω at 1 kHz. Since $P=V^2/R$, the required voltage is $V=\sqrt{(P*R)} = \sqrt{(1*9.4)} = 3.07\ V_{RMS}$. This is the effective or RMS value of the sine signal, not its peak value. The sound pressure meter should be placed in line with the speaker's center (on its "axis").

We used a Legacy combo tube guitar amp (a rebadged Epiphone Valve Junior) for our test since it had an 8" speaker. First, we measured the SPL of the existing speaker. Since we weren't interested in the absolute sensitivity values and our test bench had less than 1 meter of space, we set the SPL meter 0.6m away and chose a much lower input voltage, $0.6V_{RMS}$, because 1 Watt of input power produced a very loud and annoying sound level.

The test setup for measuring the sensitivity of an electromagnetic speaker and its SPL as a function of the excitation of its field coil.

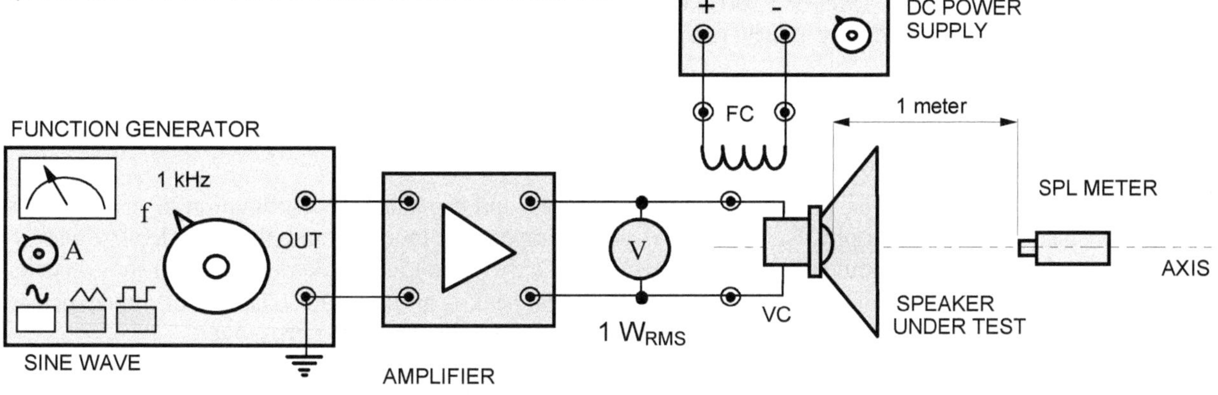

The relative comparisons still apply, of course. With such a setup, the original speaker for the Legacy combo amp produced 98 dB of sound pressure, which was very loud indeed. So, in our case, the vintage USA-made Motorola electrodynamic speaker had a DCR (DC resistance of the field coil) of 700Ω, and assuming 70mA through the coil, that would require an excitation voltage of $V_{DC}=I_{DC}*R_{DC}= 0.07*700= 49V$!

While a low voltage DC power supply of the standard 1A current capacity would be fine, most only go up to 20, 25, or 30V, you need one that can supply up to at least $50V_{DC}$! In some cases, the two DC outputs are floating and can be "stacked" or connected in series, which is a very handy capability to look for when buying a DC bench power supply.

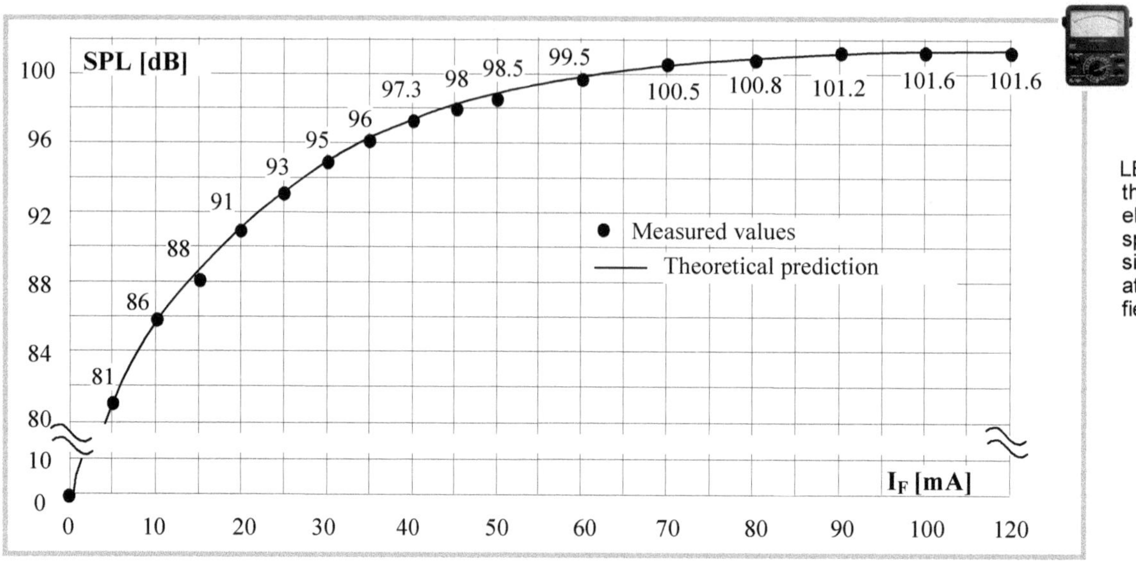

LEFT: Test results of the vintage 8" Motorola electrodynamic speaker, $0.6V_{RMS}$ signal, SPL measured at 0.6m distance versus field current I_F

The graph above shows the results of our experiment. The horizontal scale (field coil current I_F) is linear, and the vertical scale (SPL in decibels) also seems linear, but that is not the case, since dB is a logarithmic unit. The formula says that the magnetic field strength and thus the sound pressure (loudness) increase linearly with the field current, but such a linear curve on a graph with a vertical log scale has a logarithmic shape.

Even at a very low field excitation level of 5mA (the first point measured), the SPL had already reached 81dB! Doubling the current to 10mA increased the SPL by 5dB to 86 dB, and so on. Most low voltage benchtop DC power supplies have a current meter, but if that is not the case, the field current can be measured by inserting a DC ammeter into the field coil supply circuit.

The maximum SPL was around 101.6 dB, reached with about 95mA of field current. Further increase in current did not increase output loudness. Of course, you need to be careful not to exceed the nominal excitation current too much and for too long since such a high current may burn the field coil out. However, it only took a few seconds to measure each test point, so even at 120mA there was no overheating or smoke from the field coil.

The first conclusion is that by replacing a permanent magnet speaker in your guitar amp with an electrodynamic speaker of higher sensitivity, you can raise the loudness of your amp, in this case from 98 dB to 101.6 dB, an increase of 3.6dB.

Since power in dB is $P=10\log(P_1/P_2)$, we get $P_1/P_2=10^{(P/10)} = 10^{(3.6/10)}= 10^{0.36} = 2.3$ With the replacement Motorola field-coil speaker at its maximum excitation that 4 Watt amp would be as loud as a 5*2.3= 11.5 Watt amp with the original speaker! Thus, the first possible benefit of replacing the original fixed magnet speaker in your amp with a field coil speaker is higher output levels, not in terms of Watts but in terms of SPL or "loudness," which is what ultimately matters.

The second benefit is that a dynamic loudness control range of around -20dB is achievable if you decide to add variable excitation to such a speaker. A 20dB reduction in loudness is significant.

Since power in dB is $P=10\log(P_1/P_2)$, we get $P_1/P_2=10^{(P/10)} = 10^{(20/10)}= 10^2 = 100$! This means you could effectively attenuate your 5 Watt amp down to a 0.05W level or a 100W amp to a 1 Watt level (not with this tiny 8" speaker, of course, but with one or more suitable rated electrodynamic drivers in series or parallel connection).

PLOTTING SPEAKER'S IMPEDANCE VS. FREQUENCY CURVE

A speaker as a complex impedance

A typical dynamic (moving coil) loudspeaker, both drivers by themselves, and especially a combination of drivers with crossover networks in a speaker box, are complex RCL networks whose impedance modulus and phase angle vary widely with frequency. This typical impedance (modulus) vs. frequency curve of a dynamic loudspeaker (next page) shows peaks and dips, such as a sharp increase at the resonant frequency f_R=100Hz; the peak is around 20Ω. The rise of impedance at higher frequencies is due to the dominant inductance of the voice coil and the inductors in the crossover.

The "nominal" speaker impedance Z_{NOM}, usually 4, 6, or 8 ohms, "proclaimed" by manufacturers is meaningless, but, paradoxically, for simplicity's sake, we still assume such a fixed load when designing power amplifiers.

The illustrated impedance curve is very benign; many speaker boxes exhibit not just one (illustrated here at around 100 Hz), but two or even three resonant peaks and dips, some way below 2Ω or even 1Ω, meaning an amplifier will struggle to push signal currents through such a low impedance, practically a short circuit.

To understand the sonics of a certain amplifier, you have to understand the loudspeakers you are pairing it with in those listening tests, and that primarily means their impedance curve.

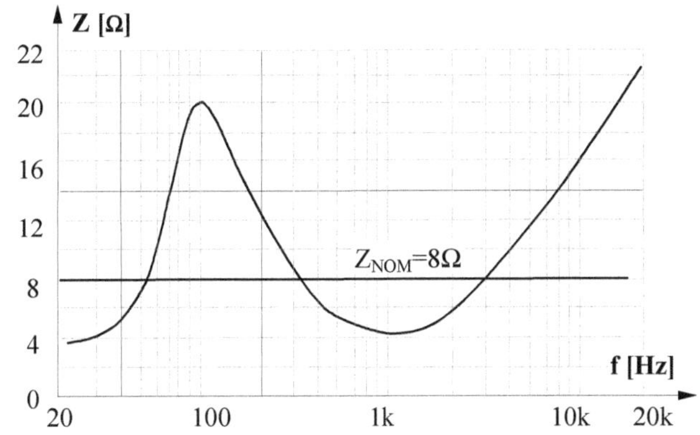

RIGHT: A typical speaker (driver only) impedance versus frequency curve bears no resemblance to its nominal (constant) impedance, a figure declared by its manufacturer!

The constant current source method of plotting loudspeaker impedance curve

Since the 220Ω series resistor has a much higher DC resistance than the impedance of the measured loudspeaker, this simple circuit approximates a CCS (Constant Current Source). That same current flows through the series resistor and the speaker's terminals.

You are measuring the difference between the readings of two AC voltmeters: V1 indicates the output voltage of a function generator (sinewave), and V2 displays the AC voltage drop across the loudspeaker's impedance at various frequencies. Had the speaker's impedance been constant with changing signal frequency, the ratio of the two measured voltages would remain constant.

However, since speakers' impedance is highly frequency-dependent, the ratio of the measured voltages will also change with frequency.

The voltmeters must be capable of operation up to at least 20 kHz, so cheap multimeters are out of the question.

If you don't have two AC voltmeters or multimeters, you can use a two-channel oscilloscope, but estimating signals' amplitude from an oscilloscope screen is much slower and less precise than using AC voltmeters.

In the experiment described below, from 10 to 150 Hz, we took measurements every 10 Hz, then from 200 to 1,000 Hz every100 Hz, and then every 1kHz. After punching in all the figures into a spreadsheet, we got a recognizable dynamic speaker impedance curve, shown on the next page.

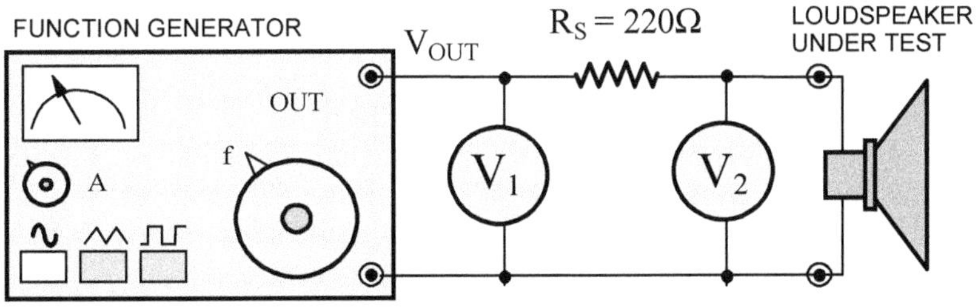

ABOVE: Test setup for manual (point-by-point) plotting of a loudspeaker impedance versus frequency curve

SPEAKER IMPEDANCE BY TWO VOLTMETER METHOD

$$Z(f) = R_S V_2(f)/[(V_1(f)-V_2(f)]$$

$V_1(f)$ and $V_2(f)$ are measured results at various frequencies, a function of that frequency, Z(f) is the calculated impedance modulus at each test frequency.

EXPERIMENT: Loudspeaker impedance curve for Opera III (Terza) loudspeaker

A pair of these speakers has served us faithfully for more than 25 years as one of our test speakers. Technically, these should be unsuitable for low-power SET amplifiers due to their complex crossovers (a 3-way design with high-order filters) and relatively low sensitivity (86-87 dB/W). Yet, for some reason, even a lowly 3 Watt amp can drive them relatively easily.

The bass drivers are small, so don't expect ceiling-collapsing bass levels, and they are not the most accurate or transparent speakers around. But, they are musical, great-looking (solid mahogany, slim and elegant), and serve their purpose well.

Our low distortion function generator kept V1 constant, so we only had to measure V2. Once the impedance curve was plotted, it became clear that Opera Terza was a very easy load to drive. Its impedance never dropped below 6.2Ω; between 3 and 20 kHz (midrange and treble), it stayed in the 12-16Ω range. The three resonant peaks are evident at around 30Hz, 80Hz, and 2,500Hz.

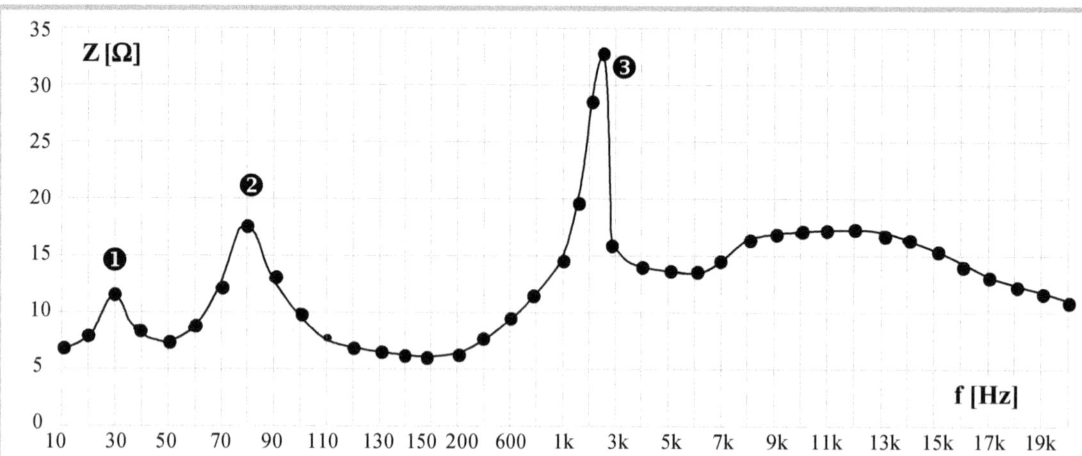

ABOVE: Impedance modulus versus frequency for Opera III (Terza) loudspeaker (ABOVE RIGHT). Dots indicate measured values.

EXPERIMENT: Amplitude-frequency response of an amplifier with a dummy load and dynamic speakers

With a dummy (purely resistive) 8Ω load, the amplitude-versus-frequency characteristic of this single-ended triode amplifier with EL156output tubes was flat over its entire bandwidth, dropping to -3dB at 15 Hz and 41 kHz. The curve is shown below and marked (4).

With Opera III speakers connected and at the same power level (1 Watt at 1 kHz, the 0 dB level), the speakers' impedance changes heavily influenced the amplitude-frequency characteristics. This is due to the relatively low damping factor of around 3.6 in the triode mode and only 2.5 in the ultra-linear mode.

The three resonant peaks of the speaker result in three amplification peaks (marked 5, 6, and 7), but the overshoots are not serious in any of the cases. Peak (6) is only 0.7dB over, and peak (7) is just 1.2 dB above the referent 0dB level. The biggest drawback is two dips, one in the lower bass region, around 45Hz, the other in the upper bass, with a minimum of -3.5 dB at around 160 Hz. These two "suckout" regions are the main reason for the relatively weak bass of these speakers when driven by most SET amplifiers. The midrange frequencies between 1 and 3 kHz will be slightly emphasized (7), up to +1.2 dB, making the bass seem even weaker in comparison.

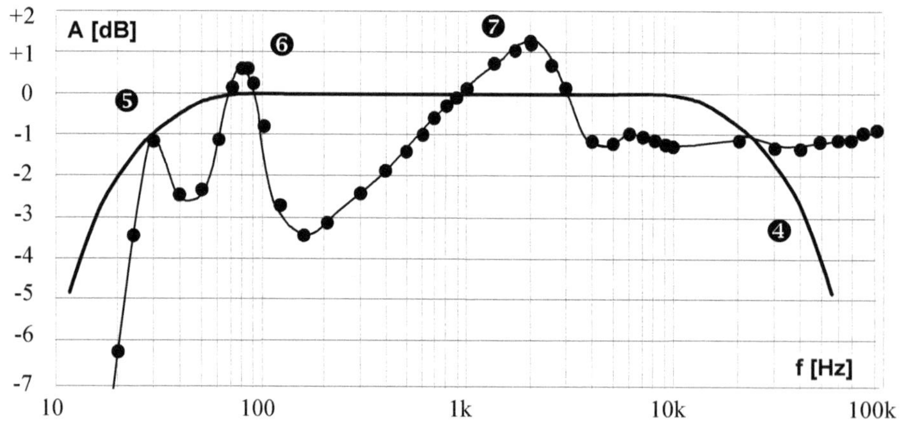

LEFT: The amplitude- versus-frequency response of EL156 SET amplifier with an 8Ω dummy load (thick line) and with Opera Terza speakers. Dots indicate measured values.

How to measure the sensitivity (dB/Watt) of speaker drivers and speaker boxes

On page 131, we've seen the test setup for measuring the sensitivity of an electromagnetic speaker and its SPL as a function of the excitation of its field coil. A similar setup can be used to measure sensitivity (dB/Watt) of permanent magnet speakers or any other loudspeaker type, such as ribbon, electrostatic, etc. The fewer Watts (power) a speaker needs to produce the same sound pressure level, or, in other words, the higher its produced SPL level for the same input power level, the more sensitive a speaker is.

You can measure speaker drivers by themselves (tweeters, midrange drivers, woofers) or whole speaker boxes. Strictly speaking, since this test should be done in a special anechoic test chamber, just as speaker manufacturers do it, your results will be different from theirs. The test room, in this case (your listening room or your workshop), will include sound pressure not just from direct sound waves but also the reflected ones. You can measure the sensitivity at one particular frequency or repeat this test at various frequencies and get the whole SPL/efficiency versus frequency curve for a particular speaker.

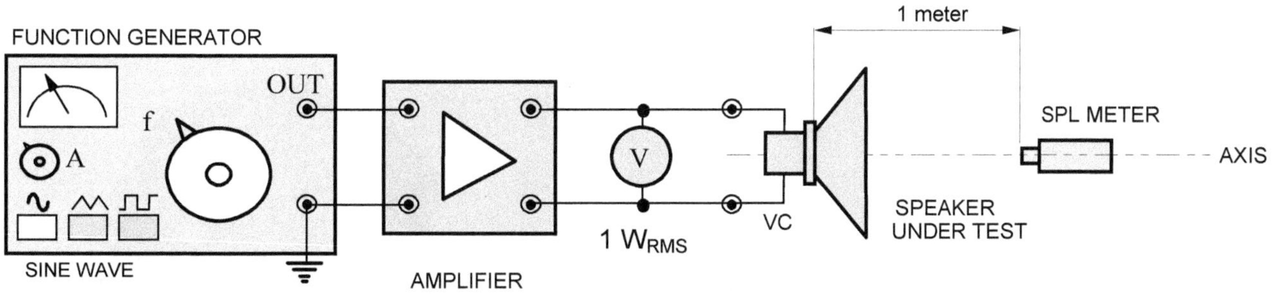

ABOVE: The test setup for measuring the sensitivity (dB/Watt) of permanent magnet (dynamic speakers) any other loudspeaker type, such as ribbon, electrostatic, etc.

Apart from the variable frequency of the test signal produced by an oscillator or function generator, all other variables must remain constant. First and foremost, the amplifier must have a flat gain-vs.-frequency characteristic over the entire measurement frequency range. If you only vary the test frequency, you will get a dB versus frequency curve, not the sensitivity curve. For that, you must measure the speaker's impedance at each test frequency, calculate the required voltage amplitude maintain the power at 1 Watt! Obviously, that is a long and tedious process.

QUICK AND SIMPLE SPEAKER CHECKS

Mechanical (manual) checks

There are four quick and simple tests to check loudspeaker drivers' mechanical/structural and electrical health. The first is the visual inspection. There should be no holes, tears, or rips in the paper cone. If there are, do not buy such a speaker; sure, re-coning can be done by experts (if a speaker is rare, valuable, or hard-to-replace) but costly.

The second test involves gently pushing the speaker cone inwards by hand. The cone should move freely and return to its original position once the force is removed. If any sudden resistance can be felt or if any "scraping" sound can be heard, that means the voice coil is rubbing against its surroundings. The "spider" or speaker's suspension has been damaged or misaligned. Such speakers may work to some extent but will sound distorted and will eventually fail.

Three speaker checks: battery, ohmmeter and LCR meter

Once a speaker passes these two mechanical tests, it is ready for electrical checks. If you have a multimeter, set it on a low resistance range and measure the resistance across the speaker's terminals. Since this is DC resistance, it will be lower than the speaker's nominal (declared by the manufacturer) impedance. For instance, an 8 ohm speaker may show a DCR of 5.7 ohms. That is fine and confirms that the speaker's voice call has continuity, i.e., that it hasn't burned out.

Take two leads with crocodile clips for this quick test and connect a 1.5V or 9V battery across the speaker's terminals. With the battery's + pole on the speaker's + terminal, the speaker's cone should quickly move forward (outward) and stay there! With the battery's + pole on the speaker's - terminal, the speaker cone should quickly move back (inward) and stay there, again, without any scraping and sticking along the way!

Finally, the best and most conclusive test is connecting a digital LCR meter (on "R" or AC resistance range) across speaker terminals. The LCR meter will send current pulses through the voice coil and show the voice coil's AC resistance on its LCD display at the selected frequency (60Hz or 1kHz). These pulses will be audible; the speaker should go "beep-beep-beep ..." about once a second.

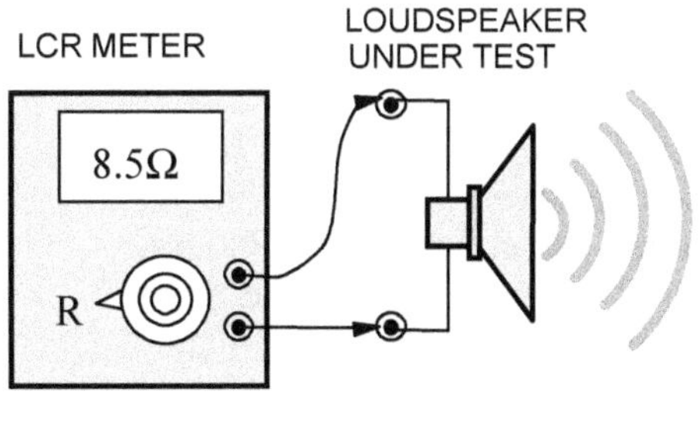

With an LCR meter set on "R" or AC resistance range, a pulsating tone should be heard in the speaker. The higher the speaker's sensitivity or SPL level (dB/W), the louder that tone will be.

SPL (SOUND PRESSURE LEVEL) METERS

The operational principle of SPL meters (the three pictured on the next page) is fairly simple. An electret condenser microphone (1) converts the sound pressure into voltage, which is then amplified, filtered and displayed on an analog or digital display (2). The range switch (3) indicates the dB figure that must be added to the meter's reading (4), the analog meter on the right indicates around -1dB on the 60dB range, meaning the reading is 59dB.

The newer models display the overall figure directly (5), although the range switch or button still must be adjusted up or down. The accuracy's in the order of ±2 dB.

Realistic (Radio Shack) and similar analog and digital SPL meters offer at least two weighting options (6), usually "A" and "C" and "Fast" and "Slow" display response (7). A-weighting has an amplitude-frequency characteristics that attenuates low and high frequencies and limits the meter's response to midrange frequencies where the human ear is the most sensitive. Thus, it is used mainly for noise checks in health & safety applications.

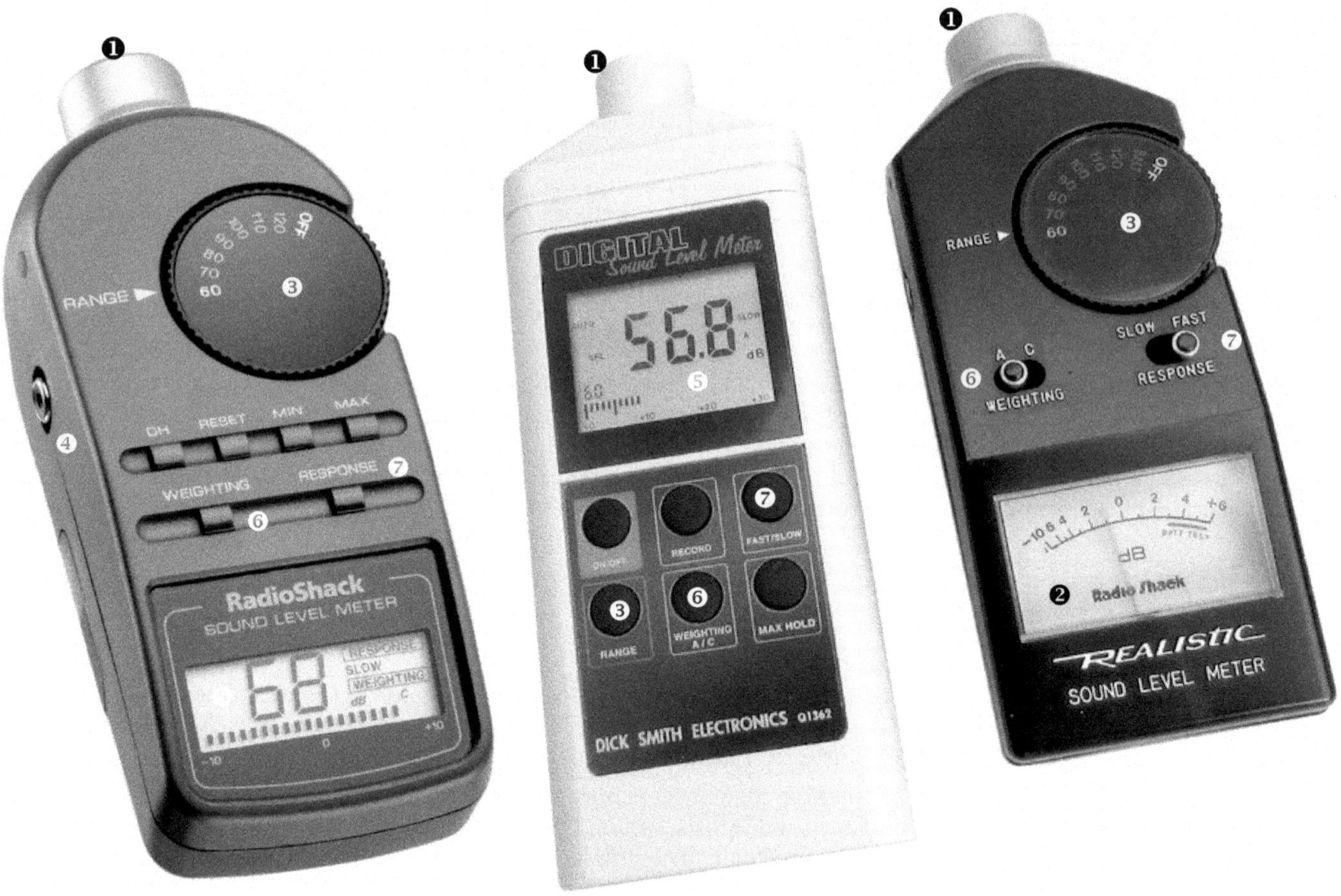

C-weighting has a much flatter frequency characteristic and is thus more suitable for audio system measurements.

There is also a signal output of 1 V_{PP} (open circuit, full scale at 1 kHz) with less than 2% distortion (4).

For the new kids on the block, there is even a NIOSH Sound Level Meter app for iPhones, which can be used with both internal (not approved for "legal" measurements) and external (calibrated and approved) microphones.

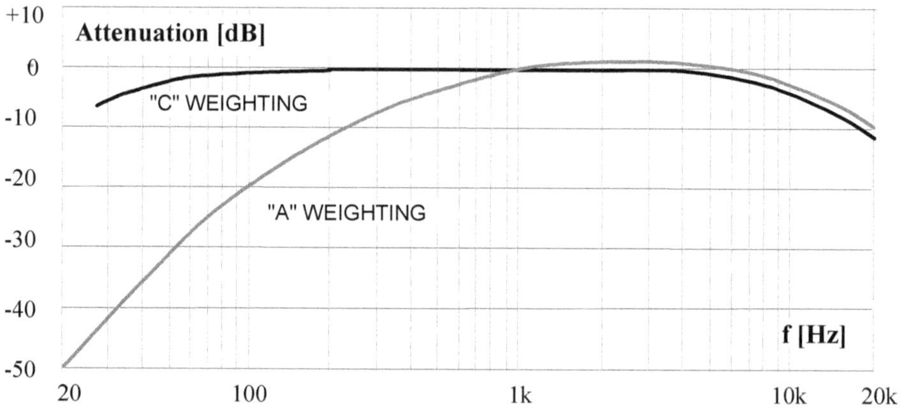

ABOVE: The A-weighting attenuation curve is mostly used for health & safety noise checks, the C-curve in audio tests

10 TRANSISTOR TESTERS AND CURVE TRACERS

- QUICK CHECKS OF SOLID STATE DIODES AND TRANSISTORS
- QUICK CHECKS OF JFET and MOSFET TRANSISTORS
- TRANSISTOR TESTERS
- SEMICONDUCTOR CURVE TRACERS

> "It is much easier to make measurements than to know exactly what you are measuring."
>
> J. W. N. Sullivan

QUICK CHECKS OF SOLID STATE DIODES AND BIPOLAR TRANSISTORS

A solid-state diode is a PN-junction, which conducts current when forward biased (P-side or anode is positive with respect to the cathode) and does not conduct when reverse-biased (anode is negative with respect to cathode). Vacuum tube diodes behave similarly but work on a different physical principle.

Germanium diodes start conducting at lower forward voltages, 0.2-0.3V, compared to 0.5-0.6V for silicon diodes. Silicon Schottky diodes also feature a low forward voltage drop (0.15-0.45V), but Schottky diodes constructed from silicon carbide have a much higher forward (1.4-1.8V at 25 °C) and reverse voltage.

While ordinary silicon diodes work in the forward biasing regime, Zener diodes (ZD) are reversely biased. This means that the cathode of the Zener diode needs to be more positive than its + electrode (anode)!

Once the reverse DC voltage $-V_Z$ across a ZD has been reached, the diode starts conducting and practically keeps the voltage across its terminal constant, a feature which makes it useful as a simple two-terminal voltage regulator.

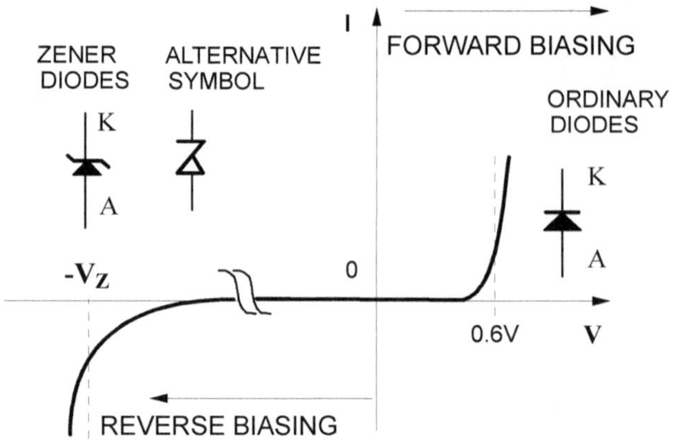

Testing semiconductor (solid state) diodes

Since ohmmeters use their internal DC battery to push current through an unknown resistor, a diode should conduct in one direction only, with the + of the battery on its anode. The measured resistance should be very low, 20-50 Ω.

When the + of the ohmmeter's battery is connected to the cathode, the diode should not conduct, and the ohmmeter should indicate a very high, almost infinite resistance. The cathode is the negative electrode, marked with a band or a line on one side of the diode's body, corresponding to the line on the symbol.

In some multimeters, the positive lead (red) is internally connected to the battery's - (negative) pole, so their readings will be the opposite.

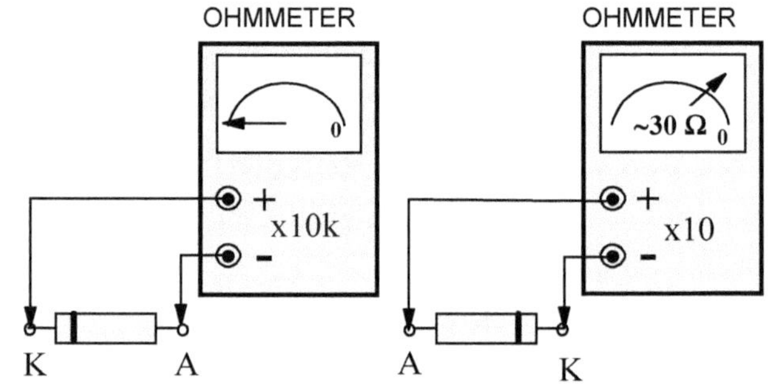

ABOVE: Ohmmeter must show diode's very low resistance one way and a very high, almost infinite resistance the other way.
BELOW: How to determine Zener diode's breakdown voltage V_Z

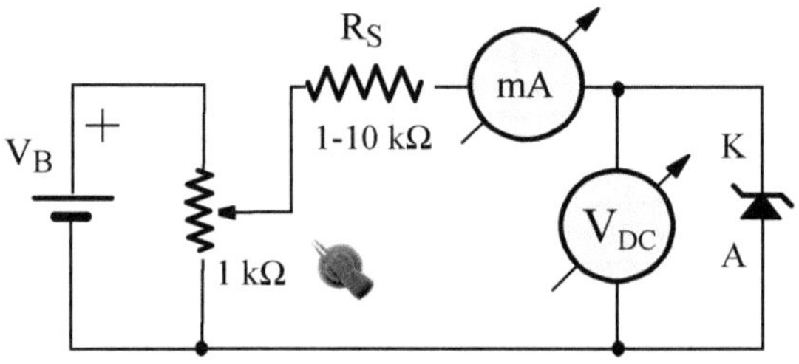

Testing Zener diodes

To determine the Zener voltage V_Z of a diode, you'll need an adjustable DC voltage source and a series resistor R_S. If your DC power supply (or a battery) is not adjustable, connect a 1 kΩ potentiometer across it (as illustrated) so you can adjust the test voltage.

Increase the voltage until there is a current jump, meaning you have reached the V_Z point, which will be indicated on the DC voltmeter. The maximum voltage V_B the DC source is capable of producing must be higher than the V_Z of the Zener diode's tested.

Testing bipolar transistors using an ohmmeter

In the first approximation, a bipolar transistor is made of two PN junctions (P stands for "Positive," N for "Negative") or "diodes." The first is the base-collector junction (test #1), and the other one is the base-emitter junction (test #2).

Using an analog ohmmeter, we should get a low resistance one way (base-collector junction is positively or forward polarized so the diode is conducting and its internal resistance is small), and when we swap the two test leads, we should get a high resistance (negatively or reverse- polarized PN junction), so the diode is not conducting and its internal resistance is high. These readings will be the opposite for a PNP transistor.

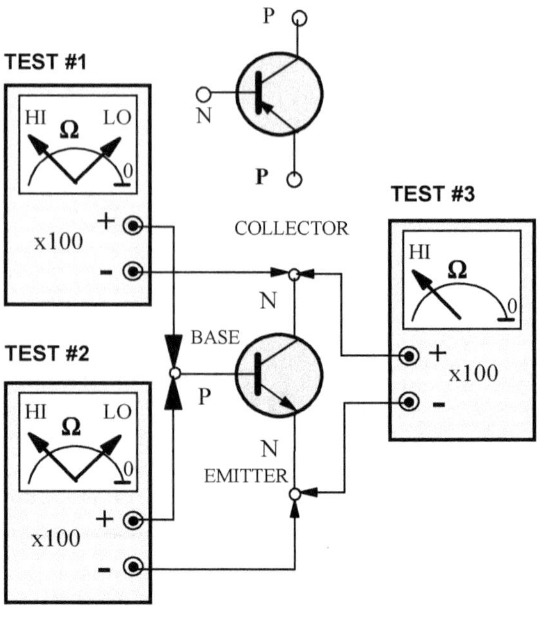

Since the emitter-collector path comprises two "opposing" PN junctions (one forward- and the other reverse-polarized), the resistance reading must always be high for both NPN and PNP transistors. What low and high means will depend on the transistor; power transistor resistances will generally be lower than those of low power ones.

However, these tests are not conclusive. Even if a transistor passes this rough test, it may still be faulty. This is called a false negative. Transistors that don't pass this quick check are definitely bad, so there are no false positives. It is not possible for transistors that fail this test to function properly.

Knowing the polarity of your ohmmeter (or multimeter on "resistance" or "ohms" function), you can easily determine if the transistor being tested is a PNP or NPN type. See page 54.

Measuring collector current in situ

Output transistors in hi-fi amplifiers are mounted either on large heatsinks or on the chassis that serves as a heatsink. Since the metal case is connected to the collector, it must be insulated from the heatsink or chassis by a mica washer. The mounting bolts must not short the case to the chassis since one is used to connect the case/collector to the rest of the circuit, marked X on the drawing (right).

To check/measure the collector current inside a working amplifier, turn the amplifier off and remove the bolt with the collector lead and spring clip attached. Connect an mA-meter in series with that lead wire and to the other mounting bolt. Do not touch the emitter and base connections. Power the amplifier back up and read the current on the mA meter.

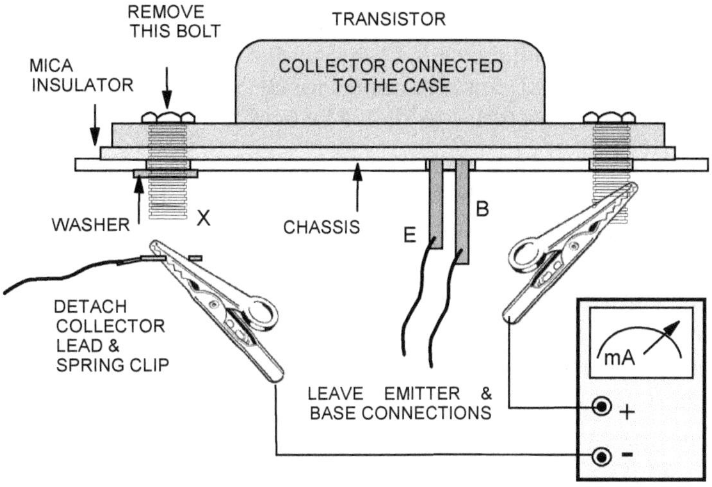

ABOVE: Measuring collector current in situ (inside a working amplifier)

QUICK CHECKS OF JFET AND MOSFET TRANSISTORS

Junction Field-Effect Transistors

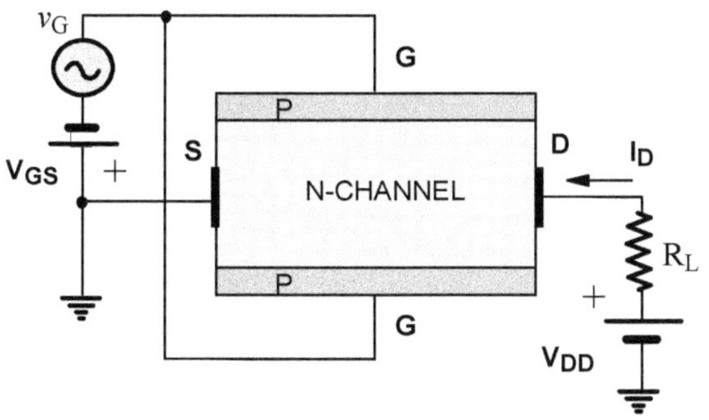

ABOVE: The physical operation of N-channel JFET

As amplifying devices, FETs behave similarly to pentodes. Both are voltage-controlled current sources, with very high input and output resistance. Unlike bipolar transistors, where there is an interaction between inputs and output circuits, unipolar FETs make interfacing with vacuum tubes very easy and are often used in hybrid circuits.

JFET works as a voltage-controlled variable resistor in one operating region (triode region). The narrow semiconductor channel (in this case N-or negative type) provides the conductive path between source S and drain D. When the gate G is negatively biased (just like a grid in a vacuum tube), the width of the top and bottom P-regions is increased, and the width of the N-channel is reduced, increasing the resistance of the channel.

FAR RIGHT: BF245B transfer characteristics for V_{DS}=15V, T=25°C

RIGHT: The output characteristics (I_D vs.. V_{DS}) for BF245B JFET

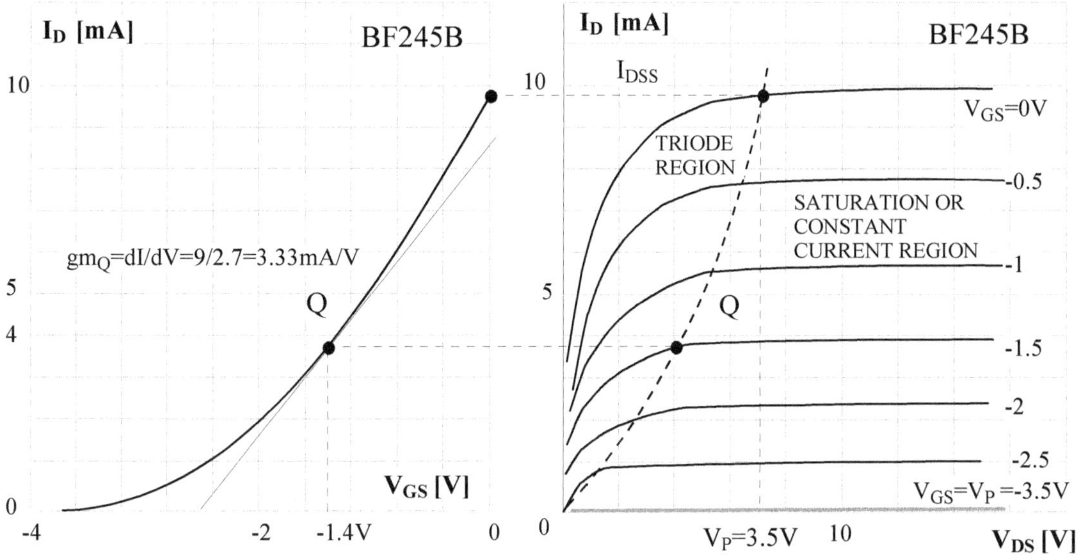

The grid-source voltage V_{GS} needed to reduce the channel thickness to zero is called pinch-off voltage V_P. The current through the channel does not cease when the drain voltage V_{DS} reaches the V_P value because a voltage equal to V_P exists between the pinch-off point and the source S, so the resulting electric field along the channel causes the current to flow.

As drain voltage is increased above the V_P value, the drain current increases only slightly, remaining practically independent of the drain voltage. In this *saturation* region, FET behaves as a voltage-controlled current source.

The drain current that flows when V_{GS}=0V and V_{DS}=V_P is called I_{DSS}, or saturated drain current with shorted input (gate-source).

FETs are semiconductors whose manufacturing tolerances are very wide (compared to vacuum tubes), so the datasheet for BF245B JFET is a typical example of a small signal JFET says that I_{DSS} should be 6-15mA.

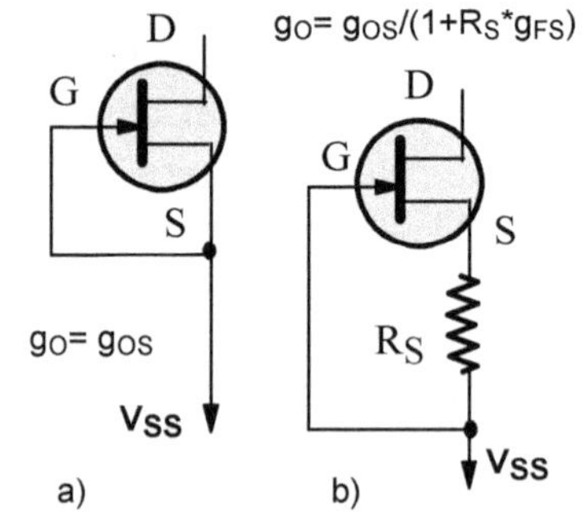

ABOVE: The simplest constant current sources with JFETs

FETs have a negative gain-temperature coefficient, which protects them from thermal runaway. The source-drain current is primarily determined by the input voltage (V_{GS}), and it stays constant with increasing load demand, maaking a FET an ideal current source. In the simplest CCS illustrated in a) V_{GS}=0 and the output conductance g_O is equal to g_{OS} (g_{OSS} is also used), the small signal common-source short-circuit *output conductance*, specified in data sheets.

The constant current is the I_{DSS} current of the chosen FET, which is the drain current when the gate is connected to the source. As g_{OS} ranges from around 1μS (microSiemens) to 50μS or more, depending on the FET type, the output impedance of this simple CCS will vary from only 20kΩ to over 1MΩ. The addition of a resistor R_S in the source circuit allows for the adjustment of V_{GS}, by which the mentioned zero coefficient of temperature drift can be achieved.

The freedom of adjusting V_{GS} allows for any constant current to be had, according to the basic FET formula $\mathbf{I_{DS} = I_{DSS}(1-V_{GS}/V_P)^2}$. The output conductance is reduced in this case by the factor $1+R_S*g_{FS}$, where g_{FS} is the *small-signal common-source short-circuit forward transfer conductance*, equivalent to Gm (mutual conductance) in a vacuum tube.

Testing JFETs with an ohmmeter

Because FETs are sensitive to static charge when not in use, their leads should be shorted by conductive foam or aluminium foil. Apart from protection, this will ensure that all residual voltage built up across the gate-channel PN junction is removed before testing.

Test #1: Due to the symmetry of the channel, there is usually no difference between the source and drain terminals, so a resistance check either way should yield the same low figure, $R_{DS}=R_{SD}$, from 100Ω to 1kΩ.

Test #2: The resistance of the PN junction between the gate and source (or gate and drain) should, just as with bipolar transistors, be very high, almost infinity in one, and very low, almost zero in the other direction (polarity). The polarity will depend on the internal construction of your ohmmeter (or multimeter on ohms) and the type of JFET, N-channel, or P-channel.

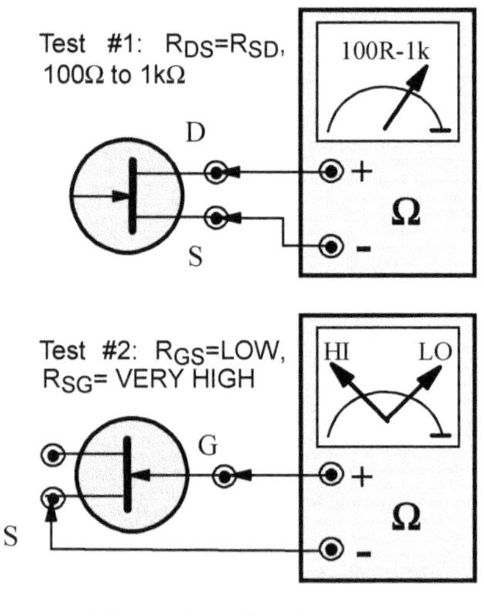

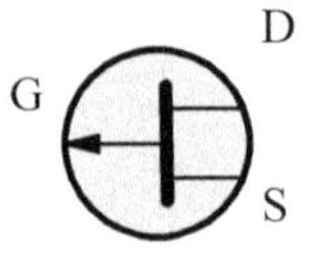

NOTE: N-channel JFET shown, meter indications for P-channel JFET should be the exact opposite

Testing MOSFETs with an ohmmeter

Checking MOSFETs requires the same preventative measures as for JFETs, but their testing can be unpredictable and inconclusive since any stray transient voltage on the gate can trigger them into conduction.

Hookup the circuit as shown (without the jumper), then remove the shorting between terminals (foam or foil). The ohmmeter should read 100-1,000 ohms for a depletion type MOSFET, as with JFETs. The resistance reading for an enhancement type should be infinity.

Then, in test #2, short the drain to the top of the upper resistor (with the jumper) as shown. The reading should drop to a lower value.

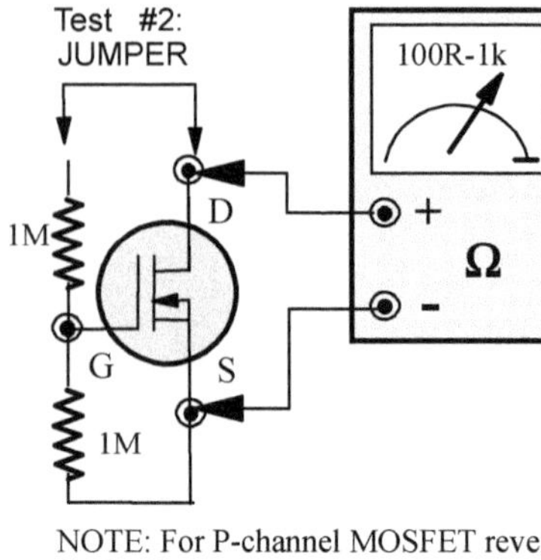

NOTE: For P-channel MOSFET reverse the leads

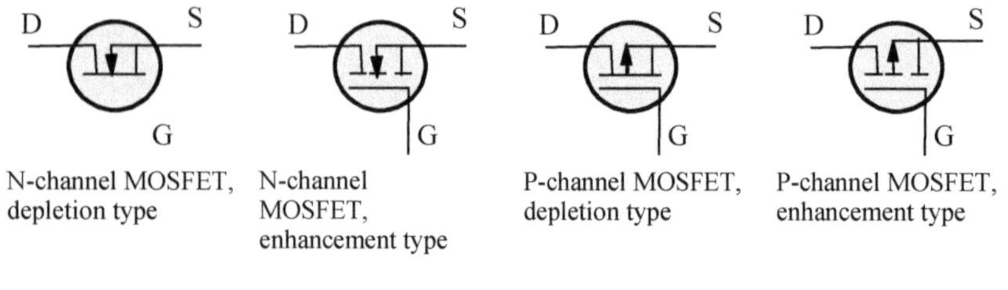

TRANSISTOR TESTERS

Bipolar transistor testing function on tube testers and digital multimeters

Some vintage tube testers also have a transistor and diode testing function, but these are very basic. Eico 666 and 667 (functionally identical) tube testers feature a 5-position switch (4), with the center position for tube testing mode and two positions each for NPN and PNP transistors.

In positions marked "1", 7.5V_{AC} transformer tap voltage is half-wave rectified and capacitor filtered and brought to the series circuit with R4 (1k) and collector-emitter transistor terminals, so the meter indicates I_{CE0} reverse leakage current (see next page).

In positions "2", a 200k resistor (R3) connects the base terminal to the supply DC voltage "E." Different shunt resistors are across the meter for the two tests (R5 and R6). The switching arrangement is not shown.

Digital multimeters are today's essential T&M tools, and since all the required "infrastructure" for transistor testing is already there, it is surprising that only a few cheaper models have the basic transistor testing capability.

The first clue (1) is the symbol used on the function selector switch and displayed on the LCD display (2), "h_{FE}". The "h" indicates hybrid parameters, "hybrid" is a mix of voltages and currents.

"Forward" stands for "forward transfer current ratio" or the current gain of a transistor (see the box below), and "E" is for common emitter connection of the transistor. There are three different sets of hybrid parameters, one for each of the three possible connections, common emitter (CE), common base (CB), and common collectors (CC).

The capital letters "FE" indicate that this is a DC current gain, in comparison to the AC gain (h_{fe}), where an AC signal of small amplitude is fed into the base, and an increase in AC collector current would be measured, with the transistor biased at some predetermined DC bias point.

The current gain of a transistor $h_{FE}=I_C/I_B$ is often also called β (beta). DMMs usually use a low collector-emitter V_{CE} testing voltage of 2.4-2.8V and push 5-10mA of base current.

Since β varies with the bias point, the multimeter h_{FE} test is limited to a single point on the curves (one current and one low voltage). Most vintage transistor testers also test in a single point; only a few better transistor "analyzers" feature adjustable bias and collector voltage.

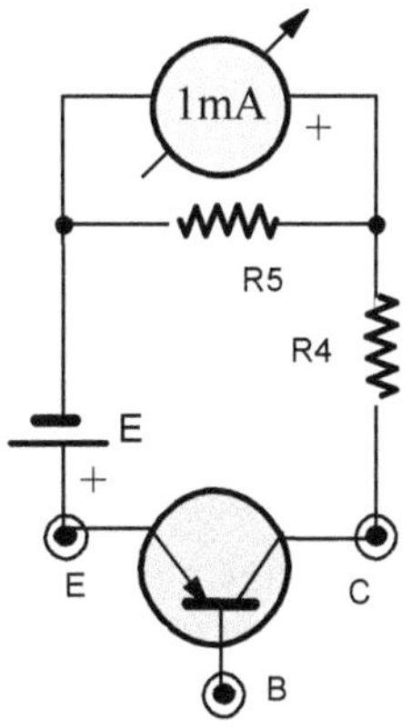

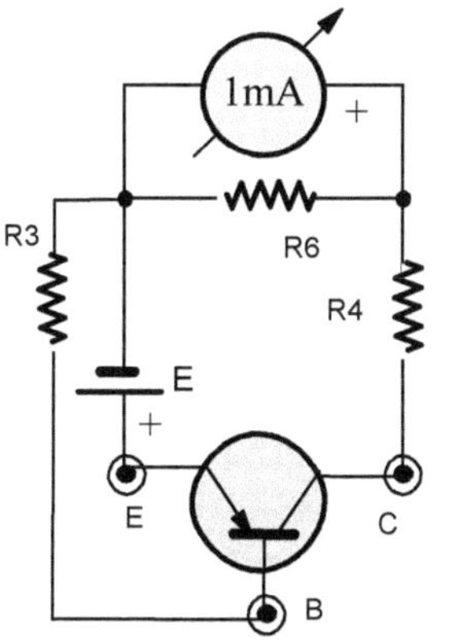

ABOVE LEFT: Measuring leakage current I_{CE0} on Eico 666 & 667 tube testers (PNP transistor shown)

ABOVE RIGHT: Measuring DC gain (DC β)

BELOW: The transistor/tube selector switch and analog meter (EICO 667). Notice that gain readout is relative only (in percentages 0-130%)

BELOW: The h_{FE} test on a true RMS DMM

A brief overview of bipolar transistor parameters

Since a transistor is a 4-pole network, two voltages and two currents are involved, as per the "black box" diagram below. The two independent variables are v_2 and i_1 and two dependent variables, v_1 and i_2:

$v_1=h_{11}*i_1+h_{12}*v_2$ and $i_2=h_{21}*i_1 1+h_{22}*v_2$

For a transistor in common emitter connection $i_1=i_B$ (base current), $v_1=v_{BE}$ (base-emitter voltage), $i_2=i_C$ and $v_2=v_{CE}$. Parameter h_{11} is the input impedance, h_{12} is the reverse transfer voltage ratio, parameter h_{21} is the negative forward transfer current ratio, and h_{22} is the output admittance (the reciprocal of output impedance). The significant letter is retained as the 1st suffix, and the letter "e" is added as the 2nd suffix to indicate common emitter mode, so they become h_{ie}, h_{re}, h_{fe} and h_{oe}.

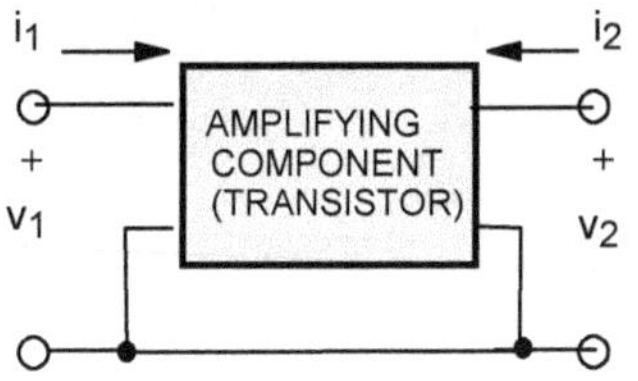

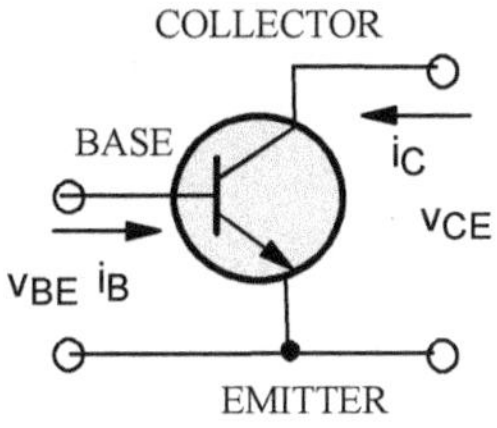

Measuring reverse leakage currents - I_{CBO} and I_{CEO}

Cheaper service type transistor testers only measure I_{CEO} leakage current that flows between collector and emitter, with base open (indicated by the "O" in the subscript). Since that current is much higher than I_{CBO} leakage current that flows between collector and base, with emitter open ("O"), less sensitive and thus cheaper analog mA meters were required.

A more sensitive meter movement (usually 50 µA) is needed to measure I_{CBO}. To measure I_{CEO} on that µA meter, better transistor checkers have a selector switch that switches in a meter shunt (R2). Most transistor testers, like Heathkit IM-30 and Eico 680, have a range selector that switches various shunt values across the meter and thus determines the current range. See the 15µA, 1.5mA, 15mA, 150mA, and 1.5A switch on the Heathkit IM-30 in the photo below (1) and the 50µA, 0.5mA, 5mA, 50mA, and 0.5A positions on Eico 680's "Range" switch (2).

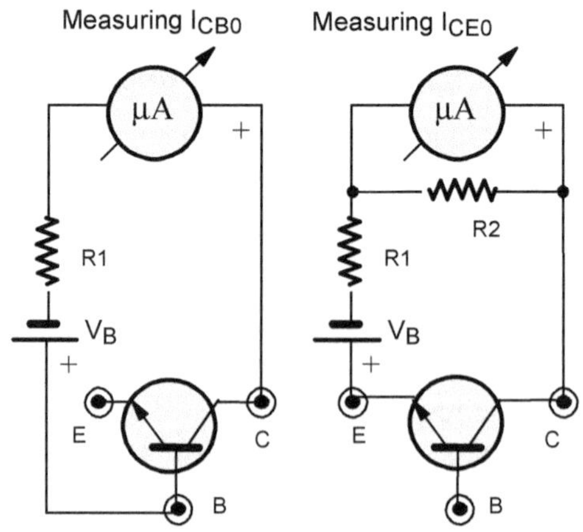

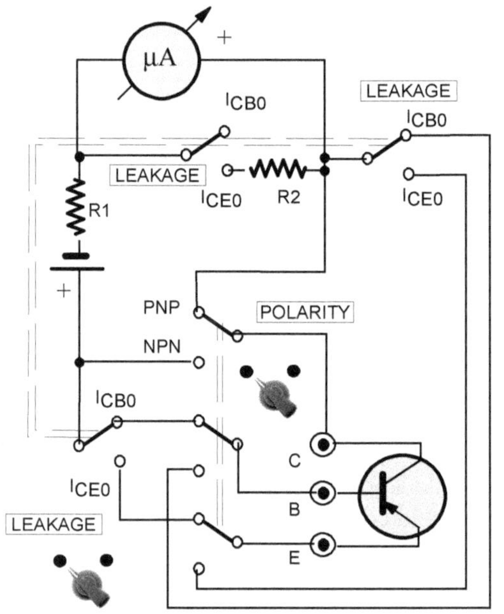

ABOVE: A combined circuit with switching to test I_{CEO} and I_{CBO} for both NPN and PNP transistors

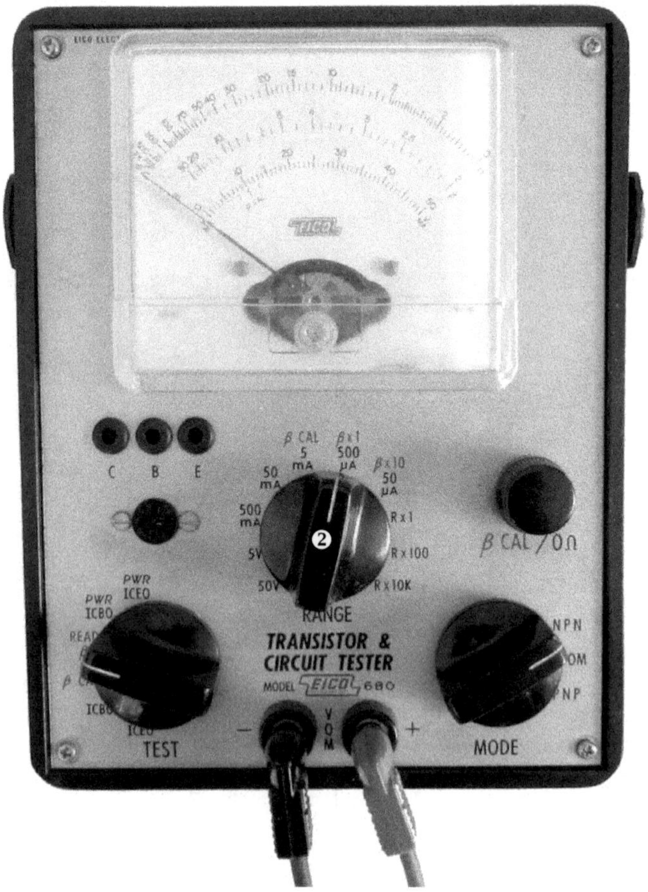

The circuit is a simple series circuit with a battery providing a DC voltage V_B and a current limiting resistor R1. Apart from the meter's accuracy class, the overall accuracy of these tests is affected by the aging (gradual reduction of its voltage) of the battery and the resistance contacts, which become dirty and oxidized over the 40 or 50 years since these checkers were built.

These checks are aimed primarily at early or vintage transistors with considerable leakage. Modern, currently produced small signal silicon transistors have a very low ICBO of only a few nano amperes, and none of these vintage checkers can measure such low currents.

Eico 680 uses five 1.5V batteries, four AA-size, and one "C" type. For leakage tests, the $4.5V_{DC}$ voltage is used. On the I_{CBO} range for "signal" transistors, the current limiting resistor is 33k, and for power transistors, a 3k9 resistor is switched in instead. On the ICEO range, the current limiting resistor is the same 3k9, and for power transistors, a 15W resistor is used.

The "circuit tester" means that 680 is also a basic analog meter that can measure resistance (ranges Rx1, Rx100, and Rx10k), DC voltage (two ranges, 5V, and 50V), and DC currents (five ranges, from 50mA to 500mA).

Apart from its educational value, the most valuable part of instruments such as Eico 680 is their sensitive analog meters (50mA in this case). These are almost impossible to buy today (or very expensive), so if, for instance, the analog meter on your expensive tube tester or another T&M instrument burns out, this is one of the candidates for its replacement (see page 49 on how to do it).

Eico 680 and similar vintage transistor checkers sell for under US$30 on eBay.

B&K model 162 (right) is in the same class as Eico 680, but it also tests FETs (1) and I_{CES} (2), which is a reverse collector to emitter leakage with the base shorted to the emitter.

During I_{CE0} leakage tests, the leakage current has to pass through the base, so it is amplified by the transistor's beta.

This means that when high I_{CE0} leakage is identified, that could result from a faulty collector-base PN junction, or the transistor could simply have normal leakage but a very high gain (since I_{CE0} is effectively I_{CB0} amplified by the transistor).

Measuring leakage currents I_{CES} and I_{BES}

To address that ambiguity, some transistor testers such as B&K 520B measure I_{CES} and I_{BES} instead. These tests are not affected by the magnitude of transistor's b yet still indicate the condition of the collector-base and emitter-base PN junctions.

The three currents are related so that I_{CES} is smaller than I_{CE0} but bigger than I_{CB0}. If any of these three currents are equal, the transistor is defective.

For instance, an avalanche breakdown would result in I_{CE0} being equal to I_{CES}.

B&K 520B is of limited value since its gain test is only qualitative, "go-no go" type, meaning the tester only indicates if a transistor has a gain or not, but not the actual value of that gain.

Otherwise, it identifies NPN versus PNP transistors, silicon versus germanium, and EBC terminals of unknown transistors (notice that terminals are marked 1-2-3 and not E-B-C (3).

The "Lo" drive is for testing small signal types, and the "Hi" drive is for power transistors (4). This is a standard feature of transistor testers of this vintage.

B&K 530 does all that plus measures gain, leakage, and gain-bandwidth (5) product (ft) for bipolar transistors, and transconductance, V_P, I_{DSS} for FETs, so it is a much more useful instrument.

ABOVE: B&K Precision 162 also features a large high sensitivity meter (0.1 mA FSD) and can also test FETs and I_{CES} , reverse collector to emitter leakage with the base shorted to the emitter.

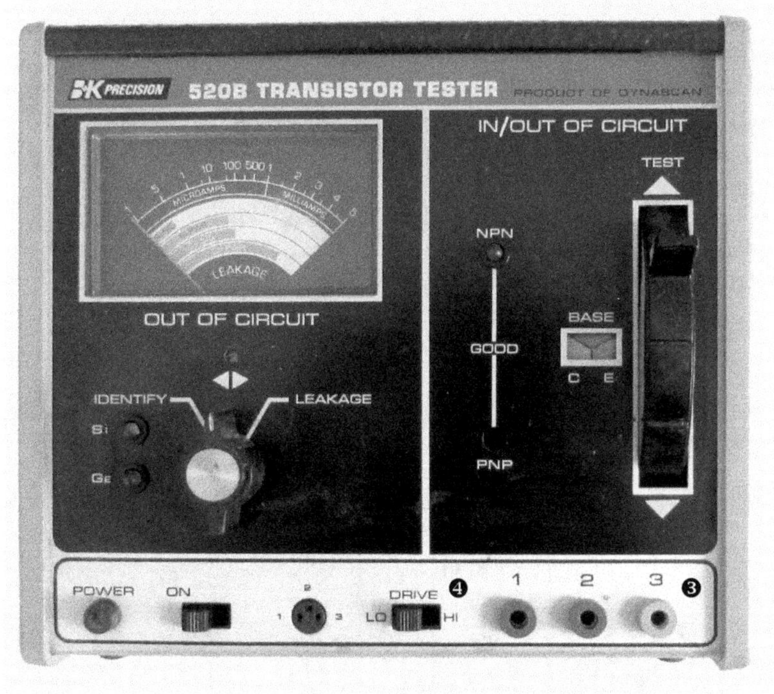

B&K Precision 520B (ABOVE) and 530 (BELOW)

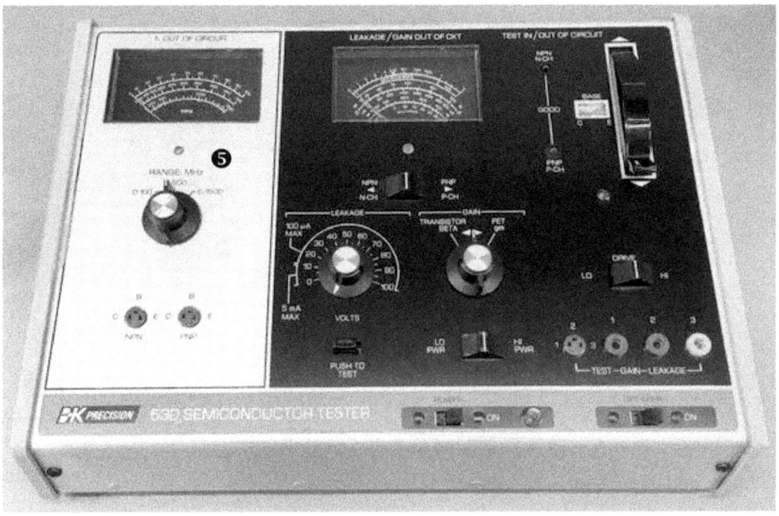

Measuring DC β

Eico 680 and most other testers can also measure β, so let's see how this is done. The operation of β-testing in Eico 680 is typical of cheaper service-type vintage transistor checkers. First, with both TEST and RANGE switches in "β CAL" position, the "β CAL/0Ω" control is adjusted until the needle is on the β CAL meter's mark. This adjusts the base current to get 1mA of collector current.

The test switch is then changed to β x1 or β x10 position, and transistor's β is read on the middle scale marked β. The meter is switched from the collector circuit into the base circuit, indicating the base current for the previously set 1mA of collector current. DC β is the ratio of the collector to base current. Since I_C=1mA, β becomes the reciprocal value of the base current in mA, which explains the nonlinear scale.

This checker can only measure DC β up to 300 (30x10), and only for signal (low power) transistors, which are its serious limitations.

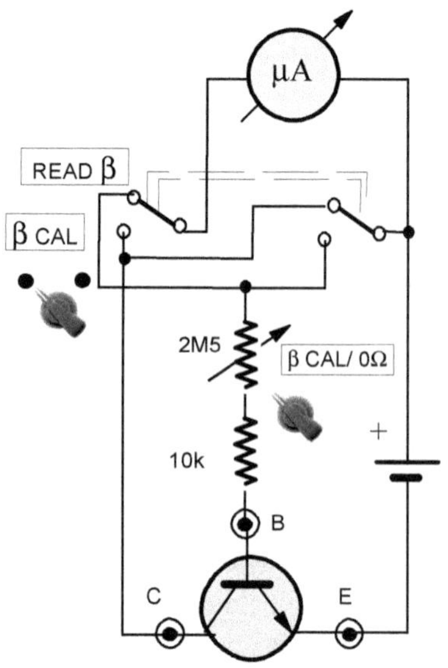

ABOVE: The DC β test on Eico 680

Measuring AC β

Sencore TR139 (photo on the next page) uses the same 2-step comparison principle as DC beta testers discussed, but it measures AC beta. As an intellectual exercise, study its circuit diagram below and try to figure out how this circuit works.

The upper section (gray frame) is simple, transformer T1, half-wave rectifier plus capacitive filter, so a DC power supply for leakage test (I_{CBO} only) and is of no interest to us here (3).

Two oppositely polarized Zener diodes (1) are on the primary (mains side) of the transformer T2. They chop the 117V (USA mains) sine wave down to an 8.2V peak in each direction or 16.4V_{PP} (peak-to-peak). Only part of that almost square-shaped voltage is then taken from the calibration potentiometer R3 and fed into T2's primary (2).

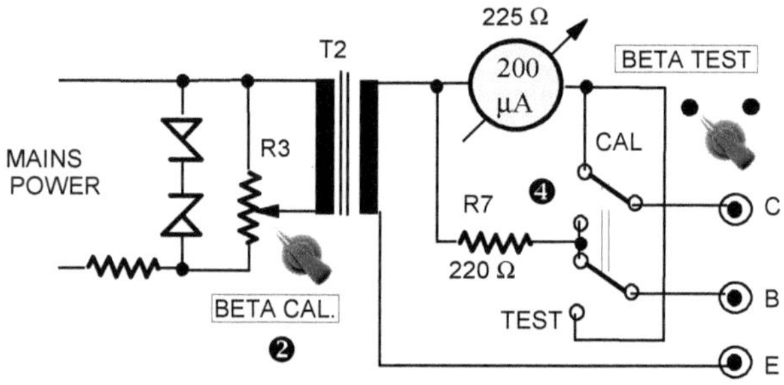

ABOVE: The in-principle schematic of Sencore TR139 tester in β testing mode

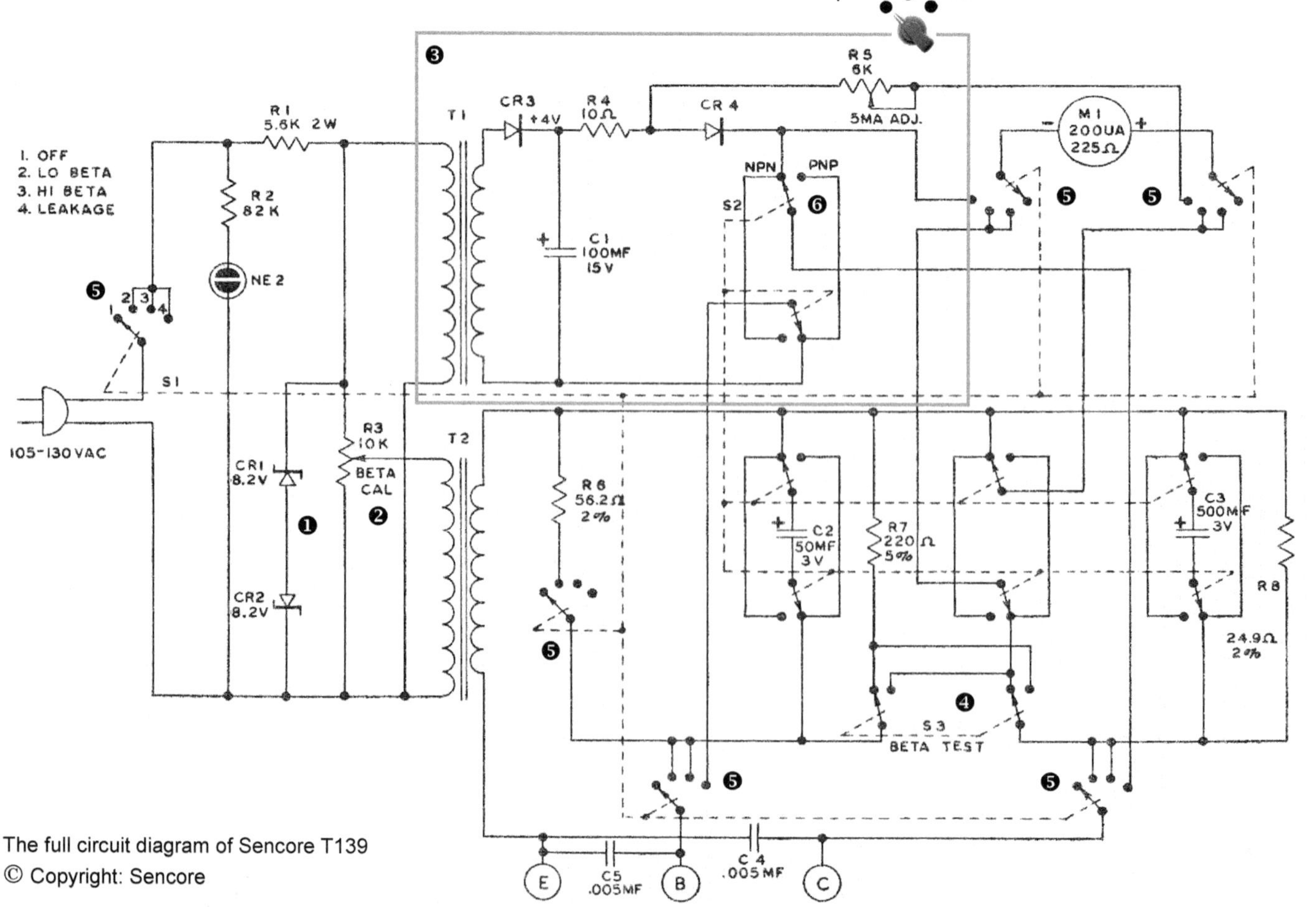

The full circuit diagram of Sencore T139

R3 sets the average value of the DC collector current pulses to 2mA. When the "Beta Test" button (4) is pressed, the meter is transferred from the collector to the base circuit to measure the average base current. The resistor R7 (220 ohms), whose resistance equals that of the analog meter, is initially in the base circuit and is now also transposed to the collector circuit.

In the "Hi beta" position, 1/10 of I_C (0.2mA) produces full-scale deflection, and in the "Lo beta" position, 1/2 of I_C (1mA) is the FSD. The ratio of I_C to I_B is the AC gain or beta.

RIGHT: Sencore TR139 dynamic in-circuit transistor tester

BELOW: Sencore TF151A uses the same testing principle as T139 but can also test FETs

BELOW RIGHT: Eico 685

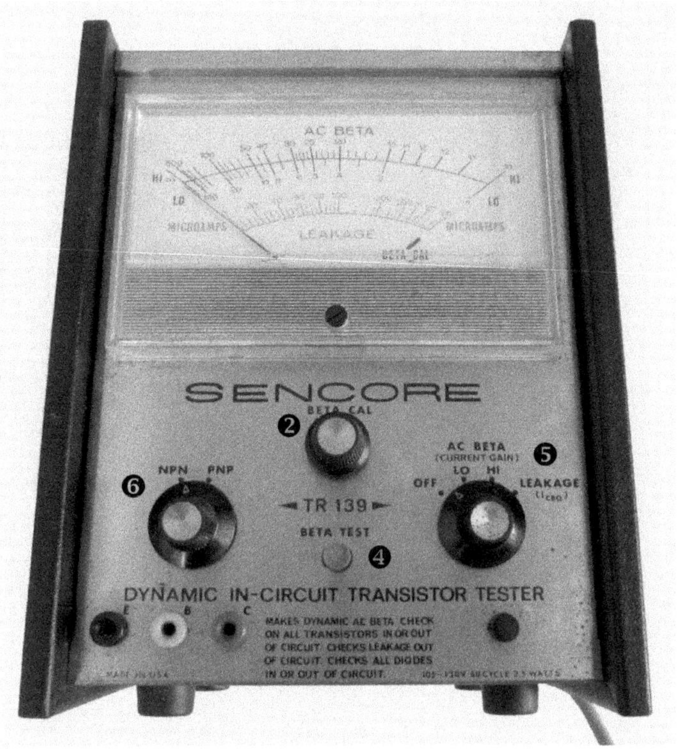

Sencore TF151A and Eico 685

We could not find the release date of the Sencore model TF151A, but Eico 685 dates back to 1970, according to its manual. They may look different (Eico is almost twice as wide), but the two instruments are functionally identical, meaning they use the same circuit for transistor testing, based on the same principle as that just discussed for Sencore TR139. They can also test field-effect transistors for transconductance Gm, for which they use a Hickok bridge with two bipolar transistors. Eico uses two NPN transistors in the bridge while Sencore opted for the PNP type, but the circuit is otherwise all but identical, even down to voltage values.

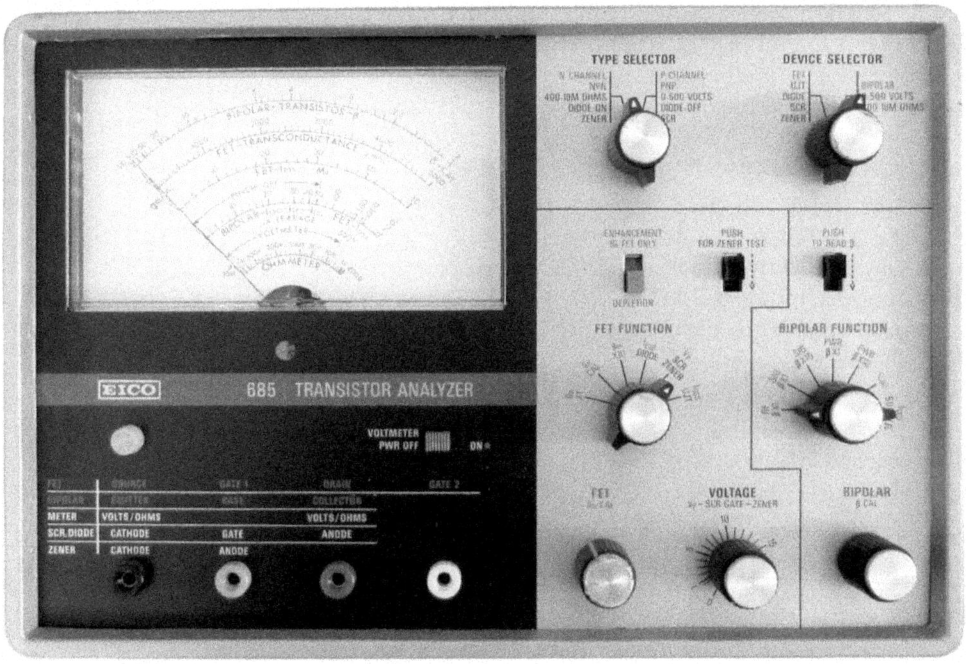

Two Gm ranges are provided, 0-5,000 and 0-50,000 micromhos. The international unit for conductance is the reciprocal of ohm or 1/Ω, called Siemens (symbol: S). However, the Americans call it "mho" (ohm spelled backward).

1S = 1 A/V, but mA/V is most commonly used in tube testing, which would be mS (milliSiemens). So, to convert micromhos to mA/V, simply divide by 1,000, meaning the ranges of these two instruments are 0-5 mA/V and 0-50 mA/V.

LEFT: The internal view of Eico 685

Internally, the tester is well laid out with easy access to most components. Notice the two transformers and two Zener diodes that clip the primary AC voltage to one of them (1). Apart from the pair of transistors mentioned, there are three calibration trimmer pots (2), almost a dozen Zener, and ordinary silicon diodes (3).

There are only a few electrolytic capacitors, and those must be replaced now since after 50 + years, they are way past their life span!

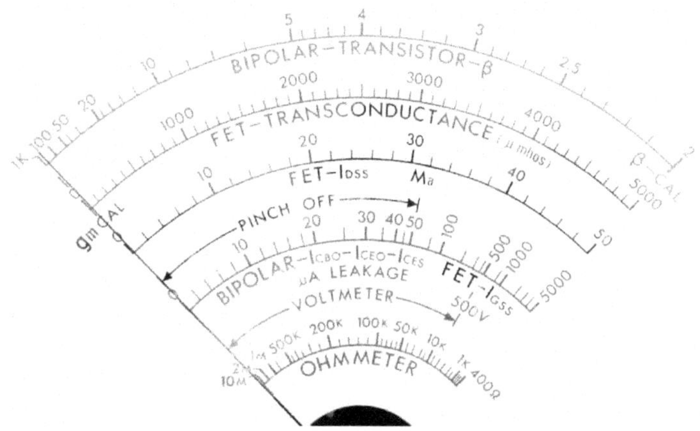

ABOVE: The meter scales of Eico 685

RIGHT: The heart of the tester, transconductance bridge and associated circuitry, most of it for FET testing

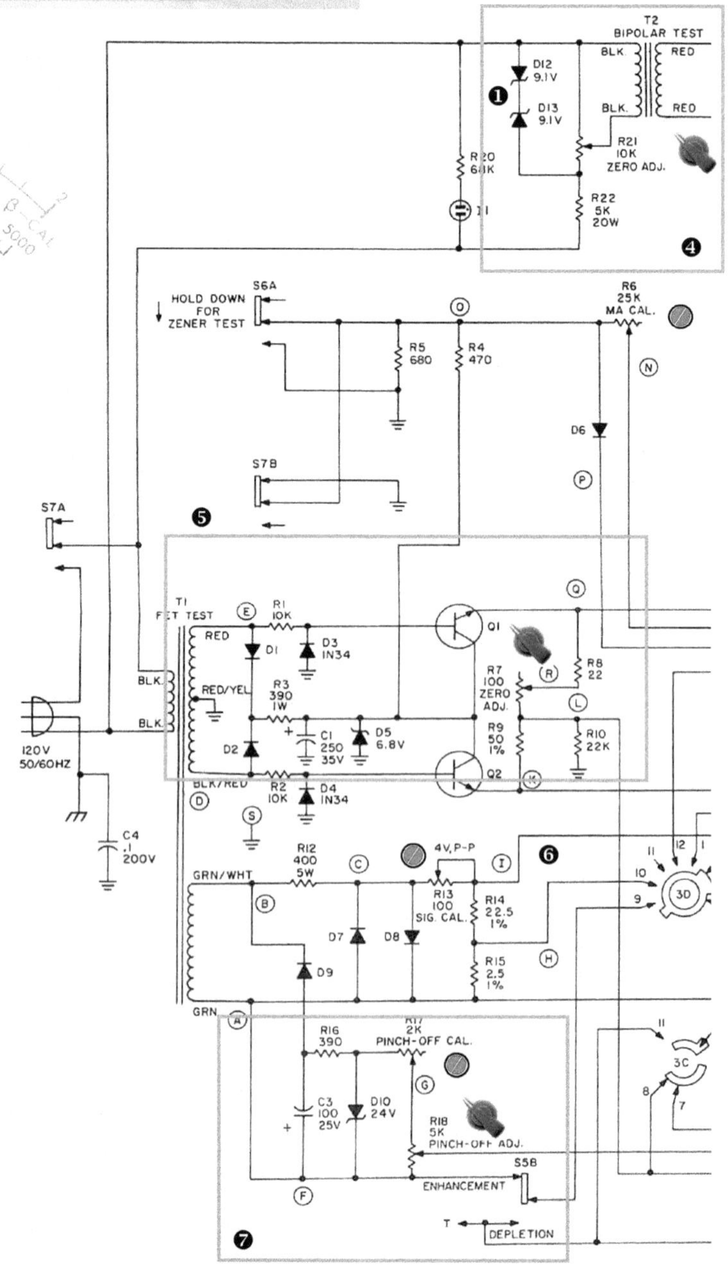

Four functional blocks are easily identified. The bipolar test circuit (4), as in Sencore TR139, the Gm bridge with two NPN transistors (5), the voltage signal for FET Gm testing (6), and the adjustable negative DC voltage (7) for FET V_P measurements and SCR and Zener testing.

Resistors R14 and R15 form a precise voltage divider, providing two test signals, $400mV_{PP}$ and $40mV_{PP}$. Notice two diodes D7 & D8 in anti-parallel connection, clipping the AC voltage down to almost a square wave of less than $1V_{PP}$, so the $4V_{PP}$ marked on the schematic is an error.

The calibration procedure and troubleshooting chart are in the operating manual, so servicing the instruments of this kind and this low level of complexity isn't hard.

The unit can be used as an analog voltmeter, but its scale is very nonlinear, with 30V at its center and 500V at its maximum, so it's suitable for quick "ballpark" service checks only.

Leader LTC-906

Despite its simple control panel layout (five 2-position and three 3-position sliding switches) and modest appearance, Japan-made LTC-906 is a complex instrument, with dozens of transistors and ICs (integrated circuits) inside. This is necessary due to its partially automated operation, where the "checker" identifies transistor terminals by itself and indicates it using LEDs (1).

The circuit enables h_{FE} to be measured in a single step, with a precise predetermined base current used (1µA).

This obviates the need for two-step manual tests where the meter is first in the collector circuit, then in the base circuit, resulting in a nonlinear beta scale (due to the calibration as the reciprocal of the base current), as we have seen in American testers.

Notice the linear h_{FE} scale, also used for the leakage current and ICEO.

DC beta or h_{FE} can be measured in 3 ranges, x1, x10, and x100 (2), and since the meter goes up to 100 (3), that means beta can be read up to 10,000, considerably higher than older American units that only go up to 300 or 500!

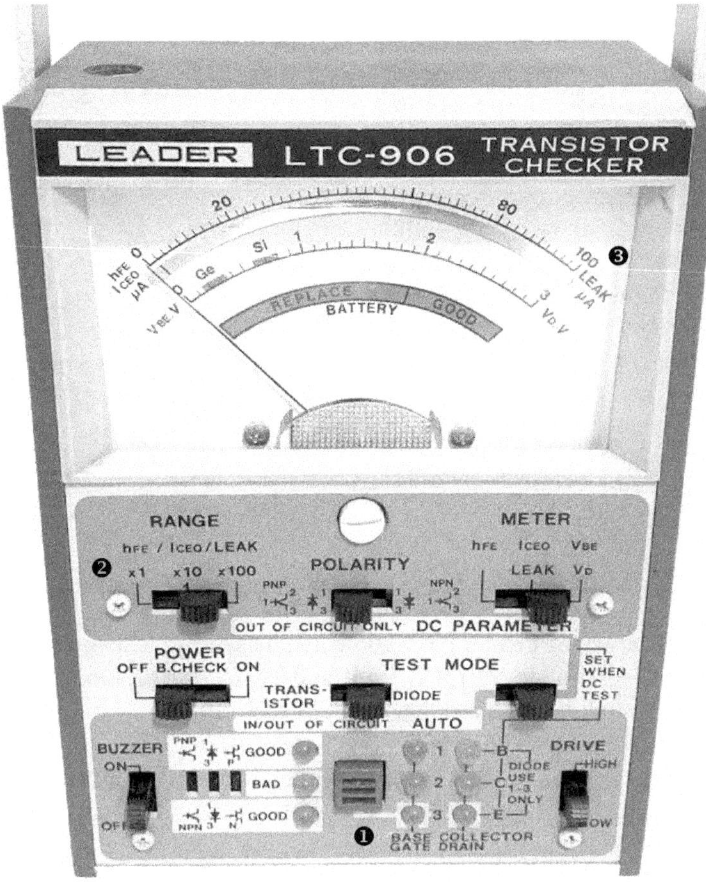

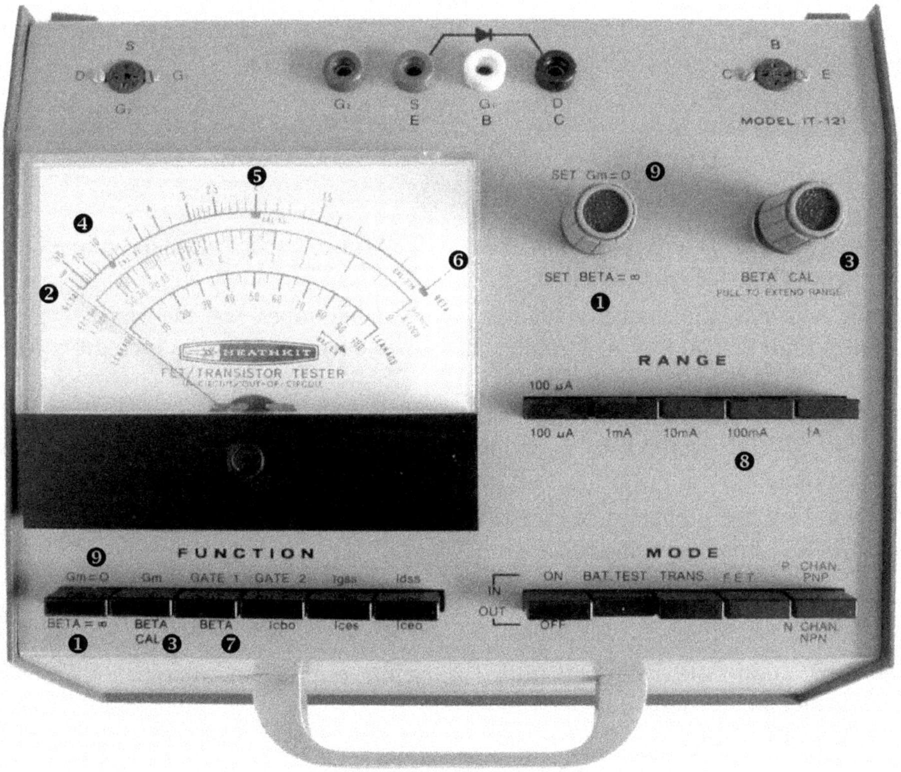

Heathkit IT-121

Heathkit IT-121 can measure beta, I_{CBO}, I_{CEO}, and I_{CES} of silicon and germanium transistors and Gm, I_{GSS}, and I_{DSS} of FETs.

For UJTs, three leakage tests may be performed (out-of-circuit only), while SCRs (silicon controlled rectifiers) and triacs may be tested for function and operation both in and out-of-circuit.

Three settings are needed to test beta. First, "SET BETA=∝" (1) adjusts the "zero" on the left of the scale (2). Then "BETA CAL" P/B must be pushed (3) and "BETA CAL" potentiometer adjusted to set the indicator to one of the three BETA CAL meter marks, x1 (4), x5 (5), or x10 (6).

Finally, the "BETA" P/B (7) is pressed, and the gain is read on the BETA scale. If the reading is greater than 50, the next lower current range should be pressed (8).

Say we were at 100mA and calibrated the reading for BETAx10, then we would press 10mA P/B, and this now means the new reading, say 3.5, must be multiplied not just by 10 (BETAx10 cal.) but also by ten, again due to 10x lower current range, so the actual beta would be 3.5*10*10 = 350!

The setup procedure for FETs is simpler; only a single calibration step is needed to set Gm=0 (9).

Compared to Leader LTC-906, IT-121 is a very simple tester, practically an elaborate switching box. There are no active components inside, and most components are on one large printed circuit board (1). Although they look like switches, the multi-terminal devices on the PCB (2) are just terminal strips.

Notice two C-size batteries that power the instrument (3).

RIGHT: The internal view of Heathkit IT-121

The accuracy of transistor testers and correction factors

The accuracy of Leader LTC-906 when measuring leakage and ICBO current is +/-6% of FS (full scale). Since there are three current ranges, 0.1mA, 1mA, and 10mA, the accuracy would then be +/-6 µA, +/-60 µA, and +/-600 µA (0.6mA). When testing h_{FE}, the claimed accuracy is +/-20.

If a tester shows a beta of, say 600, that means the real beta could be anywhere from 0.8*600=480 to 1.2*600=720! 480 to 720 is a huge range, does not instill much confidence in the whole transistor testing business, does it?

If a transistor has leakage (I_{CBO} is too large to be neglected), the indicated beta of LTC-906 and all other transistor testers mentioned will be higher than the real figure. $I_C=h_{FE}*I_B + I_{CEO}$, so the real $h_{FE}=(I_C-I_{CEO})/I_B = I_C/I_B - I_{CEO}/I_B = h_{FEM} - I_{CEO}/I_B$, where h_{FEM} is the measured value. If I_B=1µA is used to measure beta, the correction calculation becomes simple, $h_{FE}= h_{FEM} - I_{CEO}$!

If a measured value is h_{FEM}=600 and I_{CEO} =50mA, then h_{FE}=600-50=550! For testers that don't measure I_{CBO}, that current can be calculated as $I_{CBO}= I_{CEO}/h_{FE}$. For the previous case, that would be I_{CBO}= 50/550= 0.091 µA.

SEMICONDUCTOR CURVE TRACERS

Tektronix curve tracers

Cheaper curve tracers aimed at the DIY users need an external oscilloscope, but Tektronix tracers are stand-alone units with their own CRT screens and all required oscilloscope-type functions and controls.

Model 570, introduced in 1955 and designed to display the characteristic curves of vacuum tubes, was the first Tektronix curve tracer. This tube-based model is still in high demand by audiophiles and tube sellers and fetches incredibly high prices on eBay and other online shops.

Only two years later Tektronix released its first curve tracer for transistors, model 575, which remained in production until 1972. The 577-D1 (with a storage CRT) and 577-D2 (without it) were introduced in the mid-1970s and remained in production for about ten years until the mid-1980s.

370B and 371B programmable curve tracers were the pinnacles of American digital electronics in their day, with the ability to test devices up to 1,600V or up to 10A at the lowest voltage setting.

"Semiconductor Device Measurements" **FURTHER READING**

"Semiconductor Device Measurements" by John Mulvey is a 1968 Tektronix publication, part of their "Measurement Concept" series. The book is available for download online.

The principle behind the graphical display of V-I characteristics

To display I_C-V_{CE} transistor characteristics on an oscilloscope, the collector- emitter voltage must be swept from zero to some maximum value V_M. One solution is to use a full-wave rectified mains voltage from a power transformer's secondary, or, alternatively, a floating sine wave oscillator or function generator. Beware: most, especially cheaper signal sources, have single ended-outputs, with one side grounded or earthed.

The voltage drop across the measuring resistor R_M is taken straight to the scope's vertical or Y input (next page), because it is proportional to collector current I_C. The resistance of R_M must be stable and precisely calibrated.

The collector is grounded (earthed) and the emitter is connected straight to the horizontal or X-input of the scope (V_{CE}). The base circuit is simply a variable DC source (again, fully floating) to which a current limiting resistor R_B is added, and a μA or mA meter to measure the base current. This circuit only works for NPN transistors (for PNP types points A and E must be reversed, so the pulsating rectified sine wave's polarity is reversed) and only displays one characteristic curve, for one value of base current I_B.

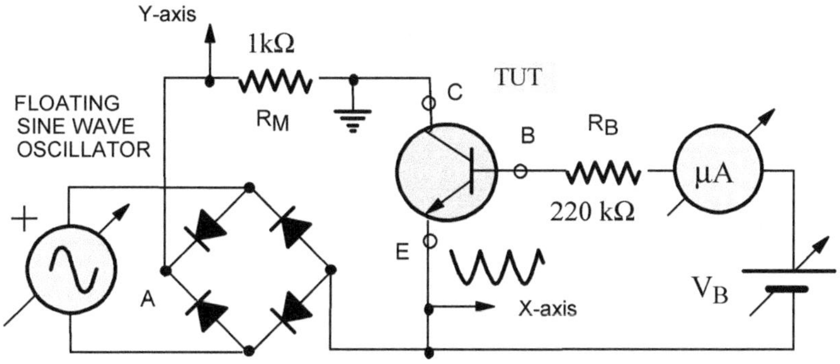

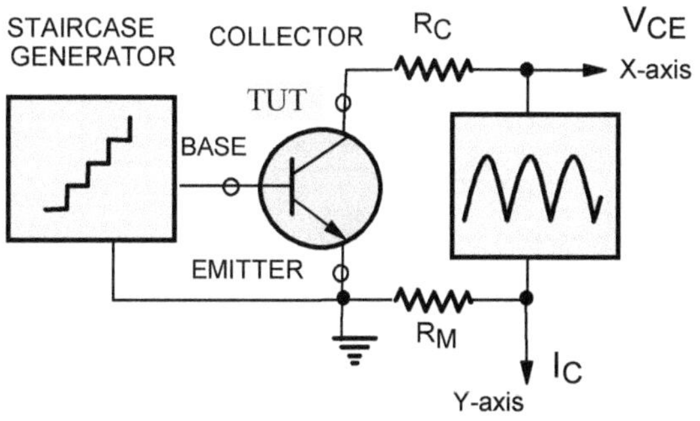

ABOVE LEFT: A floating sine wave oscillator, a bias battery V_B, two resistors, a rectifier bridge and an mA-meter are all that is needed to display bipolar transistor's I_C-V_{CE} characteristic on an oscilloscope

ABOVE RIGHT: The basic block diagram of a bipolar transistor curve tracer

BELOW RIGHT: Collector voltage sweep waveform and staircase base current as a parameter for an NPN transistor. Each step corresponds to one I_C-V_{CE} curve

Commercial transistor curve tracers

Commercial curve tracers are not fundamentally different from the basic circuit just described; all they add to this circuit is various switching options and a staircase generator with three or more base current steps so multiple characteristic curves can be displayed "simultaneously" on the screen.

Well, strictly speaking, the curves are displayed sequentially, but because the sweep is 50 or 60 times per second, our eyes cannot respond so quickly, and due to their inertia, we see all curves together on the screen.

The block diagram of a bipolar transistor tester is shown above right. For PNP transistors, the base current steps are negative (descending, instead of ascending as drawn), and the polarity of the pulsating V_{CE} voltage must also be reversed.

Field-Effect Transistors (FETs) are, just like vacuum tubes, voltage amplifiers, so instead of base current steps, gate voltage steps are needed, and of reversible polarity, so both N-type and P-type FETs can be tested and their I_D (drain current) versus V_{GS} (gate-source voltage) can be displayed.

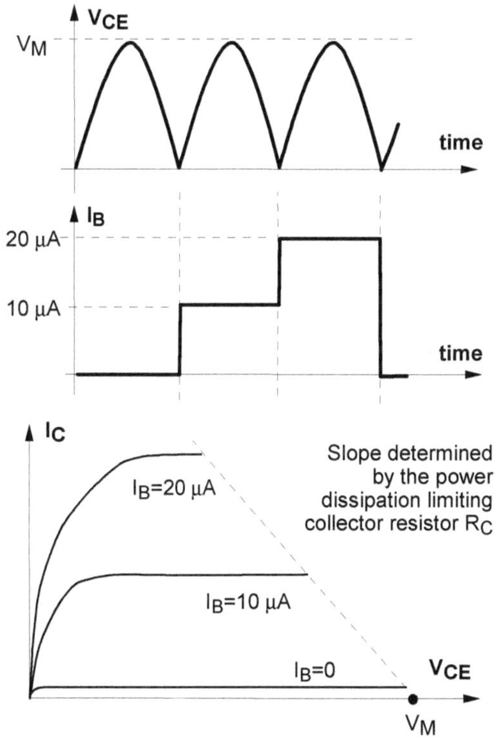

Leader LTC-905

As of this writing (late 2019), online sellers ask exorbitant sums for this vintage Japanese curve tracer, US$400-700, which is preposterous for such a simple device. Its operating and service manuals are available online and are recommended for further reading if this is your area of interest.

The manual for B&K 501A curve tracer, also posted online and available for download, is even better; it explains its operation concisely and various issues and aspects of semiconductor testing.

The sweep voltage is a full-wave rectified AC from one of the power transformer's secondary taps. The staircase generator's signal type (base current or gate voltage) and polarity are switch-selectable, depending on the type of component under test, and fed into a buffer amplifier.

Three gate voltages (0.1, 0.2 & 0.5V per step) and eight base current ranges (10, 20 &50 μA per step, 0.1, 0.2, 0.5, 1 & 2 mA per step) are switch-selectable, controlling the amplitude of the seven steps produced by the staircase generator (meaning no more than seven curves can be produced).

The "A-OFF-B" switch feeds that signal into one of the two test "bays" or disconnects them both in the center "OFF" position.

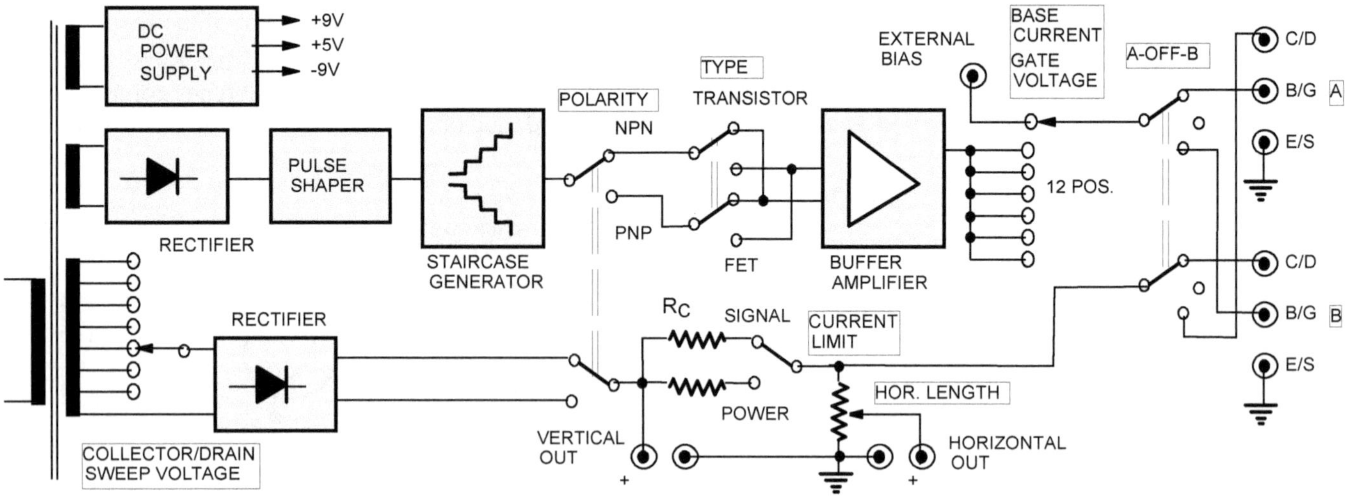

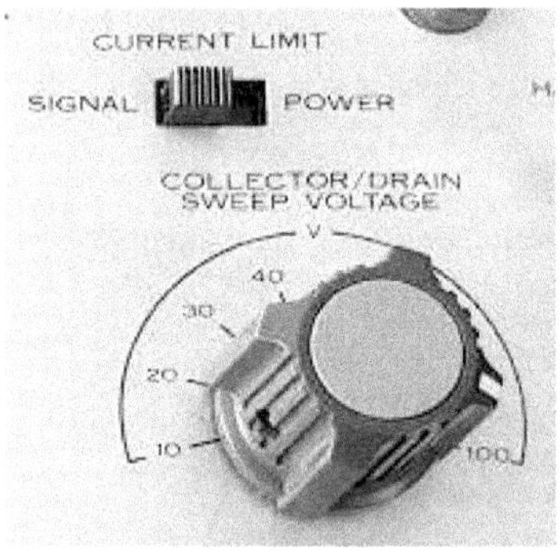

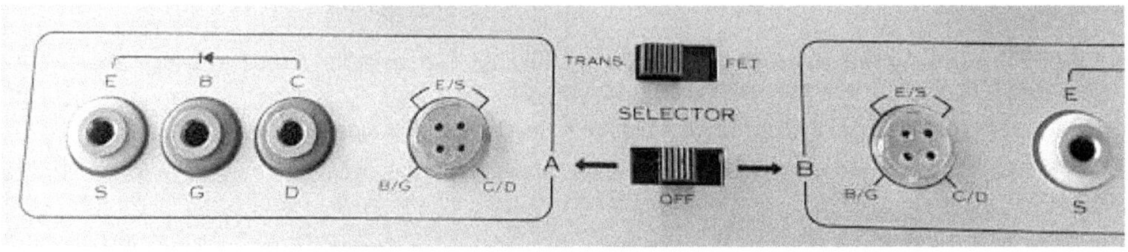

ABOVE: Functional block-diagram of Leader LTC-905 curve tracer and close-up details of its control panel

Heathkit IT-1121 and IT-3121

Heathkit IT-1121 and IT-3121 are budget curve tracers for semiconductors, capable of testing FETs and bipolar transistors. The "STEP SELECTOR" switch (7) selects gate voltage for FETs or base current for bipolar transistors.

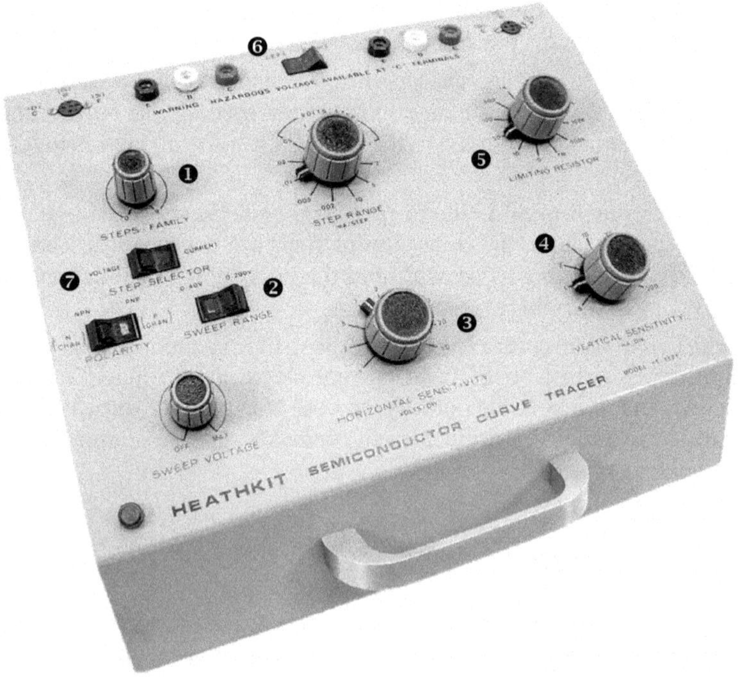

Similar models include B&K 501A and Leader LTC-905. Eico 443 cannot test FETs (has no voltage steps). These budget tracers do not include a display or vertical and horizontal amplifiers and thus require an external oscilloscope.

501A uses pulsating rectified sinewave of up to a maximum 100V peak for the collector supply, as does Leader LTC-905. The voltage steps for testing FETs are 0.05V, 0.1V, 0.2, 0.5, and 1V, with a maximum of 6 steps displayed and an accuracy of +/-4%. Leader LTC-905 has voltage steps only up to 0.5V; it can display 7 of them with +/-5%, so a mixed bag.

Heathkit IT-1121 and IT-3121 can display up to nine voltage steps (1) with 1V maximum, but its sweep voltage can be as high as 200V (2) and can supply up to 200mA.

While the horizontal sensitivity (3) is switchable between 100mV and 50V per division, the vertical sensitivity (4) ranges from 0.5 V/div to 200 V/div. Thus, Heathkit IT-1121 and IT-3121 are the most capable of all the budget curve tracers mentioned. The value of the limiting resistor is also switch-selectable (5). As with most curve tracers, there are two banks of terminal connections for two transistors (6).

11 TESTING VACUUM TUBES (VALVES)

- EMISSION TESTERS
- GRID CIRCUIT TESTERS
- DYNAMIC CONDUCTANCE TUBE TESTERS
- PROPORTIONAL MUTUAL CONDUCTANCE TESTERS
- HICKOK-TYPE TESTERS
- TRUE MUTUAL CONDUCTANCE TESTERS
- TESTING AND MATCHING TUBES WITHOUT A TUBE TESTER
- DIY CURVE TRACER FOR MATCHING TUBES
- MEASUREMENT OF TUBE COEFFICIENTS USING BRIDGE METHODS

FURTHER READING

For a more detailed, in-depth look at tube testers, tube testing and matching, please consult my book "How to Use, Calibrate, Repair and Upgrade Vacuum Tube Testers", ISBN 978-0-9806223-7-9.

Design, functionality, calibration and modifications of vintage testers by B&K, Hickok, Triplett, Mercury, Sencore, Weston, Simpson, AVO, Taylor, RCA, Precise, Precision, Eico, Jackson, Sylvania, Knight, Heathkit, Seco, Sico, Conar, Metrix and other brands are discussed.

The book is available on Amazon and all good online bookstores.

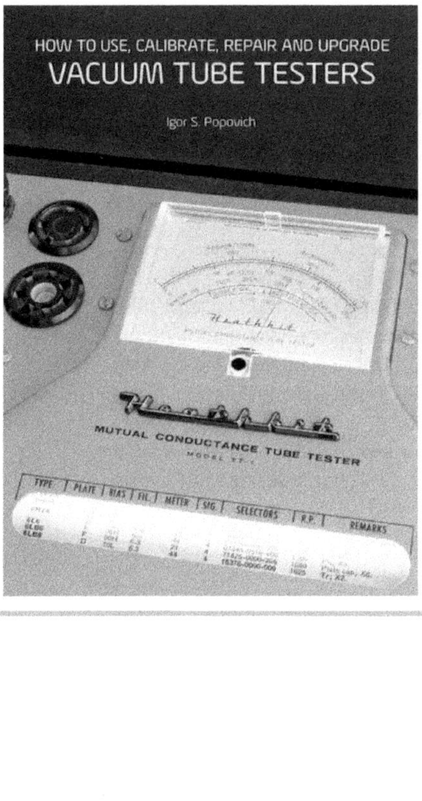

EMISSION TESTERS

How emission testers work

During the emission test, there are only two circuits. One is the heater connection between the 0V or COM terminal and the switch-selectable heater voltage (H2).

The other circuit is formed by connecting all tube electrodes together (except cathode) to a source of low AC voltage (20-40V range), which is the power transformer's secondary winding. The meter (with a series resistor and variable adjustable shunt control) is connected between the cathode and the COM terminal, and once the "Test," "Merit," or "Value" switch is closed, the circuit is closed, and current flows through the meter.

The higher the emissive capacity of the tube's cathode, the higher the current flowing. Most of that current flows from the cathode to the closest electrode, the control grid (G1). Since such grids are wound using very fine diameter (thin) wire, too much current would burn out the wire.

For that reason, some testers strap G1 to the cathode, so electrons pass through G1, and most of the current flows to the screen grid instead. The tube under test (connected as a diode) rectifies sinusoidal mains voltage into a series of sinusoidal pulses whose RMS value is 1/2 of its peak value, but moving coil DC meters respond to the average value, in this case, $V_{AV}=0.318V_{MAX}$.

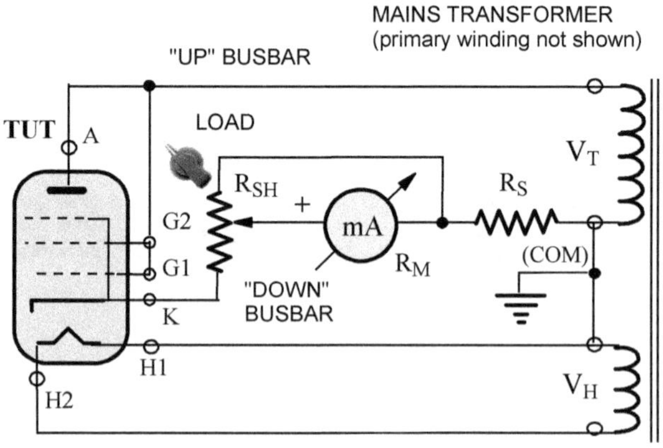

A typical tube emission test arrangement (ABOVE) and its equivalent circuit (BELOW). R_L is the total load, meter's internal resistance R_M, plus the series resistor R_S and the lower part of the "Shunt" or "Sensitivity" rheostat R_{SH}.

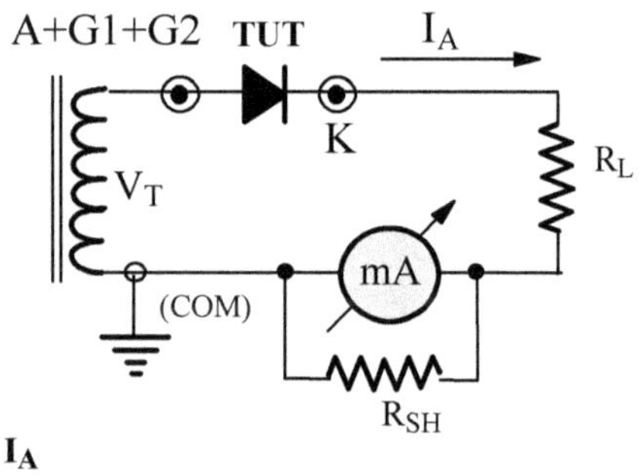

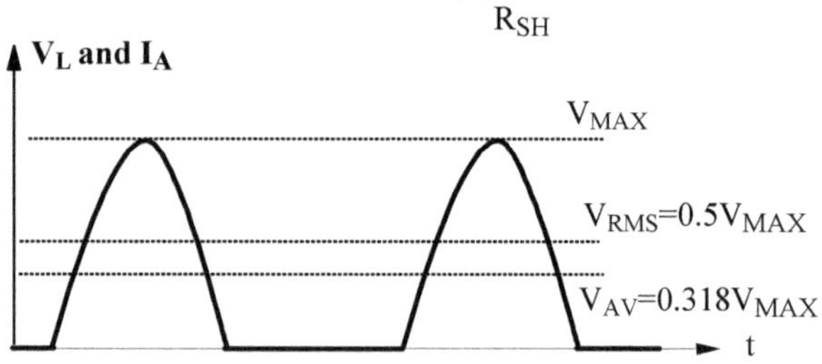

ABOVE: Tube under test (TUT) is connected as a diode and acts as a rectifier so the "DC" current through the meter is a series of half-wave rectified sinusoidal pulses, as we have already seen on page 52.

Controls of a typical vintage emission tester

Apart from Knight testers KG-600A and KG-600B, the most commonly found vintage tube checkers on the market are by Conar (models 221, 223, and 224) and Heathkit (models TC-1, TC-2, and TC-3).

Heathkit later changed to more modern units IT-17, IT-21, and IT3117. They are all very similar, in fact, functionally identical.

RIGHT: Knight KG-600A and more modern KG-600B are typical emission testers, 10 levers (one for each of the 9 tube pins and top cap) in up or down position (all electrodes connected either to cathode or anode) and a choice of four test voltages (A-circuit) selector.

1) "Line adjust" rheostat
2) On - Off switch
3) Heater (filament) voltage selector
4) "Load" potentiometer control
5) "Shorts/Line-value" test switch
6) Shorts indicator (neon)
7) "A" or "Circuit" selector switch
8) Top cap connector
9) Roll chart with settings for most common tubes

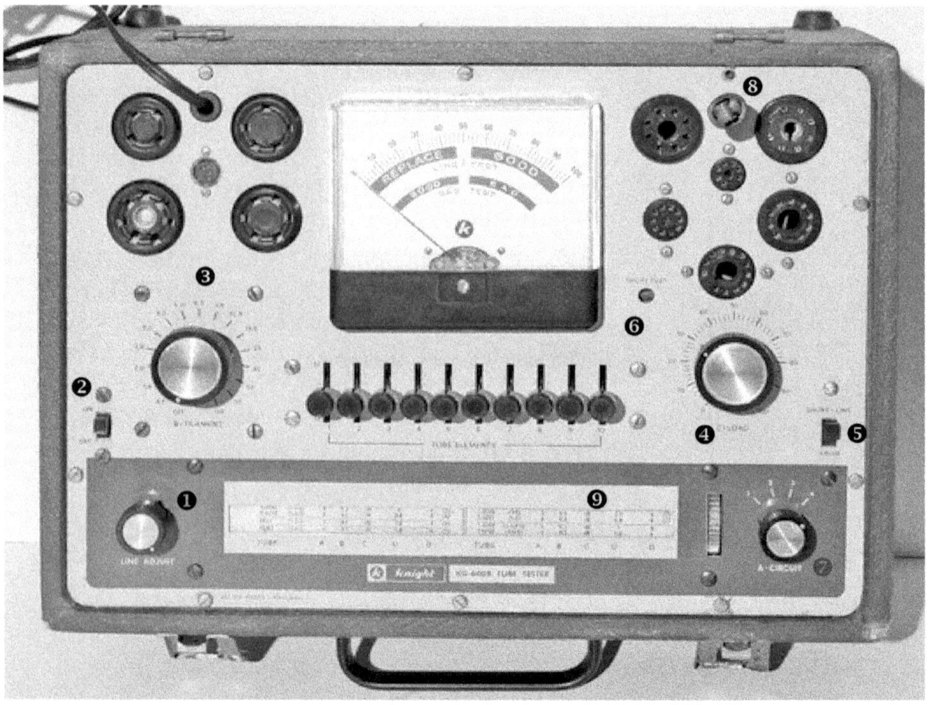

GRID CIRCUIT TESTERS

How grid circuit testers work

Instead of the cathode current flowing to the screen grid and anode, grid testers connect the anode and screen grid to the cathode, so the current only flows between the cathode, which emits electrons, and the control grid, which collects them. No current reaches the screen or the anode, so some testers such as Sico 82 leave those electrodes totally disconnected.

One faulty EL34 tube had its anode internally disconnected from its pin, so it did not work in an amplifier, yet Sico 82 tester passed it as a healthy tube! This is the most glaring inadequacy of grid testers - they only test two electrodes, the cathode, and the control grid. Even then, as with all emission testers, the current controlling action of the grid is not tested in any way; only mutual conductance testers and, to some extent, the dynamic conductance testers do that.

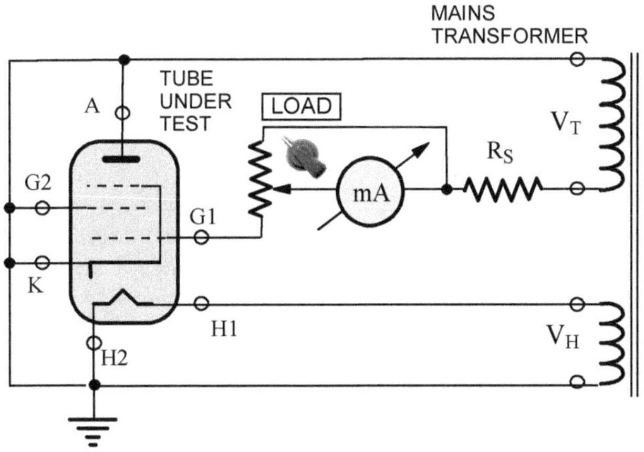

ABOVE: A typical grid circuit test arrangement.

Another problem with grid testers is their very low test voltages and currents. Most control grids are wound with superfine wire, and even a few mA of current could be enough to melt it (just like a fuse) and destroy the tube under test! This is especially so with preamplifier triodes such as 12AX7 and 6DJ8 and many pentodes.

Manufacturers liked grid circuit testers because they could save even more money on transformers, the most expensive part of any tester. Instead of 50-100 VA-rated power transformers in ordinary emission testers, tiny transformers of less than 20 VA were used with the grid testers, so further cost savings would be achieved.

Most common vintage grid circuit testers

B&K made a whole range of grid testers, starting with tube-based models 600, 606, and 625, followed by the relatively rare model 666, which was quickly superseded by solid-state models 607 and 667. Sencore's line of "Mighty Mite" tube checkers was very popular in their day; thousands were sold to DIY constructors and radio/TV repairmen over the three decades (the 1950s-1970s). TC-114, TC130, TC-136, TC-142, and TC-154 all share the same test settings and, apart from some cosmetic and minor internal differences, are the same tester for all intents and purposes. The line ended with a solid-state model, TC-162 and the TC-28 or "The Hybrider", simply a repackaged "Mighty Mite VII" or TC-162 tester with transistor testing capability.

Precision Apparatus Company or PACO made models 650 and T-62, which look different but are identical "under the hood"! Finally, we have grid checkers such as Mercury, with their models 1101, 1101C & 1101CT, and Sico, with models 78, 88, and 98, all best avoided!

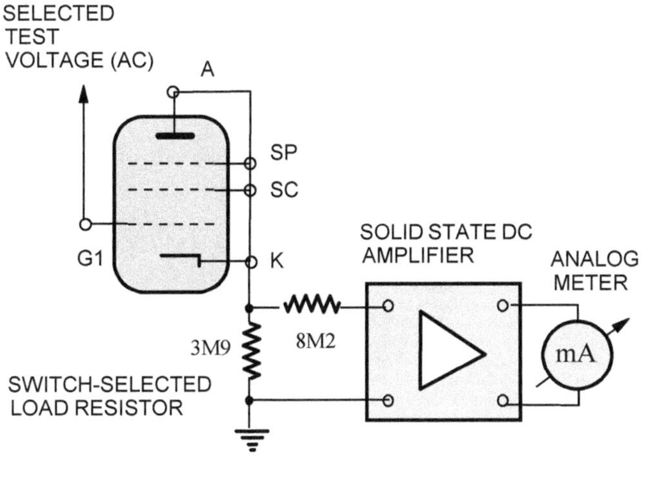

ABOVE: The principle behind B&K's 607 and 667 cathode emission test. Models 600, 606 and 625 used a tube DC amplifier.

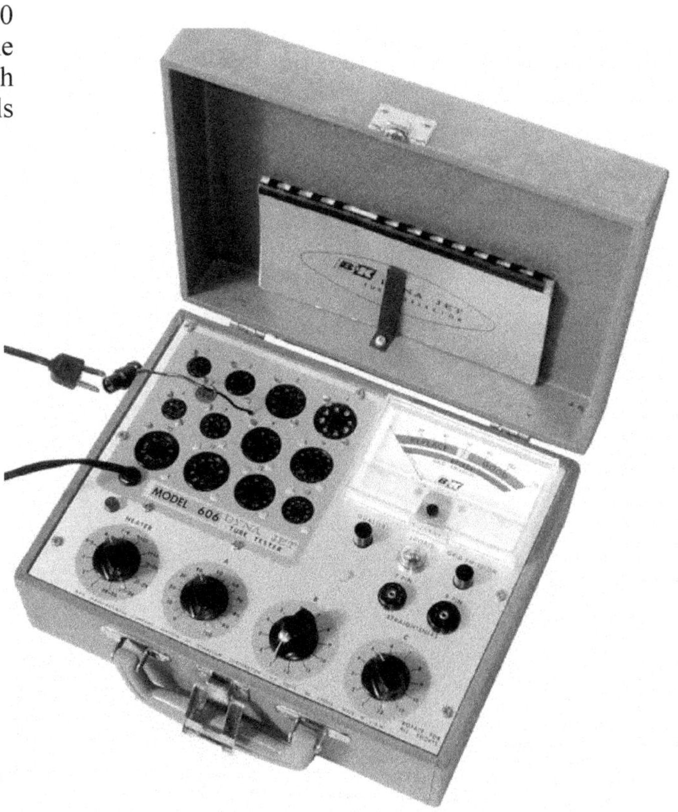

RIGHT: B&K 606 is a cute and compact quick checker, but, just like all grid testers, lacking any serious tube testing capabilities.

DYNAMIC CONDUCTANCE TUBE TESTERS

The three electrodes (control grid G1, screen grid G2, and anode A) get an AC voltage of a chosen amplitude and phase (through selector switches). The tube-under-test rectifies those three AC voltages and amplifies them. As such, this circuit doesn't only test for cathode emission (the DC anode current is proportional to emission) but also includes the effect of control grid voltage on the plate current. Therefore, Sylvania's meter reads "Composite Transconductance and Emission." Precision (PACO) touted this "electronamic" principle (a combination of 'electronic' and 'dynamic') by using this graphical explanation.

Since mains frequency alternating voltages are applied to all electrodes, the operating point of a tube under test is dynamically swept along a curved path 50 or 60 times a second. Plate voltage changes from zero to some maximum value and back, and the plate (anode) current rises and falls accordingly. Anode curves for a power pentode are shown, but the principle works for triodes, too.

However, notice that all three AC voltages are in phase, so the grid AC voltage V_{G1}, after being rectified by the control grid, biases the tube positively. Normal screen & anode voltages of 300-400V would result in very high anode & screen currents and the destruction of the tube under test, so these testers use much lower voltages to keep the currents lower not to exceed the anode power dissipation.

Notice that Precision's graph (right) shows the operating point swinging path through the negatively biased pentode, yet in these testers, the swing is above the EG=0 curve in the positively biased region, so this graph is misleading.

Sylvania also adopted the same principle in their testers, models 139/140, 219/220, and 620. Jackson, another reputable tube tester maker of its day, chose dynamic conductance testing for their 648 and 658 testers.

Simpson 1000 and its copies Eico 666 and 667 are the most commonly found dynamic emission testers to this day.

Sico (Superior Instrument Co.) 85 and TV12 are also relatively common budget dynamic conductance testers. Two different AC voltages are brought to tube electrodes, $5V_{AC}$ to control grid and $50V_{AC}$ to anode and screen grid. There is nothing "superior" about them; they are el-cheapo tube checkers of dubious value.

However, the switching matrix (10 lever switches) and the overall layout are neat and simple, and there's plenty of room in the case, so with a new power transformer and a few major improvements, TV12 can become a decent testing platform.

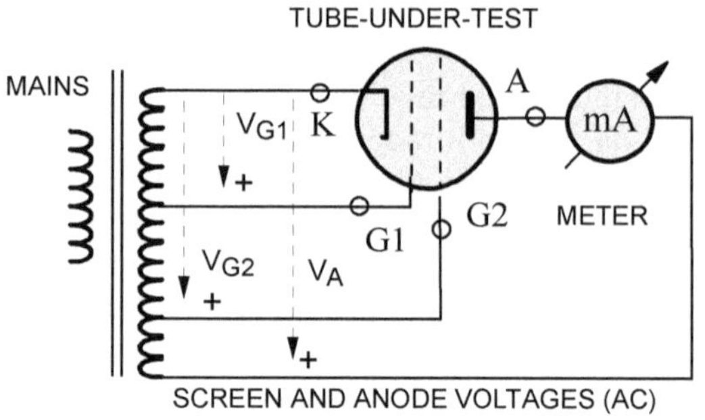

ABOVE: The operating principle behind dynamic conductance testers

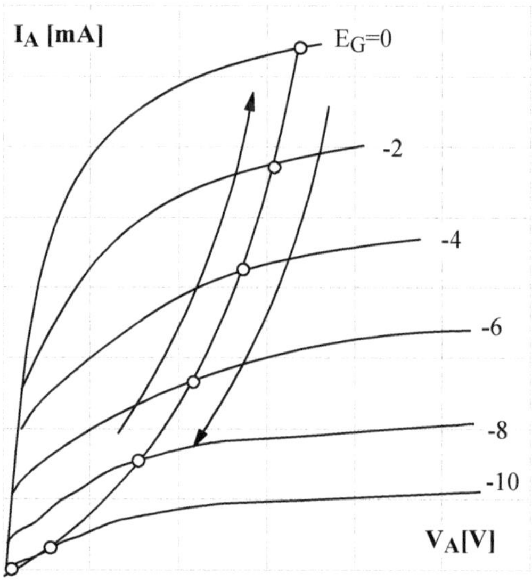

ABOVE: The illustration by PACO of the "electronamic" sweep of a power pentode used in its testers.

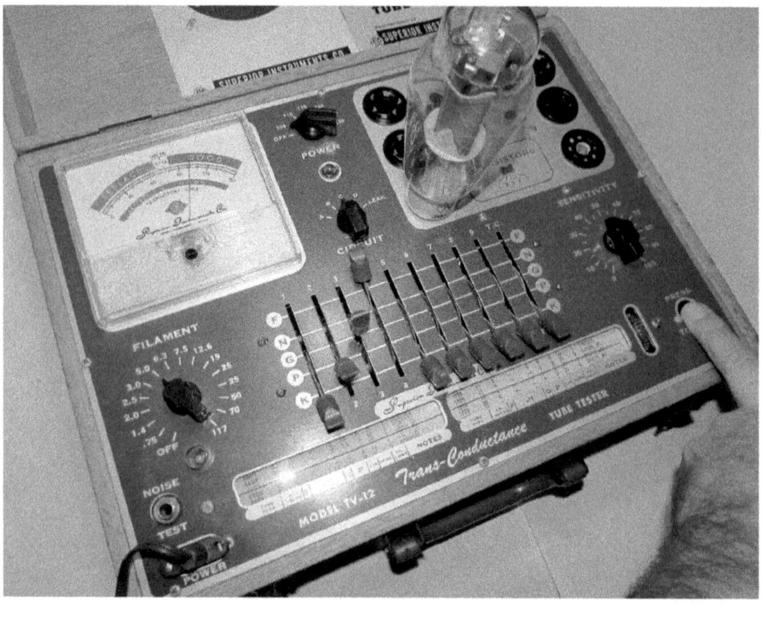

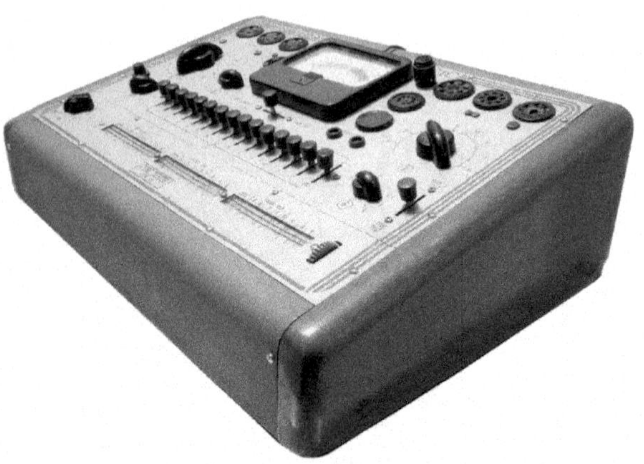

ABOVE: Great looking and solidly built, Sylvania 219 tube tester works well but isn't very intuitive to understand and operate.

LEFT: Despite SICO's claims of TV-12 being a "Trans-Conductance" tester, it is a dynamic emission tester! Shown here testing a 300B triode.

PROPORTIONAL MUTUAL CONDUCTANCE (Gm) TESTERS

The test circuit is similar to dynamic plate conductance testers, AC anode, and screen voltage, even the bias voltage is AC, but in this case, always out of phase with the rest of the secondary voltages, so the tube-under-test (TUT) is negatively biased. In dynamic conductance testers, all AC voltages are in phase, so tubes are positively biased. They are called "proportional" Gm testers because the AC voltages used get proportionally higher and higher, screen voltage V_{G2} is higher than the grid voltage V_{G1}, and the anode voltage V_A is higher than V_{G2}). Again, TUT acts as its own rectifier.

Transconductance or mutual conductance (Gm) is defined as the ratio of the anode AC current change and the grid voltage signal that caused that change: **$Gm=\Delta I_A/\Delta V_G$**. That is the slope or steepness of the transfer curve in any chosen point. Gm is related to the tube's internal resistance r_I and the amplification factor (voltage gain) μ through the Barkhousen's equation **$\mu=Gm*r_I$**.

Americans use "micromho" as a unit for Gm, "mho" being a "unit" for conductivity, a "reverse" of "ohm", a unit of resistance, but that should be "inverse" (1/Ω) and not "reverse". Prefix "micro" means 10^{-6}, so 1mA/V = 1,000 "micromhos".

An AC grid signal source and AC metering in the anode circuit are often used. When a small AC signal is applied to the grid, either from a secondary winding of the power transformer or from a dedicated higher frequency oscillator (a better option), only the AC component of anode current is measured. The meter is calibrated to read mA/V or micromhos directly.

Using mains frequency grid signal (as in Hickok, B&K, and all lower-spec testers) is considered an inferior approach because mains frequency hum affects the meter reading and because in most countries, the mains frequency isn't constant but varies all the time, often significantly.

American-made Weston 798, Triplett 3423, and Simpson 300 are the most common models on the market. French Metrix 310CTR and other models are good testers but rare outside France.

Other testers belonging to this family don't measure the AC anode current and Gm directly. Instead, they leave it to the user to perform a static Gm test or use the "backing off" method of zeroing the DC anode current through the meter. The most notable are English brands, Taylor, with their models 45A, 45B, 45C, and 45D, and AVO models Mark I, Mark 2, 3, and 4, and VCM 163.

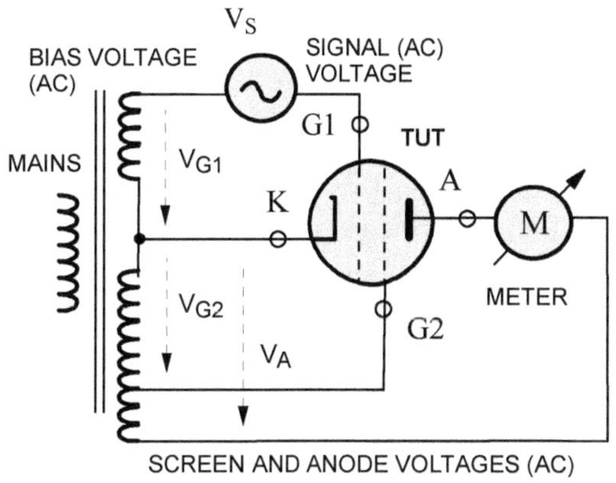

ABOVE: Test voltages in PMC testers

BELOW: Taylor 45D is a cute and compact tester, well laid out. However, it is slow and cumbersome to use.

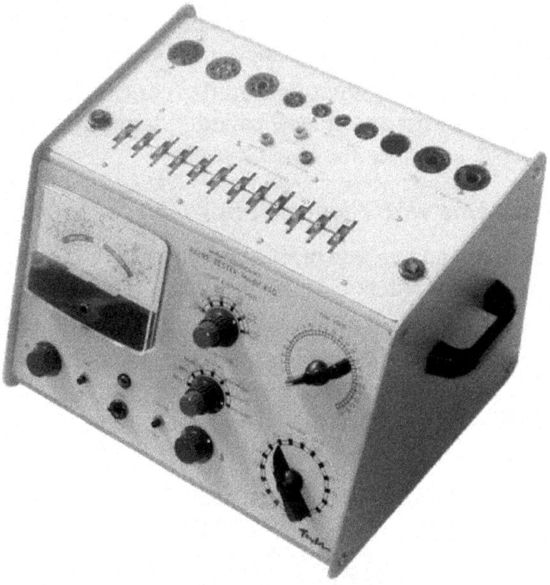

Triplett 3423 metering circuit

The anode load of the tested tube is an LC circuit tuned to the frequency of the grid signal voltage; its resonant frequency is approx. $f_R=1/(2\pi LC)) = 1/(2*3.14*0.11*20*10^{-9}) = 3.4$ kHz.

The DC component of the anode current passes only through the choke L; the 12nF film capacitor prevents it from passing through the meter.

Diode D2 provides half-wave rectification of the AC voltage across the choke, which drives the DC meter. The 5μF elco dampens the meter movement, and diode D1 protects it from reverse voltages. R20 is the internal adjustment for "calibrating" the meter circuit, although the factory calibration document does not mention it at all. Triplett's calibration procedure is not practical since it calls for eight "standard" calibration tubes such as 6AL6, 6BG6, 6CD6, 6C8, 6SN7, 2D21, 1X2, and 117N7, except 6SN7, all quite rare!

Chokes are components with an extremely wide range of parameters and tolerances, so even identical model testers will give significantly different test results, even if fully "calibrated." Apart from the generally horrible Triplett 3423 (12 internal trimmer potentiometers for "calibration"!), that applies to the flawed Heathkit TT-1 tester as well.

BELOW: Triplett 3423 metering circuit and testing principle

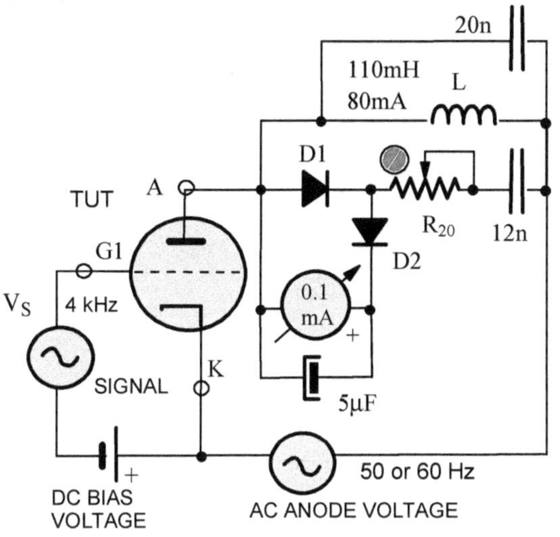

HICKOK-TYPE TESTERS

How the Hickok bridge works

Job R. Barnhart's US patent titled simply "Tube Tester" was approved on April 30, 1935, under the number 1,999,858. Over the years, it became known as "The Hickok circuit" or the Hickok bridge. When it expired, many less innovative T&M equipment makers adopted it for their tube testers, such as Mercury, B&K, and Precise.

Two identical transformer secondary windings, S1 and S2, with between 100 and 200 V_{AC} each are in two branches of a bridge, while one type 83 mercury vapor rectifier (on earlier models) or two solid-state diodes D1 and D2 on later models are in the other two branches of the bridge.

A tube-under-test is just like a fixed resistor with only DC bias on the control grid and no AC grid signal. Every half-cycle of the mains voltage, the direction of current through the meter changes, its needle pointer is subjected to equal and opposite forces, and there is no deflection. The bridge is balanced.

With the grid AC signal present, the grid voltage becomes more positive during one half-cycle and more negative during the other half. With a more positive grid voltage, the tube conducts, and its internal resistance drops significantly, so the current through the meter rises. During the period when the signal voltage makes the grid more negative, the tube conducts less current, its internal resistance rises, and the current through the meter drops.

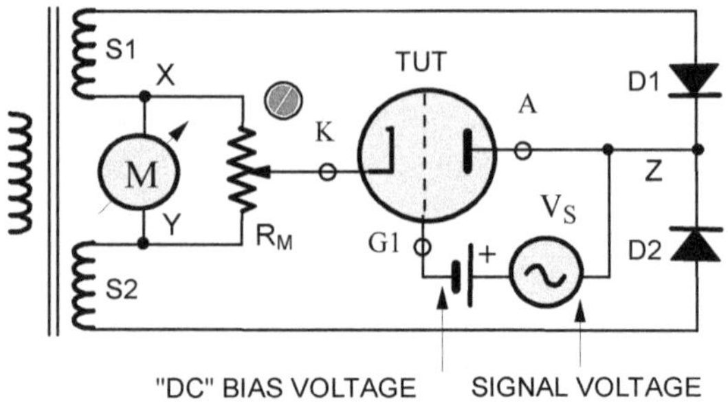

ABOVE: The Hickok bridge in principle

The bridge is unbalanced, and since due to its inertia, the meter cannot follow fast changes of current, its needle will show a deflection proportional to the average increase in current. It can be shown that the deflection of the indicator is proportional to the mutual conductance of the tube-under-test (TUT).

Another rectifier tube (5Y3GT) provides screen and bias voltages for TUT, while a separate mains transformer winding is a source of grid signal, which in this case is selectable, either $1V_{AC}$ or $5V_{AC}$, way too high for many preamp tubes.

Resistor R_M is an internal trimmer pot that balances the bridge without a signal. There is also a potentiometer connected directly across the meter, called "Shunt" or "Sensitivity" control used on Hickok-type testers by Mercury, B&K, and Precise, and on some Hickok testers.

Problems with Hickok's approach to Gm testing

There are quite a few problems with the Hickok method of Gm testing. First, the two transformer windings S1 and S2, must be balanced (identical). A significant error will be introduced into the measurement if they are not.

Second, if there is only one pair of secondary windings (as on most Hickok testers), all tubes are tested using the same anode voltage, the user has no control over the test conditions. Any tester that does not give its users such a control is a mere "quick" tube checker and not a laboratory-type instrument. Also, the test voltages are low, 150-170V in Hickok-branded models, much lower than the maximum test voltage on Triplett 3444 ($250V_{DC}$), for instance.

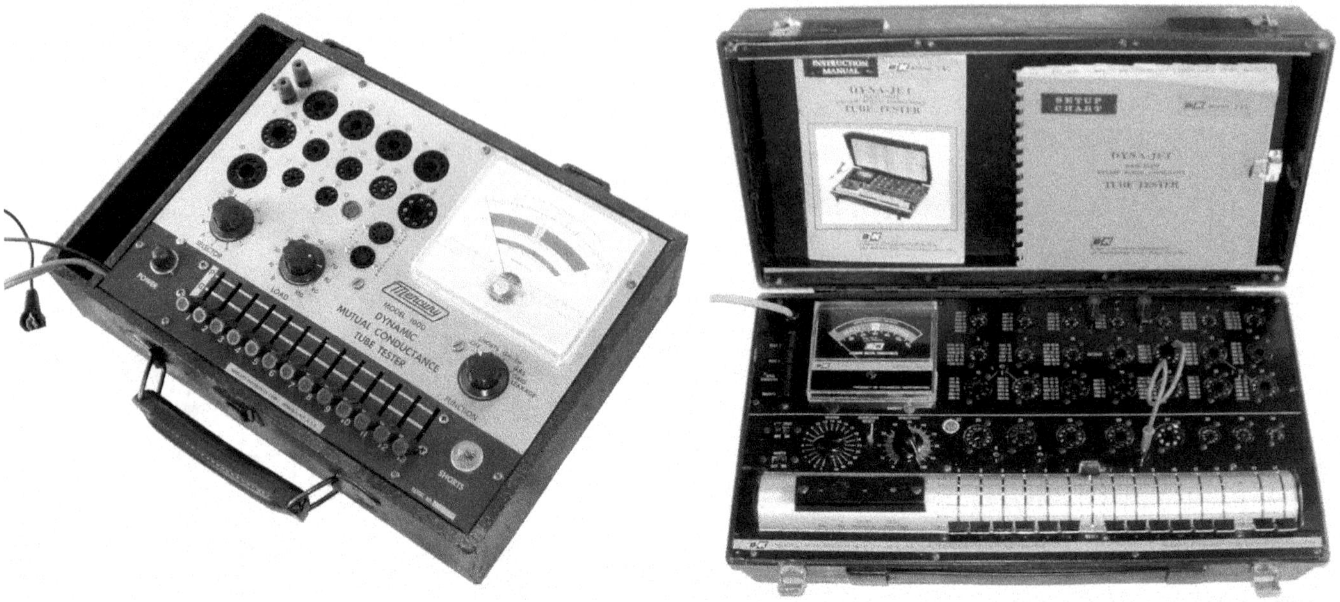

ABOVE: Mercury model 1000 (left) and B&K 747 (right) are two compact Hickok-type tube testers. Model 1000 tests all tubes as triodes but has two scales calibrated in real units, 0-5,000 and 0-25,000 micromhos. B&K 747 is a more modern solid-state unit but its meter's scale lacks fine graduations and is only in % (0-120%), not in actual Gm units. Thus, it is unsuitable for tube matching. Both use low anode and screen test voltages (resulting in low anode currents) and are compromised due to many unfortunate cost-cutting decisions.

Third, nothing is regulated or stabilized here. As the mains voltage varies, so will all secondary voltages, the signal voltage, the bias voltage, the screen voltage, and the anode voltage. Again, setting "line adjust" to one value will do nothing to prevent or compensate for these fluctuations.

Fourth, because the signal is scaled down mains AC voltage, any leakage between the heater and cathode or any other electrode of TUT will introduce hum. Such hum voltage will be added or subtracted (depending on its phase relationship with the signal voltage), thus increasing or decreasing the mutual conductance figure indicated on the meter, again seriously reducing the accuracy of the test.

Older Hickok testers use an incredibly high grid signal of 5V! Triplett 3444 uses four grid signal levels, 33mV, 100mV, 0.333V, and 1.0 V, way superior to even the highly revered Hickok 539C, which only goes down to 250mV, and only on higher Gm ranges!

For most Hickok testers, the user needs to change lots of switch settings to test dual tubes, while on Triplett, all that is needed is a simple flick of the test lever switch from T1 into T2 position. That saves significant time and prolongs the life of switch contacts and increases the reliability and longevity of the instrument.

The bias pot adjustment is not linear; for 20% of the dial, the bias voltage is -3V, while on 100% it is -40V. The only way to be sure about the bias voltage during testing is to monitor it with an external DC voltmeter or multimeter.

When Hickok's patent expired, half a dozen tube tester makers jumped on the bandwagon and started making their own versions based on the Hickok bridge. The testing speed and low production costs were the main design aims when this family of tube checkers was conceived. Many (such as later B&K models 700, 707, and 747) don't display true Gm reading; the scale is "relative micromhos" marked 0-120 or 0-130, presumable in percentage points of the "nominal" value. Testing is "blind" - no bias voltage, grid signal, or anode current is displayed, and bias voltage cannot be adjusted. Gm is tested in only one arbitrary chosen point unknown to the operator (unless voltages on tube pins are measured).

The circuitry is simple: unregulated voltages and the mains frequency used for the grid signal (no separate grid signal oscillator). Except for Hickok and Stark testers, all feature "sensitivity," "meter," "shunt," or "calibration," or similar controls which directly affect the meter reading!

TRUE MUTUAL CONDUCTANCE TESTERS

These testers approach the laboratory standard but, due to cost-cutting measures, very few, if any, reach it. The bias, screen, and anode voltages are DC and are adjustable if not continuously (that would be ideal), then at least in steps. A separate AC oscillator supplies grid signal voltage, which can be adjusted in steps. Smaller signal voltages are used for preamp tubes and larger ones for power tubes. For instance, Triplett 3444 tube tester uses four grid signal levels, 33mV, 100mV, 0.333V, and 1.0 V. The test signal's frequency is 4 kHz.

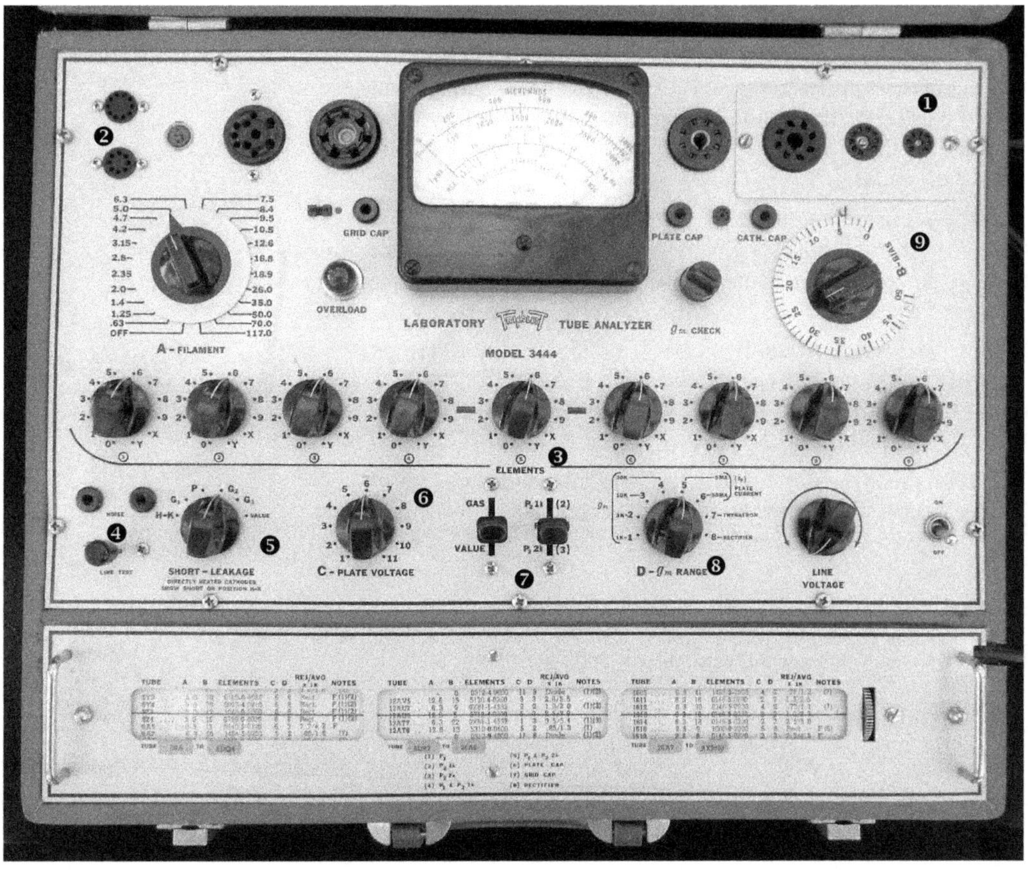

Arguably, Triplett 3444 is the best of all USA-made and designed vintage testers. Although it's called a "laboratory tube analyzer," it is far from the laboratory level of accuracy and flexibility in testing.

1. The three commonly used sockets (7-pin mini, Noval and Octal) are recessed, and socket saver module plugs into them
2. Two pin straighteners can be removed and additional sockets installed there
3. A bank of nine 12-position selector switches, each connects one tube pin to various test points
4. "Line test" push button and "Noise" terminals for headphones
5. "Short-Leakage" test switch
6. Although marked "Plate voltage", the "C" switch also selects screen and signal voltage levels (table on the next page)
7. "Gas-Value" switch and "P1-P2" switch for testing dual tubes
8. "D" switch selects one of four Gm ranges, two anode current ranges, a Thyratron or rectifier testing function
9. Bias control potentiometer

There is no range selector switch Triplett 3444 tube tester. The two bias ranges (0 to -5 V and 0 to -50 V) are automatically selected depending on the position of the "Plate voltage" switch or switch "C." The table specifies the plate, screen, and bias voltages for each position. Since the bias DC voltage is continuously adjustable, you can test a tube at any point on its transfer curve, not just in one, as basic Gm testers do.

Weston's line of testers moves from model 798 through 978 to 981-1, 981-2, and 981-3. While the looks and the switching arrangements are different, there are many similarities under-the-hood. Heathkit TT-1, a clone of Weston 981, was introduced as a DIY kit in January 1960.

Other models in this class of instruments include SECO 107, 107-B & 107-C, and RCA WT-110A, which used perforated cards instead of setup switches.

RIGHT: The 11 possible combinations of the 3 factors, the bias range, the anode and the screen voltage, selectable by switch "C" on Triplett 3444 tube tester.

Switch C ("Plate Voltage")	Plate V_{DC}	Screen V_{DC}	Bias range V_{DC}
1	250	250	0-5
2	250	250	0-50
3	250	100	0-50
4	250	100	0-5
5	100	100	0-5
6	100	100	0-50
7	100	45	0-50
8	100	45	0-5
9	30	30	0-5
10	30	12	0-5
11	12	12	0-5

EXPERIMENT: Comparison of Triplett 3444, B&K700 and Mercury 1000 testing a 6L6 power tube

To give you an idea and a feel for the test results of the same tube on different Gm testers, the table below outlines the major technical differences of these testers. Note that the voltages in the table below are as specified by the manufacturers; the actual or measured voltage may be different in the actual testers!

The B&K 550 tester was included in this table to illustrate that B&K reduced the number of bias voltages from four on the model 550 to only two on the later 700/707 models and reduced the plate voltage! The main functional difference is that models 550 and 650 display true Gm in micromhos, while later units (models 700, 707, and 747) only display on scales of 0-100 (presumably in percentages).

	TRIPLETT 3444	MERCURY 1000/1200/2000	B&K550	B&K707
PLATE VOLTAGE(S)	12, 30, 100, 250 V_{DC}	110 V_{AC}	203 V_{AC}	193 V_{AC}
SCREEN VOLTAGE(S)	12, 30, 45, 100, 250 V_{DC}	Same as plate (triode connection)	148 V_{DC}	148 V_{DC}
BIAS VOLTAGES	0-5 and 0-50 V_{DC} variable	-1.7, -5.9 V_{DC} (fixed)	-0.2 -2.5 -7.5 - 19.5 (fixed)	-0.2 - 19.5 (fixed)
SIGNAL V_{AC} (RMS)	0.033, 0.1, 0.333 and 1.0	1.0 V_{RMS} (2.82 Vpp)	1.5 V_{RMS} (4.23 Vpp)	1.5 V_{RMS} (4.23 Vpp)
SIGNAL FREQUENCY	5,000 Hz	50 or 60 Hz (mains)	50 or 60 Hz (mains)	50 or 60 Hz (mains)

We tested three pairs of 6L6 tubes, by JJ (Slovakia), Tube Art (China), and Sylvania (USA). The figures after Gm are anode currents in mA, measured with a True-RMS digital multimeter, after the anode current measuring circuits were installed in the B&K and Mercury testers since they didn't have such capability.

The specified bias voltage on Triplett 3444 was -18V, but plate currents at that setting were over 50 mA (the maximum that the tester can measure without modifications), so we increased the bias to "25" on the dial (measured as -27V).

Mercury 1000 specifies the "Load" as 95, but the readings were off the scale, so we reduced that setting to 85. Its bias was -1.55V and test voltage $92V_{DC}$. B&K 700 used a $120V_{DC}$ test voltage with -8.0V bias.

The published transfer curve at the Triplett revised test point (-27V/250V) indicates an anode current of around 20-23mA, but we measured 39-46 mA!?

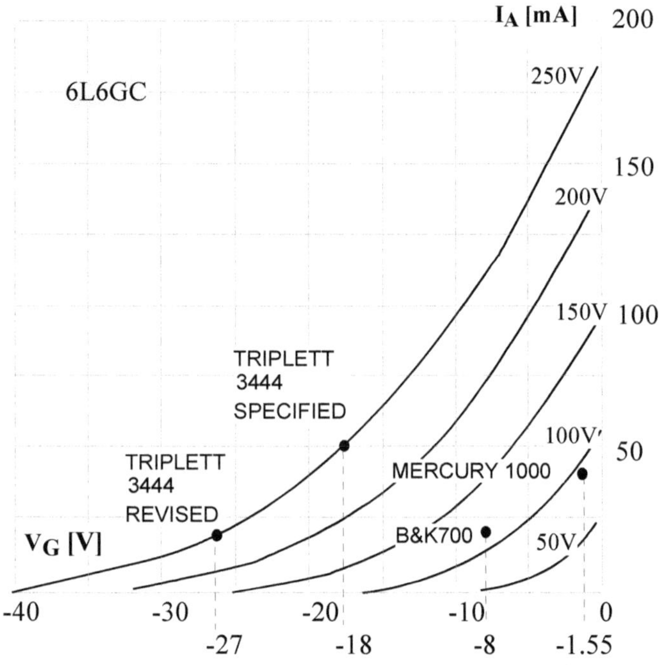

RIGHT: Published transfer curves for a 6L6 beam power tube for the five anode voltages (50-250V_{DC}) and the test points for three Gm testers, Triplett 3444, B&K 700 and 707 and Mercury 1000 and 2000.

6L6 (Gm - Ip)	#1 JJ	#2 JJ	#3 Tube Art	#4 Tube Art	#5 Sylvania	#5 Sylvania
Triplett 3444	7,000 - 42.5	7,000 - 42.5	4,600 - 39.0	4,600 - 39.0	5,500 - 46.0	5,600 - 44.0
B&K 700	107 - 34.0	105 - 34.8	74 - 26.7	78 - 27.9	91 - 33.3	90 - 34.0
Mercury 1000	4,800 - 28.3	4,750 - 28.5	3,600 - 22.0	3,700 - 23.0	4,100 - 26.2	4,000 - 26.5

TESTING AND MATCHING TUBES WITHOUT A TUBE TESTER

The ultimate test of any tube is inside its piece of equipment. Tube testers, even the more elaborate ones, suffer from both false positives (a tube tests "bad," but it works in the equipment) and false negatives when a tester passes a tube as good, but the tube then fails to work properly in an amp, transmitter, power supply, etc.

It is quite easy to test tubes without a tester if you only deal with a few tube types and test them infrequently; you don't need a tube tester at all. Here's how to do it.

Ohmmeter checks in the cold state

These cold checks are necessary but not sufficient to proclaim a tube to be OK. An analog ohmmeter should show a very low reading (a few ohms) when measuring the resistance of the tube's heater. If the ohmmeter shows infinite resistance, the heater has burned out, and you can throw that tube away. Between all other electrodes, the ohmmeter should show infinite resistance.

Pairs of electrodes adjacent to each other are more likely to show a fault, for instance, heather to the cathode, control grid to the cathode, and screen grid to the anode. However, even anode to cathode test can show low resistance due to foreign material stuck in the mechanical structure of the tube.

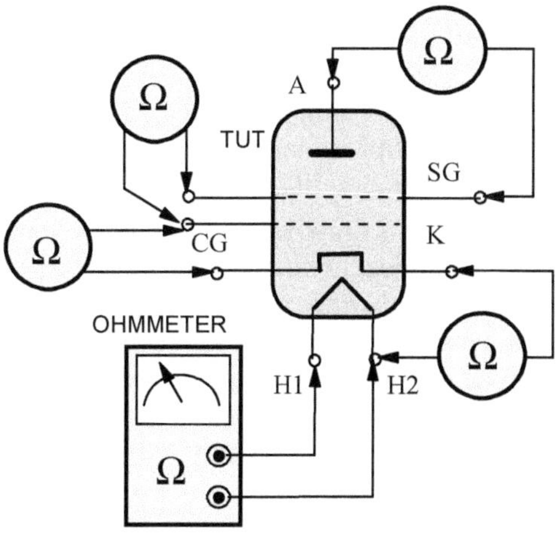

ABOVE: Apart from the very low ohm reading across the heater pins, all other ohmmeter checks should show infinite resistance.

Hot ohmmeter checks

Most cheap tube emission checkers use a similar "hot" test (calling them "testers" is a stretch of the imagination). However, indicative results of tube health can be obtained. Connect a small mains transformer to supply the heater voltage. If the heater needs 6.3V@3A, a 20-30VA transformer will be sufficient. You can use a universal adjustable DC power supply instead.

Connect an analog ohmmeter between the cathode and the control grid. As shown, you don't even have to connect the anode to the grid. Connect the + lead to the anode and control grid and the negative lead to the cathode. The ohmmeter should show some deflection, depending on the type of tube tested, its internal resistance, and other parameters. If you get no indication, reverse the ohmmeter's leads.

The typical internal circuitry of an analog ohmmeter has the internal battery and a meter in a simple series circuit.

Use the highest test range on your meter, for instance, 10k or 100k. Most multimeters use an internal 9V battery for one or two of the highest resistance ranges, but only a 1.5V battery for lower ranges. That may not be enough to check a tube. Note that digital and electronic ohmmeters, which work on a different principle (see page 54), are unsuitable for this application.

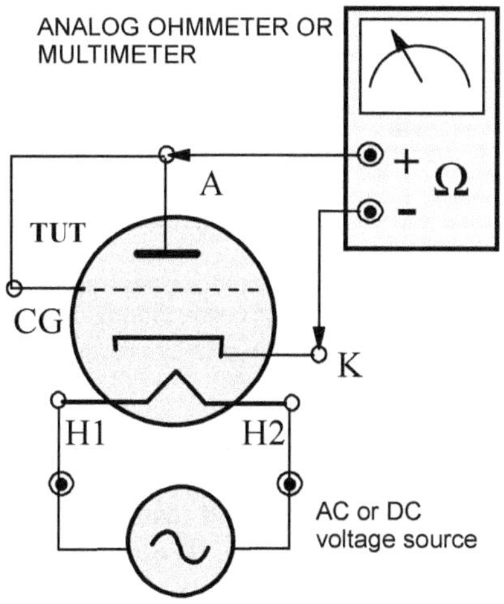

ABOVE: 9V from the ohmmeter's internal battery is enough to cause a current to flow through a hot tube, which indicates emission levels from its cathode.

Hot checks with a function generator and a multimeter

The low voltage power supply provides a DC heating voltage (adjustable and readable on the power supply voltmeter). A function generator provides a sine or square voltage connected between the cathode and the grid/anode. Again, an AC transformer of suitable voltage and current rating can be used for heating if a DC power supply is not available or is not powerful enough.

A DC milliammeter or a multimeter set on DC current range displays grid/anode current. The tube's grid should be safe at such low test voltages (most function generators provide 7-10 V_{RMS} signals). As we have seen (Gould and Philips examples), some higher-spec models may even go up to $30V_{RMS}$. If you are still worried about damaging the tube-under-test due to too much current flowing through the grid, strap the grid to the cathode instead.

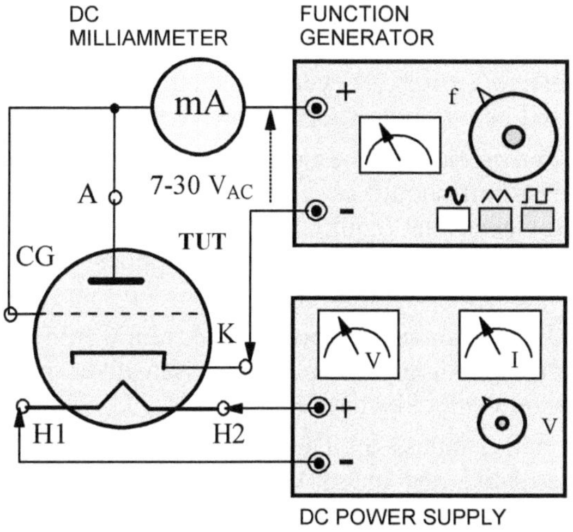

ABOVE: A function generator as a signal source plus an AC or DC heater supply, and you have a simple yet effective emission tester!

Recording transfer, anode and constant current curves of vacuum tubes

Even the best tube testers are limited in their features and range of tubes that can be tested. The most noticeable is the small range of tube socket types, but that is easily expanded through adapters or add-on sockets. The next limitation is the low heater power that their meager power transformers can supply. Tubes requiring high heater power (such as 6C33C-B, GM70, 211, 813, and 845) cannot be tested.

An external heater supply can be hooked up temporarily (crocodile clips to tube pins), or a tester can be modified (a simple switch and a couple of terminals added) to enable external heater AC or DC voltage to be brought in. During those tests, the changeover switch will automatically disconnect the tester's internal heater circuit.

The most serious drawback of all but a few laboratory testers, one that easy modifications cannot overcome, is the limited range of grid bias, anode and screen voltages available, and low anode current capabilities. Even a decent tester such as Triplett 3444 only goes up to V_A=V_S=250V_{DC} and 50mA of anode current (or 100 mA with an additional switch, shunt, and a replacement of the selenium rectifier)! The solution is to hook up your own test setup. At least two DC- and one AC-power supply (for the heater) is required for plotting curves with only negative grid bias voltages, plus two DC voltmeters and one DC milliammeter for anode current. An additional DC power supply is needed to be able to bias the grid positively unless you are happy to manually reverse the V_G polarity.

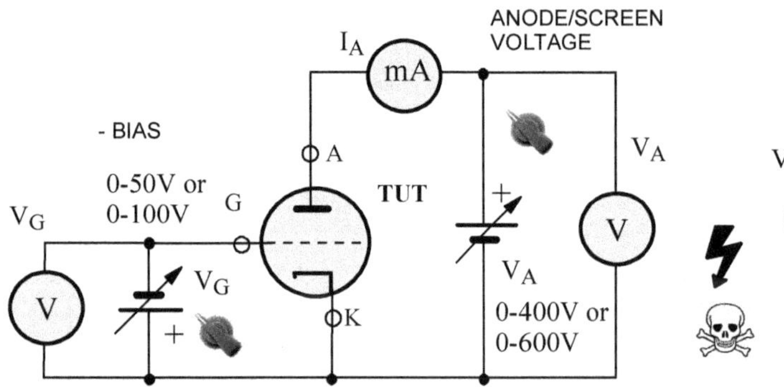

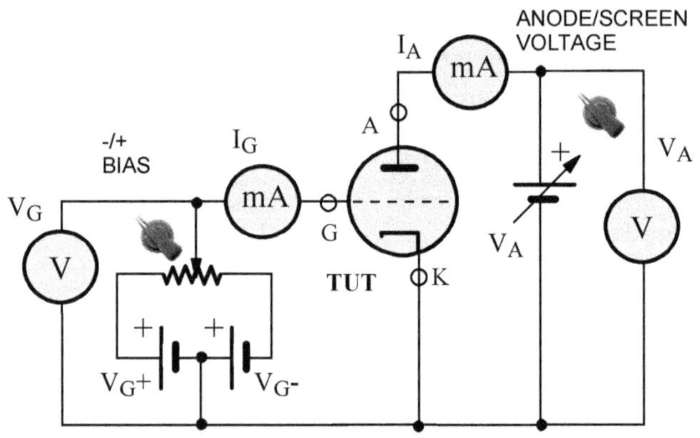

ABOVE LEFT: The circuit for recording transfer, anode and constant current curves of triodes. Only negative grid voltages are obtainable with two DC power supplies. Heater connection is not shown, a third power supply is needed for the heater circuit.

ABOVE RIGHT: A more versatile circuit for recording characteristic curves of triodes. Both positive and negative grid bias voltages are obtainable with three DC power supplies.

Measuring mutual conductance: the grid-shift method

A few vintage commercial testers used a variant of the grid-shift method as their operating principle, namely AVO and Taylor. It seems UK tester designers disliked the American way of feeding an AC test signal into the tube's control grid and measuring the AC component of anode current to arrive at the Gm figure and preferred this manual method.

A grid shift Gm tester can be hooked up in a couple of minutes. You will need a high voltage (HV) power supply, at least 0-400V_{DC} 125mA, a source of bias DC voltage 0-100V, rated at few mA only (since you will not test tubes in class A_2 with grid current flowing), a source of heater voltage and a good quality DC mA meter (ammeter) or multimeter.

The source of heater voltage could be a low voltage DC power supply, a transformer with multi-tapped secondary winding(s) and a selector switch, or even an old emission tester, used here only to supply heater voltages.

For a precise Gm measurement in one point, the bias shift should be very small, 0.1 or 0.2V. However, that requires a precise adjustment of bias, and many bias power supplies aren't capable of such fine-tuning.

Also, the change in anode DC current would be correspondingly small, and sensitive mA meters would be needed.

Finally, unless all DC supplies were closely regulated, the mains voltage would fluctuate during the test and affect results, so you would never be sure if a change in anode current is due to a change in bias or due to mains fluctuations. That makes test results of large grid shift changes more reliable and accurate!

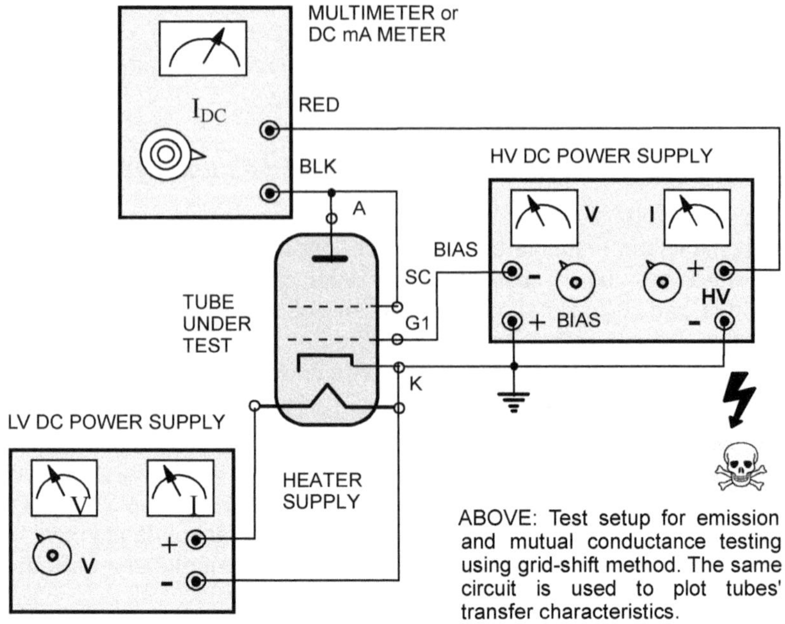

ABOVE: Test setup for emission and mutual conductance testing using grid-shift method. The same circuit is used to plot tubes' transfer characteristics.

Measuring mutual conductance: the grid signal method

The setup, in this case, is the same as for the grid shift method, with an addition of a grid audio signal of a certain frequency f_0, provided by a function generator. A True-RMS AC voltmeter or multimeter measures the voltage drop the AC component of anode current produces on the load resistor. That meter must be capable of measuring signals at f_0 test frequency. With f_0 between 1 and 5 kHz, most good quality digital multimeters will have their upper -3dB frequency limit way higher and will be fine.

You can use a decoupling capacitor in series to eliminate DC current from AC voltmeter's input, but these already have an input capacitor for that purpose; just make sure its voltage rating is at least 600V.

By the same token, to protect the output transistors of the function generator from high bias DC voltages, add a series capacitor C_S. If your test frequency is between 1 and 5 kHz, its reactance at that frequency must be negligible, less than 10Ω, meaning its capacitance should be 160-220nF.

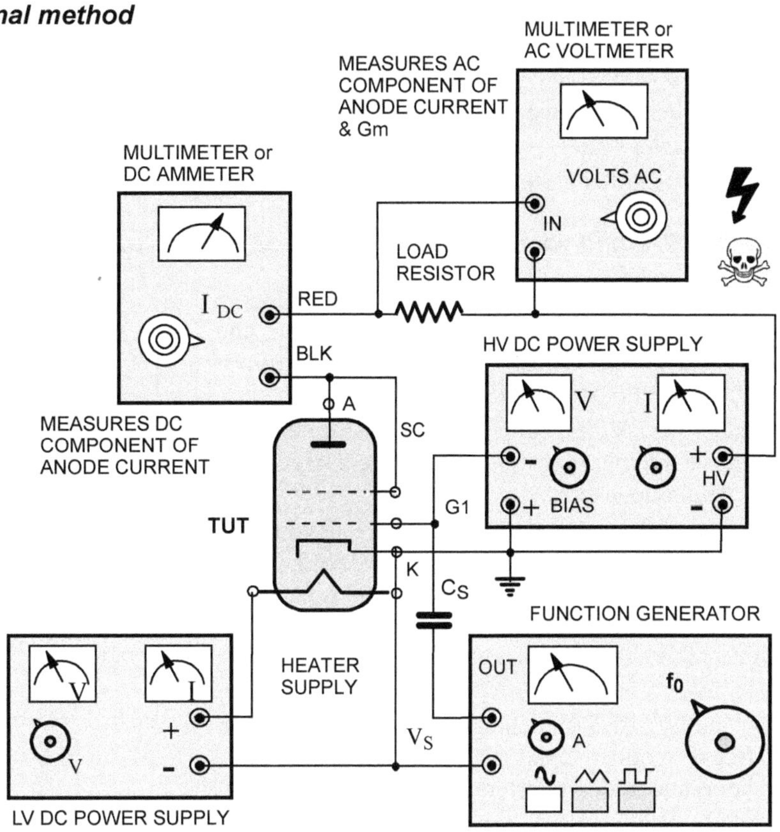

ABOVE RIGHT: Test setup for mutual conductance and emission measurement using a dedicated grid signal source (such as a function generator), a load resistor in the anode circuit and an AC millivoltmeter as an indicator of transconductance.

DIY CURVE TRACER FOR MATCHING TUBES

Option #1: Using a commercial semiconductor curve tracer

We have already discussed the operation of semiconductor curve tracers (please see the material on pages 149 and 150). Models that can test bipolar transistors (current steps) and FETs (which use voltage steps to drive the gate) such as Heathkit IT-1121, IT-3121, B&K 501A, and Leader LTC-905 can be used as a step-generator for a tube curve tracer. Eico 443 cannot test FETs (has no voltage steps) and is unsuitable for this application.

The largest step on Leader LTC-905 is only 0.5V; on most other transistor and FET testers, the largest step is 1V. Tthey are not suitable for curve tracing of power tubes where a much larger step size would be needed (5, 10, or 20 Volts).

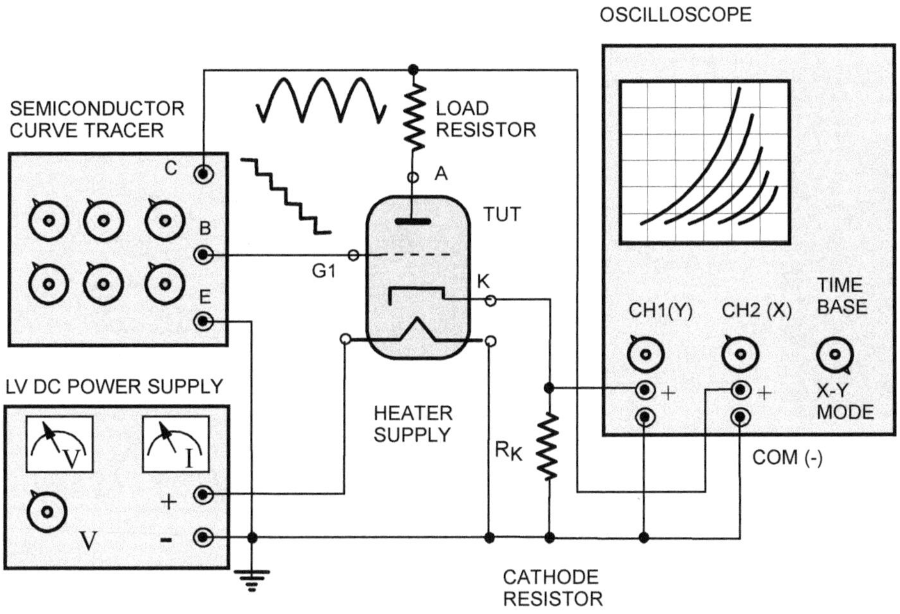

A separate power supply or an existing tube tester can supply the heater voltages.

You also need a source of sweeping plate voltage. Commercial curve tracers already have such an output (collector power supply), but if you use your own DIY staircase generator, any tube tester with full-wave rectified sine wave for plate voltage (such as Hickok or B&K models) can provide such a pulsating sweep voltage.

LEFT: Test setup for displaying triode's anode (plate) characteristics on an oscilloscope. AC heating voltage can be used instead of the DC power supply.

You can tap into plate voltage on the octal socket between pins 3 (+ or anode) and pin 8 (- or cathode). The sweep voltage goes to Channel 2 (X- or horizontal deflection) on the oscilloscope, while the input to the vertical channel (Y- or vertical deflection) is taken from the cathode resistor between the cathode and GND (COM).

The oscilloscope's time-base is switched off; it must be in "X-Y" mode.

The "Base" or "Gate" (for FETs) output of the curve tracer supplies the stepped voltage waveform to the control grid, while its "E" or "Emitter" terminal is grounded.

Option #2: Build your own curve tracer for preamp tubes

The simplest schematics for a staircase generator we could find was designed by Daniel Metzger and published in the August 1971 issue of Electronics World. It works very well.

Its steps are 1 Volt in amplitude, so it is not suitable for curve tracing of power tubes where a larger step size would be needed.

The tube biasing circuit is fairly simple; all you need is half-a-dozen resistors, a 3-pole double-throw switch (for selecting one of the two triodes you want to test), and a tube socket, Octal, Noval, 7-pin mini, RimLock, etc., depending on the tubes you want to test.

You could add a switching circuit so tubes with different pinouts could be tested and a variable cathode and anode resistors to change the value of the cathode bias resistors.

We wanted to illustrate the principle here, not to overcomplicate the schematics with multiple plate voltages, various switchable heater voltages, or similar options you can easily add yourself.

The 6.3 V winding provides heating for the tube under test. If you need more than one heater voltage, you can use a heater supply from a commercial tube tester.

The half-wave rectified, filtered, and stabilized 15V supply powers up the staircase generator.

The full-wave rectified mains is the sweep voltage for the anodes. In our case, the amplitude is 100V. With multiple secondaries, you can add a selector switch and change the sweep amplitude, say 100 - 150 - 200 - 250 Volts.

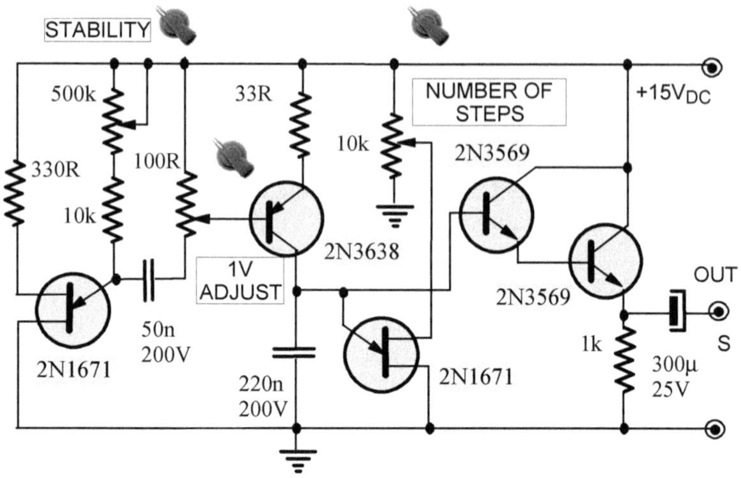

ABOVE: A simple solid state staircase generator, source & copyright: Daniel Metzger, Electronics World, August 1971

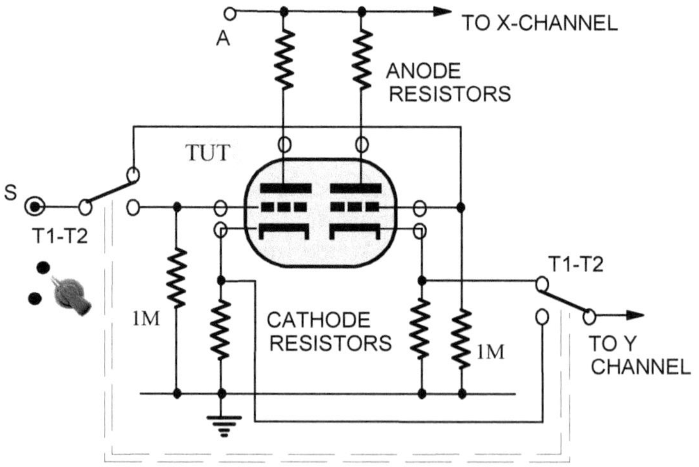

ABOVE: The main hookup indicating how the Tube-Under-Test should be connected. The values of anode and cathode resistors need to be determined experimentally for each tube type.

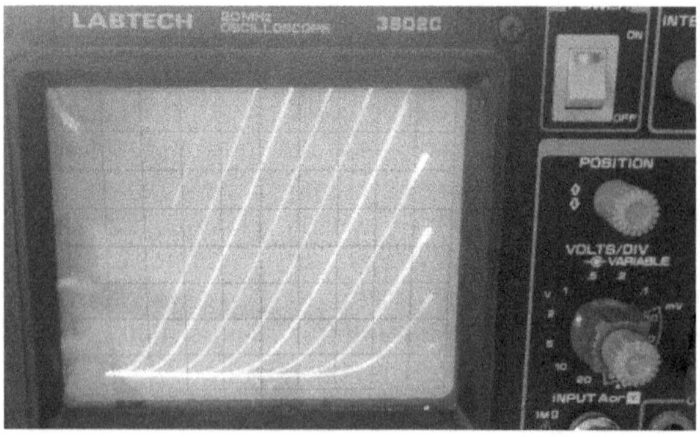

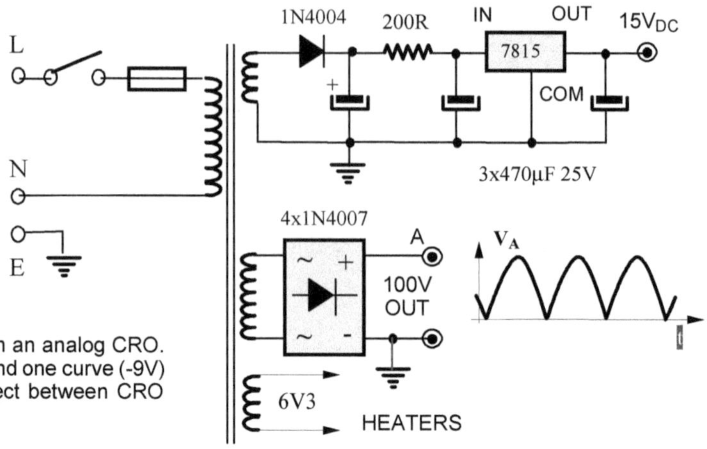

RIGHT: One possible power supply configuration for the curve tracer

ABOVE: A family of ten plate curves for 6DJ8 duo triode displayed on an analog CRO. Each VG step (one curve) is -1V. Half of the 0V bias curve on the left and one curve (-9V) on the right is missing on the photo due to the synchronization effect between CRO sweep frequency and digital camera operation.

MEASUREMENT OF TUBE COEFFICIENTS USING BRIDGE METHODS

Measuring internal resistance r_I

An ordinary Wheatstone AC bridge can be used for determining the internal resistance r_I of a tube, providing the signal voltage amplitude V_S is kept small. R1 and R2 are used to balance the bridge, in which case there is no sound in the headphones or vertical deflection on an oscilloscope if used as a null-detector. Transformer TR can be any small audio transformer salvaged from a tube radio or amplifier. Its impedance ratio is not critical, although, depending on the headphones' impedance, it will affect the sound volume in the headphones.

Any audio frequency can be used, but it should be kept low so the tube's capacitances do not affect the accuracy of the measurement. A 1 kHz signal is recommended. If anode DC voltage V_{BB} and grid bias V_{CC} are adjustable, as they almost always are if a commercial regulated HV power supply is used, r_I can be measured at any operating point. At balance $r_I/R_1=R_3/R_2$ so $\mathbf{r_I=R_1R_3/R_2}$

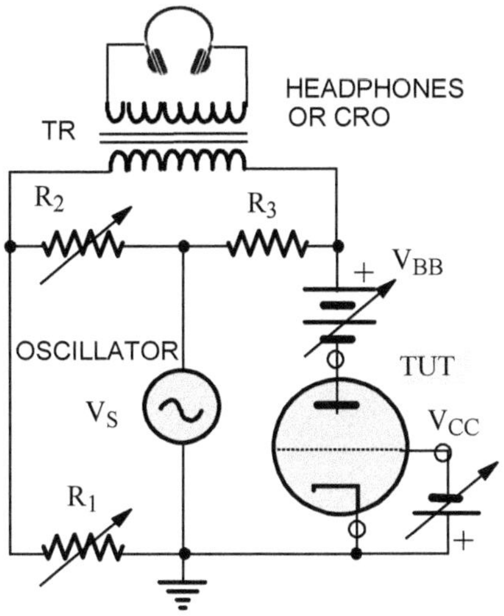

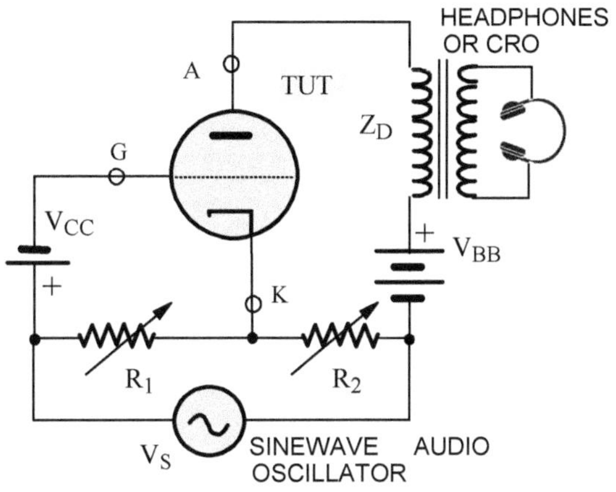

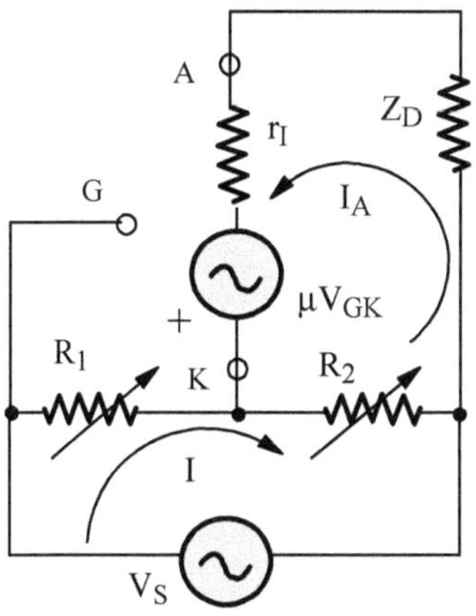

LEFT: Measuring tube's internal resistance ABOVE: Measuring tube's amplification factor

Measuring amplification factor μ on Miller bridge

The small-signal voltage equivalent circuit will help us analyze this bridge. The cathode-to-grid voltage is equal to the voltage drop on R_1: $V_{GK}=IR_1$ For the anode current loop we have $\mu V_{GK} = \mu IR_1 = (I+I_A) R_2 + I_A(Z_D+r_I)$ and if we express I_A we get $I_A= I(\mu R_1-R_2)/(R_2+Z_D+r_I)$

If R_1 and R_2 are adjusted so that the bridge is balanced and the signal anode current is zero, there will be no sound in the headphones or vertical deflection on the oscilloscope screen (if used as a null-detector). In that case, the impedance Z_D (reflected headphones impedance onto the primary side of the matching transformer) is irrelevant and does not feature in the balance equation: $\mu R_1-R_2=0$ and so $\boldsymbol{\mu=R_2/R_1}$.

Measuring mutual conductance Gm

For a balanced bridge and no sound in the headphones, $GmV_{GK}=(r_I||R_3)=I_1R_2$ and since $V_{GK}=I_1R_1$, we get $Gm= R_2(r_I+R_3)/(r_IR_3R_1)$.

The measurement should not depend on another tube coefficient, namely r_I, so we should make $R_3 \ll r_I$. In that case, r_I in the nominator cancels r_I in the denominator and $\mathbf{Gm \approx R_2/R_3R_1}$. For small-signal triodes such as 6SN7 and ECC82, R_3 should be kept below 1 kΩ, but for testing of power triodes such as 2A3 or 300B, it should be under 100Ω!

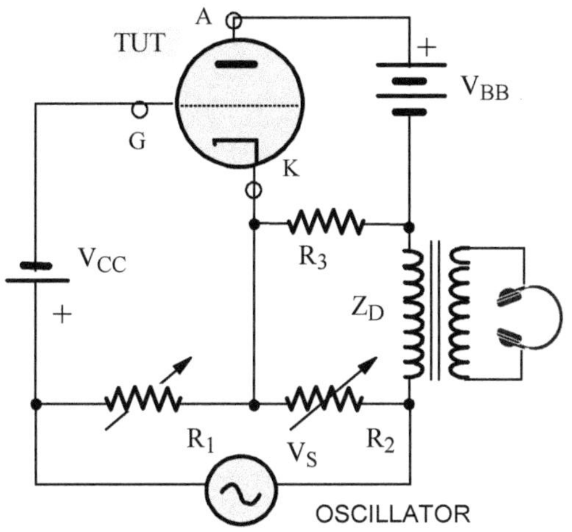

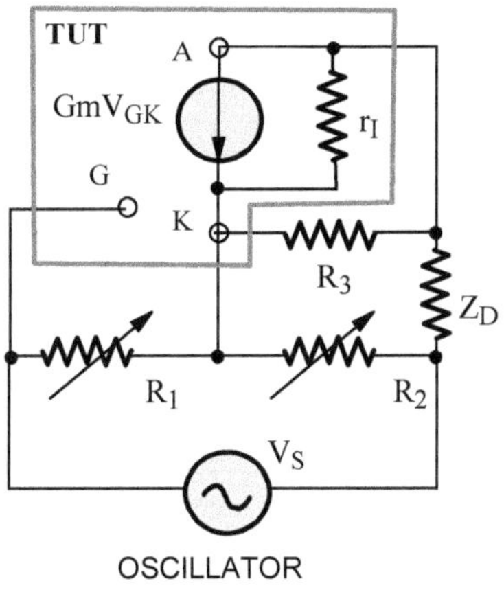

FAR LEFT: The bridge for determining mutual conductance Gm of a tube

LEFT: The tube has been replaced by its simple model (gray box), a current source in parallel with its internal resistance

THIS PAGE WAS DELIBERATELY LEFT BLANK

12 THE END MATTER

- INDEX
- DIY RESOURCES: DVD PROGRAMS
- AUDIOPHILE TUBE AMPLIFIER BOOKS BY IGOR S. POPOVICH
- OTHER TECHNICAL BOOKS BY IGOR S. POPOVICH

> "All the measurement in the world is useless if you don't make any changes based on the data.
>
> Amber Naslund"

INDEX

A

B

C

D

E

F

G

H

I

J

L

M

N

O

INDEX, continued

O, cont.

P

Q

R

S

S, cont.

T

V

W

X

Z

DIY RESOURCES: DVD PROGRAMS

While a technical book such as this one presents information in a visual way, nothing beats the power of a video. If a picture is worth a thousand words, than video instructions are worth a thousand! This unique video series, produced, filmed and manufactured in Australia, has been selling all over the world since 2005. The feedback has been overwhelmingly positive.

You can see amplifier building, transformer winding, testing of tube amplifiers and many other operations as they are performed by Dr. Bob and me live on camera, with commentaries and discussions. Most DVD programs come with accompanying manuals, containing all the schematics and diagrams used, plus additional technical information.

FOR BEGINNERS - THEORY & BASIC CONCEPTS

FUNDAMENTALS OF ELECTRONICS - 2 DVD set

Learn fundamentals of DC and AC circuits, inductance, resistance, capacitance, current and voltage laws, and much more. AU$59.-

SECRETS OF VACUUM TUBES & TUBE AMPLIFIERS - 4 DVD set

Learn how tubes & amplifiers work. Covers the operating principles of electron tubes and the secrets behind their successful practical use in designing & building hi-fi amplifiers & preamplifiers. AU$99.-

BUILDING HI-FI AMPLIFIERS AND PREAMPLIFIERS

DESIGN & BUILD TUBE AND SOLID STATE PHONO PREAMPLIFIERS - 4 DVD set + MANUAL

The secrets of building a world-class phono stage, many proven designs you can choose from! MM & MC designs, MC input transformers, active & passive equalization designs, battery-powered low voltage designs, tried & tested audiophile designs using a dozen various tube types. AU$129.-

DIY AUDIOPHILE QUALITY VALVE (TUBE) AMPLIFIER - 4 DVD set + MANUAL

The first DVD workshop in the world that shows you how to make your own hi-fi amp, using a SET 300B amp as an example. How to convert a circuit diagram into a wiring diagram and construct & test a tube amplifier. AU$129.-

REPAIRING AND UPGRADING HI-FI AMPLIFIERS AND PREAMPLIFIERS

REPAIR & RESTORE VINTAGE TUBE HI-FI AMPLIFIERS & PREAMPLIFIERS - 4 DVD set + MANUAL

Vintage gear is now at least 40-50 years old. To bring it back to operational state you need knowledge of troubleshooting methods and repair principles. Earthing unsafe old equipment, checking & replacing components, cold checks, DC & AC voltage checks, signal tracing and other troubleshooting methods & examples. AU$129.-

UPGRADE & IMPROVE VINTAGE TUBE HI-FI AMPLIFIERS & PREAMPLIFIERS - 4 DVD set + MANUAL

The most advanced program of all. Discusses components, designs and circuits and how to convert an old $300 amp into a beast that will put modern $4,000 amps to shame. Lots of well kept secrets revealed here! Dozens of easy modifications that will add functionality, increase reliability & tube life, and improve the sound of any vintage amplifier or preamplifier. AU$129.-

HOW TO MODIFY & IMPROVE TUBE & SOLID STATE PHONO STAGES - 4 DVD set + MANUAL

Makie any phono stage better sounding. Lots of examples, tips and techniques! Ten different budget phono preamplifiers analyzed and improvements suggested, from simple component changes to major redesign or rewiring. AU$129.-

MODIFY & UPGRADE LINE-LEVEL TUBE PREAMPLIFIERS - 3 DVD set + MANUAL

Buy a cheap tube preamp and improve it for a much better performance. Four line-level preamps reviewed & analyzed, tests & measurements performed, various problems identified and rectified, power supply, safety, reliability and sonic improvements that can be applied to any tube preamplifier. AU$129.-

MODIFY & UPGRADE TUBE HI-FI AMPLIFIERS - 4 DVD set + MANUAL

Buy a cheap made-in-China tube amp, identify & fix its problems & weaknesses, and improve it so it works & sounds like an amplifier many times its price! Buy a cheap tube amp and improve it for a much better performance. We show you how to get a $5,000 sound from a $600-$900 amp! AU$129.-

AUDIO TRANSFORMERS

DIY TRANSFORMERS FOR TUBE AMPLIFIERS - 3 DVD set + COLOR MANUAL

How to Make (Wind) Chokes, Power & Output Transformers for Valve (Tube) Amplifiers. Step-by-step instructions and examples, all wound, assembled and tested on camera! AU$129.-

TESTS & MEASUREMENTS

TESTS & MEASUREMENTS IN ELECTRONICS - 4 DVD set + COLOR MANUAL

For all amp builders, repairers and enthusiasts! How to use multimeters, LCR meters, oscilloscopes & function generators to measure voltage, current, power, phase shift, resistance, capacitance, inductance, gain and other parameters. AU$129.-

TESTS & MEASUREMENTS IN TUBE AUDIO - 4 DVD set + COLOR MANUAL

How to measure gain, output power, THD & IM distortion, S/N ratio, crosstalk, input and output impedance, damping factor, frequency range, AC & DC voltages & waveforms, phono stage accuracy & much more! How to measure gain, output power, harmonic and IM distortion, damping factor, frequency range, test tubes without a tube tester! AU$129.-

TRANSFORMER TESTS & MEASUREMENTS - 1 DVD + COLOR MANUAL

How to measure transformer's voltage & impedance ratios, power rating, (idle) magnetizing current, primary & leakage inductance, interwinding capacitances and frequency range, phasing of transformer windings, and much more! AU$69.-

TUBE GUITAR AMPLIFIERS

DIY TUBE GUITAR AMPLIFIER - 3 DVD set + COLOR MANUAL

Build your own boutique tube guitar amp. Four projects to choose from, from a small single-ended baby amp, through a 50 Watt push-pull beast, to hybrid designs. Keep the preamp and controls of your solid state amp and replace the output stage with a tube power amplifier. AU$129.-

TROUBLESHOOT & REPAIR TUBE GUITAR AMPLIFIERS - 4 DVD set + COLOR MANUAL

Learn how to fix tube guitar amps. Two real-life examples, with lots of upgrading ideas as well. Test & measurements, transformer, rectifier & capacitor checks, cold checks and hot (power) tests, unique testing jigs & tools, safety rules and protective measures, no sound, radio interference, hum, excessive power draw, blown fuse and other faults. AU$129.-

TUBE TESTERS

A VIDEO JOURNEY THROUGH THE WORLD OF VINTAGE TUBE TESTERS - 2 DVD set

See 22 different tube testers in operation, learn about their capabilities and limitations, before you decide which tester to buy! AU$89.-

REPAIR, UPGRADE & BUILD VACUUM TUBE TESTERS - 4 DVD set + COLOR MANUAL

Add features, improve accuracy of old testers, or make your own tube tester, way superior to most old designs from the 50s and 60s! Test and replace capacitors, resistors, diodes, transistors, analog meters and transformers, convert 115 V testers to 220V or 240V mains, how to calibrate various testers, add plate current measuring terminals, bias meters, socket adapters, or, if truly ambitious, how to build a true mutual conductance tester and curve tracer for matching tubes. AU$149.-

Ordering information

Please e-mail us the titles you want to order and your full address so we can advise you on the total cost, including registered airmail postage worldwide. All prices are in Australian Dollars (AU$ or AUD).

PayPal transfer and bank transfer are the only payment methods accepted.

Our e-mail is **tubeampdvds@yahoo.com.au**

AUDIOPHILE TUBE AMPLIFIER BOOKS BY IGOR S. POPOVICH

Available from Amazon, Barnes & Noble, Book Depository and all other major online bookstores

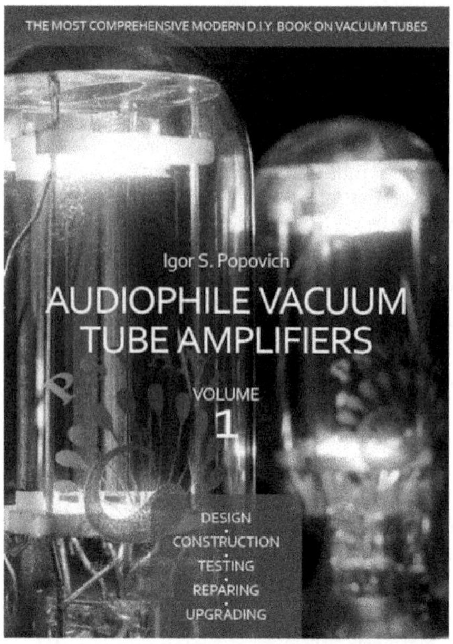

Audiophile Vacuum Tube Amplifiers, Vol 1

ISBN: 978-0-9806223-2-4

- BASIC ELECTRONIC CIRCUIT THEORY
- ELECTRONIC COMPONENTS
- AUDIO FREQUENCY AMPLIFIERS
- PHYSICAL FUNDAMENTALS OF VACUUM TUBE OPERATION
- VOLTAGE AMPLIFICATION WITH TRIODES - THE COMMON CATHODE STAGE
- OTHER VOLTAGE AMPLIFICATION STAGES WITH TRIODES
- TETRODES AND PENTODES AS VOLTAGE AMPLIFIERS
- FREQUENCY RESPONSE OF VACUUM TUBE AMPLIFIERS
- IMPEDANCE-COUPLED STAGES AND INTERSTAGE TRANSFORMERS
- NEGATIVE FEEDBACK
- TONE CONTROLS, ACTIVE CROSSOVERS AND OTHER CIRCUITS
- PRACTICAL LINE-LEVEL PREAMPLIFIER DESIGNS
- PHONO PREAMPLIFIERS
- SINGLE-ENDED TRIODE OUTPUT STAGE
- PRACTICAL SINGLE-ENDED TRIODE AMPLIFIER DESIGNS
- PRACTICAL SINGLE-ENDED PSEUDO-TRIODE DESIGNS
- SINGLE-ENDED PENTODE AND ULTRALINEAR OUTPUT STAGES

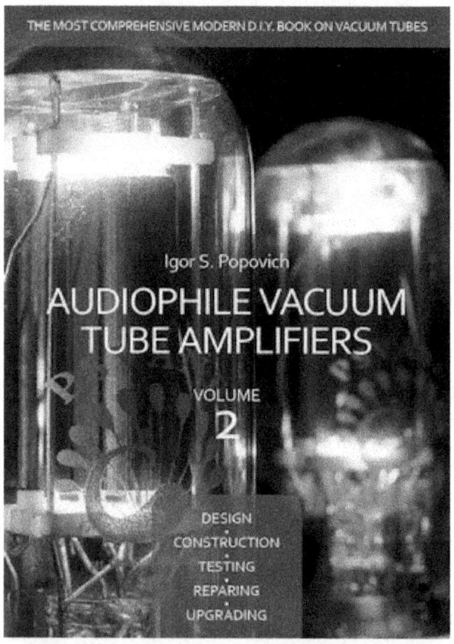

Audiophile Vacuum Tube Amplifiers, Vol 2

ISBN: 978-0-9806223-3-1

- PRACTICAL SINGLE-ENDED PENTODE AND ULTRALINEAR DESIGNS
- PUSH-PULL OUTPUT STAGES
- PRACTICAL PUSH-PULL AMPLIFIER DESIGNS
- BALANCED, BRIDGE AND OTL (OUTPUT TRANSFORMERLESS) AMPLIFIERS
- THE DESIGN PROCESS
- FUNDAMENTALS OF MAGNETIC CIRCUITS AND TRANSFORMERS
- MAINS TRANSFORMERS AND FILTERING CHOKES
- POWER SUPPLIES FOR TUBE AMPLIFIERS
- AUDIO TRANSFORMERS
- TROUBLESHOOTING AND REPAIRING TUBE AMPLIFIERS
- UPGRADING & IMPROVING TUBE AMPLIFIERS
- SOUND CONSTRUCTION PRACTICES
- AUDIO TESTS & MEASUREMENTS
- TESTING & MATCHING VACUUM TUBES

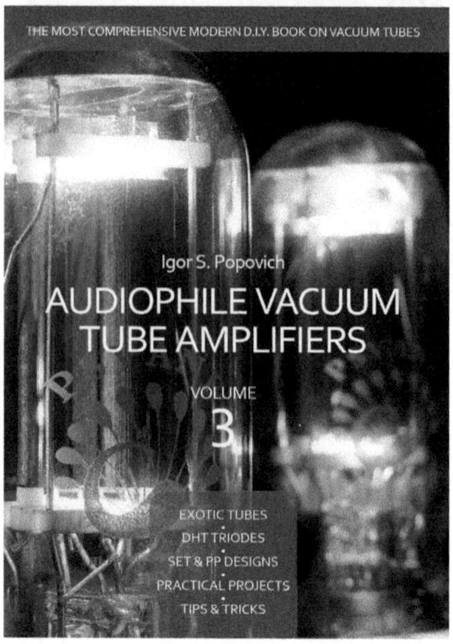

Audiophile Vacuum Tube Amplifiers, Vol 3

ISBN: 978-0-9806223-4-8

- THE FRONT-END: SUPERIOR INPUT & DRIVER STAGES
- FROM SHOCKING TO SUBLIME: LESSONS FROM COMMERCIAL LINE STAGES
- DIY LINE-LEVEL PREAMPLIFIERS: $10,000 SOUND ON $500-$1,000 BUDGET
- THE STARS OF THE AUDION ERA: ANCIENT TUBES IN MODERN AMPS
- CHEAP & CHEERFUL: PREAMP & DRIVER TUBES FOR AUDIO EXPLORERS
- SLEEPING GIANTS: OUTPUT TUBES FOR THOSE WHO WANT TO BE DIFFERENT
- THE QUEEN OF HEARTS: SINGLE-ENDED AMPLIFIERS WITH 300B TRIODES
- TRIODES, PENTODES AND BEAM TUBES: MORE SINGLE-ENDED DESIGNS
- BIG BOTTLES: SET AMPLIFIERS WITH HIGH VOLTAGE TRANSMITTING TUBES
- THE WAY IT USED TO BE: VINTAGE PUSH-PULL AMPLIFIERS
- NEW? IMPROVED? MODERN PUSH-PULL AMPLIFIER DESIGNS
- CUTE, CLEVER OR CONTROVERSIAL? INTERESTING IDEAS FROM TUBE AUDIO'S PAST AND PRESENT
- THRIFTY TIPS & TRICKS: TIME & MONEY SAVING IDEAS
- OUTPUT AND INTERSTAGE TRANSFORMERS: FROM COMMERCIAL BENCHMARKS TO YOUR OWN DESIGNS
- MEASUREMENTS VERSUS LISTENING AND OTHER AUDIO DESIGN DILEMMAS

OTHER TECHNICAL BOOKS BY IGOR S. POPOVICH

Available from Amazon, Barnes & Noble, Book Depository and all other major online bookstores

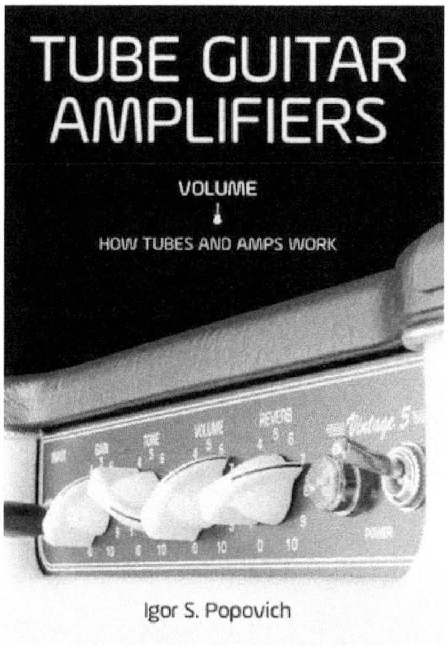

Tube Guitar Amplifiers Volume 1: How Tubes and Amps Work
ISBN: 978-0-9806223-5-5

- BASIC ELECTRONIC CIRCUIT THEORY
- AUDIO AMPLIFIERS
- ELECTRONIC COMPONENTS
- PHYSICAL FUNDAMENTALS OF VACUUM TUBE OPERATION
- TRIODES AS VOLTAGE AMPLIFIERS
- TETRODES, PENTODES AND BEAM-POWER TUBES
- INPUT CIRCUITS AND STAGES
- TONE CONTROLS
- ANALOG EFFECTS (TREMOLO, VIBRATO, REVERB) AND EFFECTS LOOPS
- POWER SUPPLIES FOR TUBE AMPLIFIERS
- SINGLE-ENDED TRIODE, PENTODE AND ULTRALINEAR OUTPUT STAGES
- PHASE SPLITTERS OR INVERTERS
- PUSH-PULL OUTPUT STAGES
- NEGATIVE FEEDBACK
- TRANSISTOR AND HYBRID GUITAR AMPLIFIERS

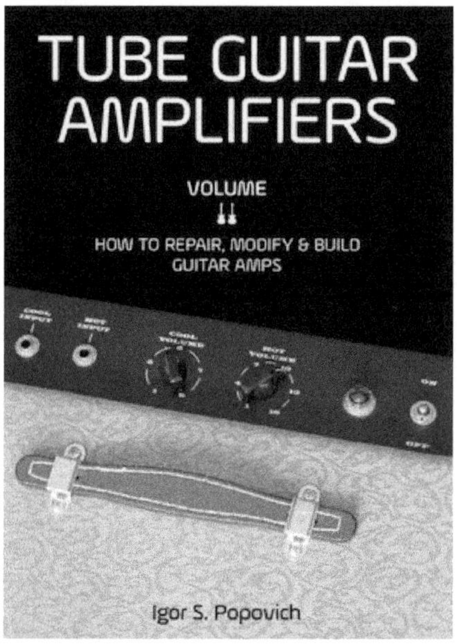

Tube Guitar Amplifiers Volume 2: How to Repair, Modify & Build Guitar Amps
ISBN: 978-0-9806223-6-2

- OUTPUT AND INTERSTAGE TRANSFORMERS FOR TUBE GUITAR AMPS
- LOUDSPEAKERS, OUTPUT ATTENUATORS & HEADPHONE CIRCUITS
- TROUBLESHOOTING AND REPAIRING TUBE GUITAR AMPLIFIERS
- WIRING, SOLDERING & MODIFICATION PRACTICES
- POWER SUPPLY MODIFICATIONS AND IMPROVEMENTS
- TONE TWEAKS
- MODERN PUSH-PULL AMPS
- DIY PROJECTS: CONVERTING SOLID STATE GUITAR AMPS TO TUBES
- DIY PROJECTS: ULTRA-SMALL AMPS
- REBUILDING COMMERCIAL AMPS IN A HANDWIRED (POINT-TO-POINT) FASHION
- DIY PROJECTS: QUIRKY & UNUSUAL DESIGNS
- CONVERTING VINTAGE TUBE GEAR INTO GUITAR AMPS

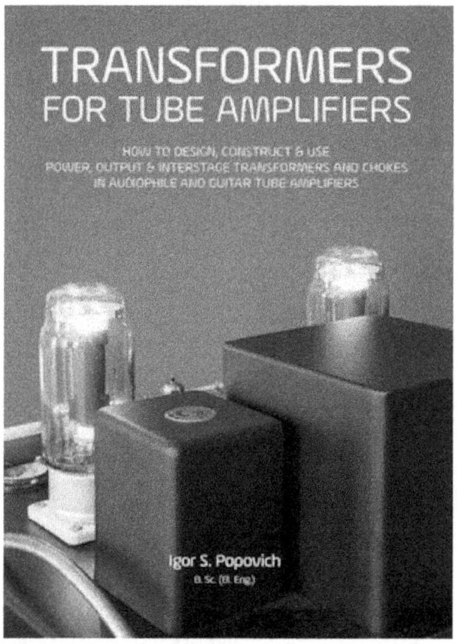

Transformers For Tube Amplifiers: How to Design, Construct & Use Power, Output & Interstage Transformers and Chokes in Audiophile and Guitar Tube Amplifiers

ISBN: 978-0-9806223-8-6

- PHYSICAL FUNDAMENTALS OF MAGNETIC CIRCUITS AND TRANSFORMERS
- FILTERING CHOKES (INDUCTORS WITH DC CURRENT)
- TRANSFORMER MATERIALS, CONSTRUCTION METHODS AND ISSUES
- MAINS (POWER) TRANSFORMERS
- PHYSICAL FUNDAMENTALS OF AUDIO TRANSFORMERS
- SINGLE-ENDED OUTPUT TRANSFORMERS
- PUSH-PULL OUTPUT TRANSFORMERS
- SPECIAL MAGNETIC COMPONENTS: LOW POWER INPUT, PREAMP OUTPUT & DAC OUTPUT TRANSFORMERS, TRANSFORMER VOLUME CONTROL
- INTERSTAGE TRANSFORMERS, GRID & ANODE CHOKES
- OUTPUT AND INTERSTAGE TRANSFORMERS FOR TUBE GUITAR AMPS
- TRANSFORMER TESTS & MEASUREMENTS

Lightning Source UK Ltd.
Milton Keynes UK
UKHW051113181022
410640UK00003B/73

9 780980 622393